NON–INVASIVE
PHYSIOLOGICAL
MEASUREMENTS

MEDICAL PHYSICS SERIES

P. N. T. WELLS. Physical Principles of Ultrasonic Diagnosis. 1969
D. W. HILL and A. M. DOLAN. Intensive Care Instrumentation. (Second edition) 1982.
P. N. T. WELLS. Biomedical Ultrasonics. 1977
P. ROLFE. Non-invasive Physiological Measurements. Volume 1. 1979
P. ATKINSON and J. P. WOODCOCK. Doppler Ultrasound and its Use in Clinical Measurement. 1982
A. R. WILLIAMS. Ultrasound: Biological Effects and Potential Hazards. 1983
G. LASZLO and M. F. SUDLOW. Measurement in Clinical Respiratory Physiology. 1983
P. ROLFE. Non-invasive Physiological Measurements. Volume 2. 1983

NON-INVASIVE PHYSIOLOGICAL MEASUREMENTS

VOLUME 2

Edited by

Peter Rolfe

*Bioengineering Unit, Department of Paediatrics,
University of Oxford, John Radcliffe Hospital,
Oxford, England*

1983

Academic Press
A Subsidiary of Harcourt Brace Jovanovich, Publishers

London · New York · Paris · San Diego
San Francisco · São Paulo
Sydney · Tokyo · Toronto

ACADEMIC PRESS INC. (LONDON) LTD.
24–28 Oval Road
London NW1 7DX

U.S. Edition published by
ACADEMIC PRESS INC.
111 Fifth Avenue
New York, New York 10003

British Library Cataloguing in Publication Data
Non-invasive physiological measurements.—(Medical
  physics series, ISSN 0076-5953)
  Vol. 2
1. Human physiology—Measurement
I. Rolfe, Peter   II. Series
612   QP54

Library of Congress Catalog Number: 78-72551
ISBN: 0 12 593402 5

Typeset by Bath Typesetting Ltd., Bath
and printed in Great Britain by
Thomson Litho Ltd., East Kilbride, Scotland

CONTRIBUTORS

B. H. BROWN, *Department of Medical Physics and Clinical Engineering, Royal Hallamshire Hospital, Glossop Road, Sheffield S10 2JF, Yorkshire, England.*

E. C. BURDETTE, *Biomedical Research Division, Engineering Experiment Station, Georgia Institute of Technology, Atlanta, Georgia 30332, USA.*

K. R. GREENE, *Department of Obstetrics and Reproduction, University of Southampton, Princess Anne Hospital, Oxford Road, Southampton SO9 4HA, England.*

G. A. HOLLOWAY, Jr, *Associate Professor, Center for Bioengineering, 310 Harris (WD-12), University of Washington, Seattle, Washington 98195, USA.*

K. LINDSTRÖM, *Department of Biomedical Engineering, University Hospital, S-214 01 Malmö, Sweden.*

R. L. MAGIN, *Assistant Professor of Electrical Engineering and Bioengineering, Bioacoustics Research Laboratory, University of Illinois at Urbana-Champagne, 1406 West Green Street, Urbana, Illinois 61801, USA.*

K. MARŠÁL, *Department of Obstetrics and Gynaecology, University Hospital, S-214 01 Malmö, Sweden.*

J. NAGEL, *Zentralinstitut für Biomedizinische Technik, Friedrich-Alexander-Universität, Erlangen-Nürnberg, West Germany.*

M. J. O'BRIEN, *Department of Developmental Neurology, University Hospital, Oostersingel 59, Groningen, The Netherlands.*

H. F. R. PRECHTL, *Department of Developmental Neurology, University Hospital, Oostersingel 59, Groningen, The Netherlands.*

M. SCHALDACH, *Zentralinstitut für Biomedizinische Technik, Friedrich-Alexander-Universität, Erlangen-Nürnberg, West Germany.*

Y. SHIMADA, *Assistant Professor, Department of Anesthesiology, Osaka University Hospital, Fukushimaku, Osaka 553, Japan.*

R. H. SMALLWOOD, *Department of Medical Physics and Clinical Engineering, Royal Hallamshire Hospital, Glossop Road, Sheffield, S10 2JF, Yorkshire, England.*

L. A. VAN EYKERN, *Department of Developmental Neurology, University Hospital, Oostersingel 59, Groningen, The Netherlands.*

P. N. T. WELLS, *Department of Medical Physics, Bristol General Hospital, Guinea Street, Bristol BS1 6SY, England.*

A. YAMANISHI, *Minolta Camera Co. Ltd., Technical Center, 3-91, Daisen-Nishimachi, Sakai-Shi, Osaka 590, Japan.*

I. YAMANOUCHI, *Children's Medical Center, Okayama National Hospital, Minamigata 2-13-1, Okayama 700, Japan.*

I. YOSHIYA, *Department of Anesthesiology, Osaka University Hospital, Fukushimaku, Osaka 553, Japan.*

PREFACE

The subjects covered in this volume complement those presented in Volume 1, and together they demonstrate the breadth of the now well-established field of non-invasive physiological measurement.

There has been much discussion recently regarding the precise definition of "non-invasive", and indeed this was of some importance to me in deciding which topics should, and could, be included in this book. Although one might regard the matter as one of remote academic interest, it does seem appropriate to clarify the terminology which, after all, is the *raison d'être* for the work contained within these covers.

The verb "to invade" means, amongst other things, "to intrude upon, to enter or penetrate" (*Shorter Oxford English Dictionary*). It seems straightforward to apply these descriptions to certain measurement techniques, such as those involved in puncture of the skin and blood vessels for insertion of mechanical objects of any kind, and thereby evolve categories of invasive and non-invasive techniques. However, it has been argued that "invasion" also occurs when energy of any form is made to enter the subject, and of course use of this criterion would mean that some techniques commonly regarded as non-invasive should be called invasive. For example, ultrasound for blood flow and imaging purposes requires high frequency pressure waves to penetrate the subject. Even the apparently innocuous technique of skin surface gas measurement based on electrically heated transducers would be categorized as invasive, since it requires thermal energy to pass into the skin in order to produce local vasodilatation. For a method to be accepted as non-invasive under this definition, it would need to be based on what might be regarded as passive principles. This could include the detection of any form of energy actually generated by the subject, such as biopotentials (EEG, ECG, EMG), thermal energy, pressure etc., or straightforward dimensional measurement.

Consideration of energy exchange between a measuring instrument and the subject is of interest for a number of reasons, perhaps the most important being safety. But is it useful as the basis of differentiation between invasive and non-invasive measurement techniques? It will be apparent from the contents of this book that I think not. I prefer to regard the primary criterion for non-invasiveness as being that of the absence of the need to insert an object into the subject. There is a tendency to assume that "non-invasive" must mean "safe", but some non-invasive methods are not yet proven to be entirely safe, and others are not entirely unencumbering to the subject. Therefore, even when using a technique which is agreed to be non-invasive, we must ask whether it is acceptably safe and non-disturbing to the subject.

Some anomalies do seem to exist whichever definitions are employed, because of the common assumption that non-invasive techniques are inherently safe. For example, it seems to be difficult to categorize those measurement techniques which involve the inhalation of radioactive gases. According to one definition, these methods are non-invasive since they do not involve insertion of an object into the patient, but nobody would ignore the possibilities of hazards when using radioactive gases. The alternative definition based on energy transfer would suggest that these methods are invasive, since radioactive energy enters the subject, although use of very low doses can ensure acceptable safety.

It seems clear from this reasoning that when assessing the acceptability of a measurement technique for a particular application, we need to consider quite separately the issues of, firstly, invasiveness and, secondly, safety.

A significant proportion of this volume is concerned with the very real clinical problems of assessing the condition of the fetus during pregnancy, labour and delivery. Of course, during pregnancy our subject, the fetus, is entirely inaccessible by straightforward mechanical contact, and this situation is perhaps one of the clearest examples of why we need non-invasive measurement techniques. The three chapters on what might be called "fetal monitoring" provide a useful indication of the ways in which the continuing developments in bio-engineering have brought important improvements to this difficult area of measurement. This is particularly relevant at a time when the widely practised procedure of assessing fetal status on the basis of heart rate measurements is being brought into question.

In addition to the elegant utilization of EMG recording to assess uterine activity, there are important contributions describing the ways in which recordings of muscle activity may be used to study respiratory activity in infants and gastric activity. Optical measurement techniques are already being used widely in medical diagnosis, particularly in the biochemistry laboratory. It is intriguing to consider how such *in vitro* principles may be adapted and employed for direct measurement within the subject, and the three chapters in this volume concerned with optical techniques are good examples.

Ultrasound imaging has had enormous impact on the ways in which a variety of diagnoses may be made. Following the early introduction of large and immobile ultrasound scanners, there has been rapid technological development to produce small, portable low-cost instruments which, in many situations, may be transported easily around the hospital to bring the measurement procedure to the patient. This is particularly true in neonatal intensive care where portable ultrasound scanners are now used for the clinical diagnosis and research study of brain haemorrhage in preterm infants. There has long been an interest, and indeed fascination, in the electrical properties of tissue, and the final chapter provides an introduction to one aspect of this subject which could ultimately lead to the evolution of entirely novel diagnostic techniques. These could provide convenient means for the detection of abnormal tissue in a way which could complement existing imaging procedures.

Broad as the contents of Volumes 1 and 2 are, they do not encompass the entire field of non-invasive measurement. The subject continues to evolve to produce entirely new approaches, and refinements of old techniques, through both technological advances and improved understanding of the physical and physiological processes involved.

August, 1983   PETER ROLFE

CONTENTS

1. FETAL CARDIAC ASSESSMENT

Keith R. Greene

*Department of Obstetrics and Reproduction,
University of Southampton, Princess Anne Hospital,
Southampton, England*

1.0 INTRODUCTION

The first written reference to "fetal cardiac assessment", by auscultation of the fetal heart, is contained in a poem written by Philip le Goust in 1650, who chides a Dr Marsac for claiming to have heard the fetal heart "beating like the clapper of a mill". By the early nineteenth century (de Kegaradec, 1822; Kennedy, 1933) auscultation had established that the fetal heart rate was faster in early pregnancy and that it varied from occasion to occasion. At the beginning of this century, Von Winkel first drew attention to the association of a slowing of the fetal heart with a poor fetal outcome and, in his wards, a few years later, Cremer (1906) recorded the first fetal electrocardiogram. Since the work of Hon in the 1960s we have become used to recording the instantaneous fetal heart rate continuously, both antepartum and intrapartum. Clinically, however, the fetal heart rate is still used as it was at the beginning of the century, to judge the condition of the whole fetus. The taking of a pulse to reach a diagnosis is a little naive, but is done largely because of a lack of other information. Over recent years the application of new technology, particularly ultrasound, has allowed us to examine anatomical and physiological cardiac function so that we really are now approaching the goal of comprehensive fetal cardiac assessment.

All of the information gained from human observation and fetal animal experimentation is of clinical relevance but as yet not much is of clinical

NON-INVASIVE MEASUREMENTS: 2   *Copyright© 1983 by Academic Press Inc. (London) Ltd.*
ISBN 0 12 593402 5   *All rights of reproduction in any form reserved*

value. An attempt will be made to indicate what is likely to be useful in the future. Any workers new to the field should gain a basic understanding of developmental cardiac anatomy, developmental physiology and the different fetal animal preparations to judge for themselves the significance of published work. These areas are covered briefly in the following sections of the introduction.

1.1 Fetal Animal Studies

Our knowledge of fetal cardiovascular dynamics is mostly derived from physiological measurement and experiment in fetal animal preparations (principally the sheep, goat and monkey). The fetuses of these animals are comparable in size to the human fetus and the configuration of their circulations is similar. It is unlikely that there are any major differences in cardiovascular physiological response between the species, apart from those associated with developmental maturity. To review these data a broad understanding of the different animal preparations is needed.

There are basically two types of fetal experiment, termed the acute and the chronic preparation. True chronic preparations are uncommon in the monkey as the discoid placenta is more liable to separation and the uterus is more irritable, so that labour is often initiated by surgery. In the acute preparation experimental manoeuvres are performed at the time of surgery, either with the fetus exteriorized or remaining *in utero*, and usually anaesthetized though such procedures can be performed under local anaesthesia. However, the preparation may be compromised because exteriorization can produce a fall in umbilical blood flow (Heymann and Rudolph, 1967), and high levels of circulating catecholamines (Jones and Ritchie, 1978) and anaesthesia (Assali *et al.*, 1974) have varying effects on cardiovascular reflexes. In the chronically instrumented preparation the fetus is returned to the uterus after surgery and allowed to recover for some days prior to any experimental procedure, or it may be left undisturbed for developmental changes to be studied. Whilst offering undeniable advantages over the acute preparation, infection, haemorrhage or asphyxia can occur, so that an experimental response may not be that of a healthy animal. Displacement of catheters and electrodes can also occur, due to poor fixing, fetal movement or growth. Whilst this would make little difference to a cathether for blood-pressure measurement, displacement of an electromagnetic flowmeter on a major vessel may alter axial symmetry and give false blood-flow measurements.

Thus data from both acute and chronic fetal animal preparations need to be interpreted with care, with particular attention paid to the normality of values in the control period.

1.2 Developmental Anatomy

By the end of the eighth embryonic week the fetal heart is fully formed as a four chamber structure and no major changes occur until the revision of the circulation which takes place with the first breath at birth.

An awareness of the broad features of cardiac differentiation is helpful in understanding the mechanisms of congenital heart disease. By the sixth week the fetal heart tube has undergone lengthening, coiling and rotation to assume the general external shape of the adult heart. The atria are divided between the sixth and eighth week by the parallel septa (septum primum and septum secundum) which are imperforate in different areas to form the foramen ovale. The ventricular septum forms by an upgrowth of the common ventricle from the fourth week and fuses with the ridges dividing the common ventricular exit (bulbus cordis and truncus arteriosus) so that an aortic channel empties the left ventricle and a pulmonary channel the right ventricle by the seventh week. During this process endocardial cushions of cardiac jelly condense to form the aortic and pulmonary and atrioventricular valves to complete the four chambered structure.

Initially the heart beats as a contractile tube about the fourth week at a rate of 60 beats per minute. High in the right atrium some of the developing musculature soon becomes specialized to form the sino-atrial node with a rhythm of about 130 beats per minute which spreads to the rest of the heart. At this time there is continuity between the musculature of the different chambers so that this contractile impulse spreads over the whole heart. As the atrial muscle becomes separated from the ventricular muscle with the formation of the fibrous atrioventricular ring a specialized conducting system is formed by the persistence of a band of muscle between them. This ordinary heart muscle at first accumulates glycogen and a connective tissue insulation to form the conducting bundle of His with right and left branches passing to the apex of the ventricles to form a Purkinje network which ramifies through the ventricular muscle mass. At the point of origin of the bundle a nodule of atrial muscle differentiates to form the atrioventricular node. The adult form of ECG can be detected by direct contact with the fetus from the eighth week.

It is apparent that various degrees of non-differentiation may give a single atrium, single ventricle, single ventricular outflow tract or a ventricular septal defect; and that all of these could be associated with abnormalities of the conducting pathway.

1.3 Functional Anatomy

The course of the fetal circulation is now well known (see Dawes, 1968) and is illustrated in Fig. 1, in which the numerals indicate the mean oxygen saturation in each vessel. Well oxygenated blood from the placenta is diverted preferentially to the heart and brain before supplying the abdominal viscera, lower trunk and limbs. The principal mechanism of achieving this, elucidated by cine-angiography (Barclay *et al.*, 1944; Barcroft, 1946), is the diversion of a large proportion of oxygenated blood from the inferior vena cava through the foramen ovale into the left side of the heart without mixing with the relatively deoxygenated blood returning to the heart by the superior vena cava. In the resting state all blood returning via the superior vena cava passes into the right atrium. Blood returning via the inferior vena cava divides on the crista dividens (the lower free margin of the septum secundum); the larger proportion passes directly through the foramen ovale and the smaller proportion is directed into the right atrium where it mixes with the superior vena caval blood. Under certain circumstances of asphyxia and vagal stimulation associated with a

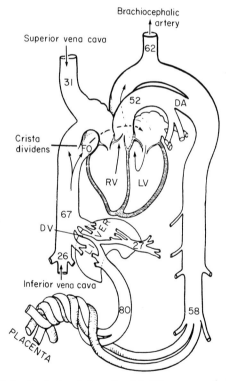

Fig. 1   Diagram of the fetal circulation (lamb). The numerals indicate the mean O_2 saturation (%) in the great vessels. RV: right ventricle; LV: left ventricle; FO: foramen ovale; DA: ductus arteriosus; DV: ductus venosus. (From Born, Dawes, Mott and Widdicombe (1954). *Cold Spring Harb. Symp. Quant. Biol.* **19**, 102, with permission.)

bradycardia (Dawes, 1968) blood returning via the superior vena cava regurgitates to the upper end of the inferior vena cava around the crista dividens and into the left atrium through the foramen ovale. Thus the distribution of the venous return to the two sides of the heart is dynamic, depending not just on the gross anatomy of the great vessels but also on the physiological forces affecting relative flow.

The ductus arteriosus, which is as large as the arch of the aorta, directs the bulk of the blood from the right side of the heart away from the lung to the descending aorta, whilst the bulk of the output from the left ventricle passes to the upper body. The ductus arteriosus and the foramen ovale thus allow the two sides of the fetal heart to pump in parallel and not in series as in the adult (Fig. 2).

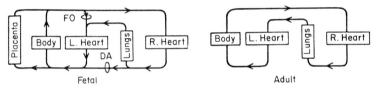

Fetal   Adult

FIG. 2   Diagrams to illustrate the changes from fetal to adult types of circulation. (After Born *et al.* (1954) *Cold Spring Harb. Symp. Quant. Biol.* **19**, 102, with permission.)

At birth the arrest of the umbilical circulation with the separation of the placenta, and the increase in pulmonary blood flow, with ventilation of the lungs, alter pressure and flow on the two sides of the heart to close the foramen ovale and this, together with increased oxygen tension, closes the ductus arteriosus (see Dawes, 1968). Occasionally some part of this mechanism fails in the neonatal period, with shunting of blood either right-to-left or left-to-right, across these fetal channels.

1.4 Developmental Physiology

Although many of the mechanisms which operate to control the circulation in the adult at rest and during stress have been extensively studied, there is limited and sometimes conflicting information in the fetus. Just as there are large differences in the functional anatomy of the fetal circulation from that of the adult, so there are differences in physiological control. Further, the degree of control depends very much on developmental maturity, the rate and timing of which vary in different species.

There are also different stages in establishing autonomic control to the heart. In mammalian fetuses parasympathetic innervation precedes sympathetic innervation and in both situations the effector organ (the cardiac or smooth muscle cell) has the ability to respond to the neurotransmitter substance before the nervous endings have the ability to release it. Thus circulating catecholamines will have an effect on the heart earlier in gestation than direct stimulation of the sympathetic nerves. These considerations emphasize the importance of considering gestational age when examining cardiovascular responses.

The fetal circulation is to a large extent dominated by the high blood flow, low vascular resistance and relative unreactive nature of the umbilical–placental circulation. This receives 57% of the combined output of both ventricles in contrast to 10% to the fetal lungs, and 15% and 18% respectively to all the foreparts (excluding lungs) and all the hindparts of the sheep fetus (Dawes, 1968). The lack of autoregulation within the placental circulation means that placental blood flow and thus fetal oxygenation are largely dependent on heart rate and arterial and venous pressure. Their control will depend on the maturity of the reflexes (e.g. baroreceptor, chemoreceptor and mechanoreceptor) and on functional sensitivity to circulating hormones. Some reflex responses may also be modified by neurological sleep state (Dawes *et al.*, 1980).

Certainly there is a well-organized response to moderate hypoxia (induced by maternal hypoxaemia) in later gestation, with intense vaso-constriction in the gut and lower limb beds to redistribute an increased proportion of cardiac output to the heart, brain and adrenals. The in-creased systemic vascular resistance causes a rise in arterial blood pressure and thus a proportionate rise in umbilical blood flow, with a presumed increase in oxygen extraction at the placental bed.

Myocardial metabolism also differs in the fetus compared with the adult. All oxygen consumption can be accounted for by carbohydrate rather than free fatty acid metabolism. Cardiac glycogen stores are three to four times greater than in the adult (Dawes *et al.*, 1959) and the fetal myocardium is more dependent on energy supplied by glycolysis than aerobic metabolism (Su and Friedman, 1973). This also increases the heart's ability to maintain the circulation during hypoxia.

For further reading on these aspects of developmental cardiac physiology, the reader is referred to Dawes (1968), Rudolph and Heymann (1973), Kirkpatrick *et al.* (1973, 1976), Goodwin (1976) and Rudolph *et al.* (1981).

1.5 Physical Methods

The physical methods used will be discussed in detail under the heading of the particular cardiac function being measured. The physical principles of ultrasound in its various modes of operation and phonocardiography have been discussed in other chapters and the basics of electrocardiography will be assumed. Figure 3 summarizes the physical methods used to measure

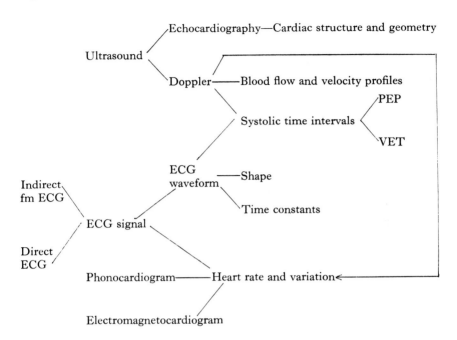

FIG. 3   Summary of the physical methods and measurements available in fetal cardiac assessment. PEP = pre-ejection period. VET = ventricular ejection period. fm = fetomaternal.

the various cardiac assessments discussed. The fact that some methods provide information on more than one variable may prove to be important clinically in the future.

2.0 CARDIAC STRUCTURE AND HAEMODYNAMICS

2.1 General

The dramatic advances in two-dimensional real time ultrasound scanning equipment (Chapter 9) have now permitted clear visualization of fetal cardiac structure at different stages in the cardiac cycle (Sahn *et al.*, 1980). This is now being used to exclude congenital cardiac abnormalities, measure cardiac geometry and output and, in combination with Doppler ultrasound, measure aortic artery and umbilical vein flow, and peripheral resistance of the placental circulation.

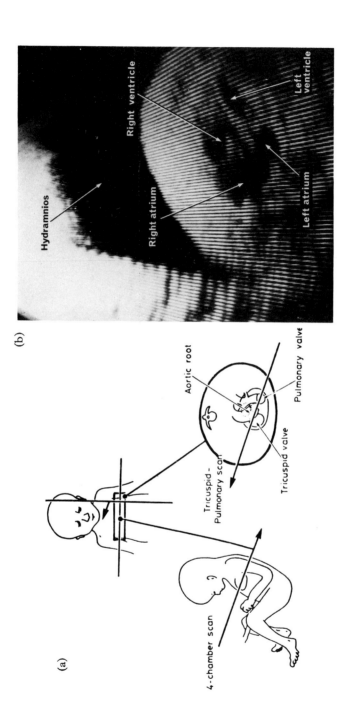

FIG. 4 Real-time scanning of cardiac anatomy. (Reproduced from Allan *et al. Prenatal Diag.* **1**, 131, 1981, with permission.)
(a) Left-hand panel shows the plane producing the four chamber scan. The right-hand panel is a cross-section of the thorax as shown in the upper panel. The scan plane across this cross-section illustrates the angle needed to produce the tricuspid pulmonary plane. (b) A four chamber scan from a 35-week fetus with idiopathic right ventricular dilatation and hypertrophy showing dilated right heart chambers and hydramnios.

2.2 Cardiac Structure

Most fetal cardiac abnormalities are now capable of being detected by ultrasound. After studying eight ultrasound scan planes in 200 pregnancies (Allan *et al.*, 1980), with anatomical correlation in a number of fetuses, Allan *et al.* (1981) have suggested that two echocardiographic sections are sufficient to exclude abnormality. The tricuspid–pulmonary and four chamber planes (Fig. 4) are easy to obtain with no obstruction by the fluid-filled lungs and demonstrate the four cardiac chambers, atrioventricular and semilunar valves and their outflow tracts. Normality appears to be readily identifiable; an inability to recognize all the normal structures should lead to the suspicion of a cardiac malformation. Thereafter M-mode studies (Kleinman *et al.*, 1980) may assist in diagnosis. Routine screening has been suggested for mothers with diabetes or a previous history of congenital heart disease, and for fetuses with cardiac arrhythmias or ascites.

2.3 Cardiac Geometry and Output

Wladimiroff (Wladimiroff and McGhie, 1981) has concentrated on measuring cardiac geometry and output from changes in structural shape, with the cardiac cycle. Two-dimensional real time scanning is used to locate the largest longitudinal cross-section of the left ventricle and six landmarks are noted to ensure continuous visualization of the same section through systole and diastole. After recording several cardiac cycles on videotape during fetal apnoea measurements are made of the transverse and vertical axes of the ventricles from still frames in the end-diastolic and end-systolic phases. The end-diastolic and end-systolic volumes are calculated from these measurements, assuming the ventricle to have the shape of a prolate ellipse, and the difference between them gives the calculated left ventricular cardiac output.

A significant increase in the transverse axis diameter was found with advancing gestational age, associated with an increase in "left ventricular cardiac output' from 210 ± 48 ml min^{-1} (mean $\pm$ SD) between 27 and 33 weeks to 336 ± 35 ml min^{-1} between 34 and 41 weeks gestation. Values of 114–123 ml kg^{-1} min^{-1} were calculated in four fetuses shortly before delivery, which is rather less than that of 250 ml kg^{-1} min^{-1} found in the fetal lamb at term (Kirkpatrick *et al.*, 1973).

Although this technique makes a number of assumptions and needs to be critically examined by other workers it nevertheless illustrates the new capacity for non-invasive measurement of cardiac function, which may prove useful in the future.

2.4 Flow and Flow Profiles

A combination of pulsed echo and pulsed Doppler ultrasound has been used to measure volume flow in the umbilical vein (Gill and Kossoff, 1979) and fetal descending aorta (Eik-Nes *et al.*, 1980). The Doppler transducer is set at a fixed angle on a real time transducer which is used to locate the vessel so that its longitudinal axis is parallel to the transducer (Fig. 5). The vessel diameter is measured by a time/distance measurement and the B-mode scanner switched off whilst the Doppler velocity measurement is made. The blood flow Q is calculated from the formula $Q = V \times A/\cos \alpha$, where V is the mean or maximum velocity, A the cross-sectional area of the vessel and α the fixed angle between the Doppler and B-mode beams. This means of measurement of blood flow has been found to have a high correlation ($r = 0.98$) with an electromagnetic flow transducer in *in vivo* experiments, examining flow in the aorta of an adult pig (Eik-Nes *et al.*, 1981).

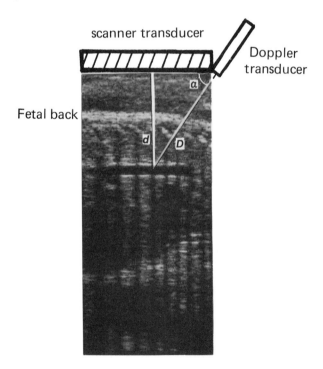

FIG. 5 Diagram to illustrate measurement of fetal blood flow, showing real time image of the fetal aorta in a longitudinal body section. D is distance from Doppler transducer to vessel; d is distance from real-time transducer to vessel; α is fixed angle between real-time and Doppler transducers.

Gill and Kossoff (1979) show a striking consistency in umbilical vein mean flow of 105 ml kg^{-1} min^{-1} in their few measurements at gestations from 25 to 40 weeks. The values of Eik-Nes *et al.* (1980) measured near term based on maximum velocity are 110 ml kg^{-1} min^{-1} for mean umbilical vein flow and 191 ml kg^{-1} min^{-1} for mean flow in the fetal thoracic aorta. Though lower than those reported for the mature fetal lamb (Dawes, 1968), these values appear credible but further studies to observe changes, are needed to substantiate them.

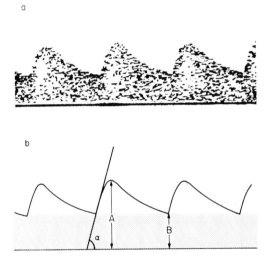

FIG. 6   (a) The sonogram displays frequency on the ordinate plotted against time on the abscissa. (b) The outline of the sonogram yields a blood velocity waveform. The angle α is measured to calculate the acceleration slope which represents the rate of acceleration of blood during cardiac systole. The highest systolic velocity (A) and the end diastolic velocity (B) are measured as height above the baseline. The shaded area represents the velocity of blood continuously perfusing the placenta. (Modified from Stuart, Drumm, Fitzgerald and Duignan, *Br. J. Gynaecol.* **88**, 865, 1981, with permission.)

Doppler ultrasound can also be used to study flow profiles of the circulation by examination of the fetal blood velocity waveform (Fitzgerald and Drumm, 1977; McCallum *et al.*, 1978). Fetal blood velocity signals from the umbilical arteries are obtained by a continuous-wave directional Doppler system after localization of the umbilical cord by B-mode scan.

The shape of the blood velocity waveform obtained after audiofrequency analysis of these signals (Fig. 6) reflects the pulsatile nature of blood flow and contains information about flow dynamics both distal and proximal to the point of measurement. Measurements from the umbilical artery should enable characterization of fetal cardiac and placental vascular dynamics (Stuart *et al.*, 1980, 1981), and may be particularly informative in the future combined with measurements of flow.

3.0 FETAL HEART RATE

3.1 General

Heart rate is the clinical representation of the interval between successive cardiac cycles. It may be recorded traditionally as a pulse count over a period of time (conventionally 1 min) or more accurately as the reciprocal of the measured interval (heart period) between repetitive events in the cardiac cycle. If measured as a pulse count over one minute (e.g. via a fetal stethoscope) heart periods are effectively averaged over that time period and information is lost, for, if measured as individual heart periods, it is apparent that there are small differences in interval length which, plotted as an instantaneous rate over time reveal beat to beat differences. This difference in data acquisition is illustrated in Fig. 7.

The variation in heart period is described as heart rate variation or variability. It may obviously be expressed as variation between adjacent heart periods or between heart periods considerably displaced in time and this has given rise to terms such as short-term and long-term irregularity (de Haan *et al.*, 1971). Thus it is important when talking of heart rate variation to express clearly, either as time or number of beats, the epoch being considered. Beat-to-beat variation should be reserved for just that, i.e. measurement of differences between successive heart periods.

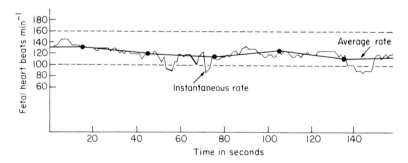

Fig. 7   To illustrate the difference between instantaneous heart rate measurement (——) and the same rate counted and plotted every 30 seconds (——). (Reproduced from Hon (1975), *An Introduction to Fetal Heart Rate Monitoring,* with permission.)

It is now clear that variation in fetal heart rate is as important a physiological variable as the heart rate itself.

3.2 Physical Methods

The essential differences between a pulse count and an instantaneous heart rate recording have been discussed.

Changes in instantaneous heart rate are easily recognized by changes in rhythm on auscultation, but very much more information is available if the heart rate is plotted against time on paper records, enabling both measurement and analysis of patterns of change. There are a number of different trigger signals which may be used to obtain these continuous heart rate recordings and these are summarized in Fig. 3. Though we will consider the role of the fetal ECG in particular it is important to discuss briefly the other signal modes as they produce slightly differing fetal heart rate (FHR) recordings and are used both clinically and in research. All use a repetitive event in the cardiac cycle to produce a trigger so that heart period is measured as the interval, t, between successive triggers. This is measured electronically and a potential which is inversely proportional to t is computed for each successive interval. The series of potentials is then displayed in terms of heart rate against time by plotting it on moving paper. Obviously horizontal (paper speed) and vertical scaling (beats min^{-1} scale) must be considered in assessing patterns, for variations in either alter pattern shape.

3.2.1 System Logic

In practice, even with the direct fetal ECG, noise will occasionally cause false triggering to produce an abnormally short and erroneous interval, while a real trigger can be missed to produce an erroneously long interval. Logic systems are used to prevent these generated potentials from producing wild excursions of the plot but the logic used differs between different monitors both in terms of the acceptable limits and the pen response. Some monitors lift the pen whilst others continue to write a straight line at the last accepted rate. These editing systems give a tidier looking trace but may also edit out the abrupt changes in rate which occur with some fetal cardiac arrhythmias. Simple auscultation and/or the use of an oscilloscope, often incorporated in the monitor, should clarify the situation.

Doppler ultrasound and phonocardiographic signals have two valvular components each cardiac cycle to the signals and a refractory period is incorporated as part of the logic to prevent both components being counted.

Occasionally the FHR exceeds the limits of a particular system and this logic is overwhelmed with sudden doubling (e.g. 70 to 140 beats min^{-1} or halving (e.g. 200 to 100 beats min^{-1}).

The appearance of FHR recordings obtained by Doppler ultrasound or phonocardiography is often improved by using a logic system which calculates an average FHR over two or three beats. Whilst this technique would normally obscure beat-to-beat changes these changes are obscured anyway by the non-discrete nature of these signals (particularly ultrasound).

It is apparent that any research worker or clinician using such systems must be aware of the logic employed in his particular instrument.

3.2.2 *Fetal Electrocardiography*

Apart from the stethoscope the fetal electrocardiograph is the earliest reported form of fetal heart monitoring (Cremer, 1906). In its simplest form the instrumentation for fetal electrocardiography requires the interposition of a high gain, low noise preamplifier between the patient and a standard ECG machine. A continuous fetal heart rate is obtained by using the QRS complex, or some derivative, as the ratemeter trigger. It may be obtained after suitable processing *indirectly* from the fetomaternal ECG using maternal abdominal electrodes or *directly* as a pure fetal ECG by application of an electrode to the fetal presenting part after rupturing the membranes. It has also been obtained directly by abdominal percutaneous puncture of the fetal buttock (Figueroa-Longo et al., 1966). The direct ECG, whilst providing the clearest signal and most precise ratemeter trigger, can only be used during or immediately prior to labour.

(a) *Indirect fetal electrocardiography*. The amplitude of the fetal ECG complex from the abdominal fetomaternal ECG is 5–50 μV and though its presence can be recognized on most patients the signal-to-noise ratio is frequently too poor to enable processing for continuous heart rate monitoring. It has been recorded as early as the eleventh week of gestation but is more easily obtained after the sixteenth week with a maximum R wave amplitude at 21 to 24 weeks. Many studies (e.g. Bolte, 1961; Caughey and Krohn, 1963) document the decrease in signal strength between 26 and 34 weeks gestation, with an improving signal from 37 weeks to term. This variation has been attributed to the electrical insulating properties of vernix and the amplitude of the QRS complex has been correlated with the amount of vernix at birth (Roche and Hon, 1965). These same authors were unable to correlate positional variations in QRS amplitude with the placental site.

The placement of electrodes for the best signal is largely by trial and error but it is noticeable that midline positions avoid much EMG noise. Good electrical coupling of the electrodes to maternal skin is also essential.

Before the fetomaternal ECG can be used to derive an FHR the maternal ECG needs to be eliminated. Methods have included filtering, blanking and subtraction. The maximum power in the ECG occurs at about 40 Hz in the fetus and at 25 Hz in the adult (Smyth, 1953), but there is considerable overlap so whilst filtering can be used to improve the fetal signal-to-noise ratio it cannot eliminate the maternal signal.

In the blanking technique (Offner and Moisand, 1966) a second maternal ECG only signal is obtained from a chest lead and this signal used to switch off the combined signal during the maternal R wave. This also removes coincident fetal complexes which leads to loss of 20–40% of fetal beats. Similarly, techniques based on distinguishing fetal from maternal complexes by comparison of frequency content, amplitude and repetition rates (van Bemmel, 1968) eliminate coincident complexes. The missing data are either interpolated or omitted from display. Most commercial monitors use such blanking systems.

The subtraction technique, first described by Hon and Hess (1957) and developed further by Sureau and Trocellier (1961), has the particular advantage of preserving all the fetal complexes for measurement of heart rate. Again a fetomaternal signal and maternal only signal are required but to achieve satisfactory subtraction the waveform of the latter must resemble the maternal QRS of the combined signal in all respects. A suitable waveform may be found by a search of many abdominal sites using a multi-electrode switchable array (Sureau and Trocellier, 1961) or by careful placement of electrodes (Curran and MacGregor, 1970) then matching the amplitudes of the maternal QRS complexes by varying the signal gain.

Wheeler *et al.* (1978) have developed a system to avoid this often difficult search for a matching maternal waveform by using analogue delay lines to enable subtraction of the last clean maternal QRS complex on the combined signal from a subsequent fetomaternal complex on the same signal. The combined signal is first categorized into pure fetal, pure maternal and fetomaternal complexes (Fig. 8). Pure fetal complexes are identified by synchrony with a signal of pure fetal complexes produced by blanking the original signal with a separate maternal channel. A 'fetal expectation period' is derived from this blanked signal and fetomaternal complexes are presumed to occur on the abdominal signal when a fetal complex is absent but a maternal complex is present within this period. Pure maternal complexes are those which have synchrony with the QRS complex of the separate maternal ECG and are not already identified as

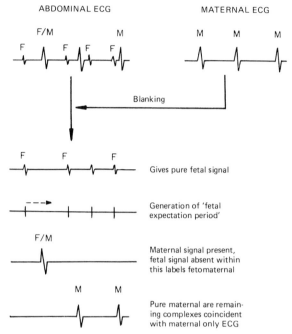

ABDOMINAL ECG   MATERNAL ECG

Blanking

Gives pure fetal signal

Generation of 'fetal
expectation period'

Maternal signal present,
fetal signal absent within
this labels fetomaternal

Pure maternal are remain-
ing complexes coincident
with maternal only ECG

FIG. 8   Illustrating the method of Wheeler *et al.* (1978) for characterizing the fetomaternal ECG prior to the subtraction process.

fetomaternal. After the subtraction (Fig. 9) trigger pulses are generated for the ratemeter from the now continuous fetal ECG signal. This system probably offers the best antenatal measurement of beat-to-beat heart rate changes.

However, apart from research work, the use of such signal processing methods remains limited by the frequently poor initial abdominal ECG recording. Schuler *et al.* (1968) describe an interesting system which offers multichannel cancellation of the maternal ECG but might be more useful for improving the quality and success of recording the fetal abdominal ECG. The object is to simulate the ideal electrode position within the maternal abdomen. A ring of eight electrodes is placed circumferentially about the maternal abdomen at the level of the fundus with a common electrode above the symphysis pubis. The signal from each electrode and the common one is passed to the appropriate point on a resistance mat cut to an approximate cross-section of the pregnant abdomen and the best fetal signal is found on this by a hand-held exploring electrode (Fig. 10). In theory this system should work, and Curran (1975) confirms that it does, but there would need to be compelling reasons for wanting an antepartum ECG to use it!

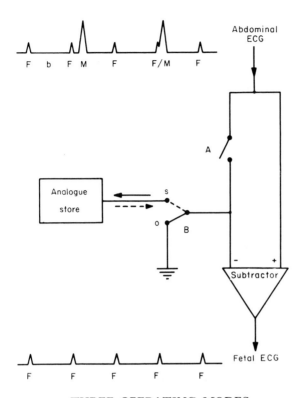

THREE OPERATING MODES

Signal	Switch A	Switch B	Effect
Baseline (b) or fetal (F)	open	o	signals passed without alteration
Pure maternal (M)	closed	s	signal stored and subtracted from itself
Combined fetomaternal (F/M)	open	s	previously stored pure maternal subtracted from combined complex

FIG. 9   The analogue subtraction method of Wheeler, Murrills and Shelley. (Reproduced from *Br. J. Obstet. Gynaecol.* **85**, 12, 1978, with permission.)

(b) *Direct electrocardiography.* The direct application of a clip or spiral electrode to the fetal presenting part after membrane rupture provides the best and easiest signal to process for continuous heart rate assessment and is widely used clinically. The clip or spiral is one electrode of a bipolar pair—the second electrode, situated a few centimetres away on the insula-

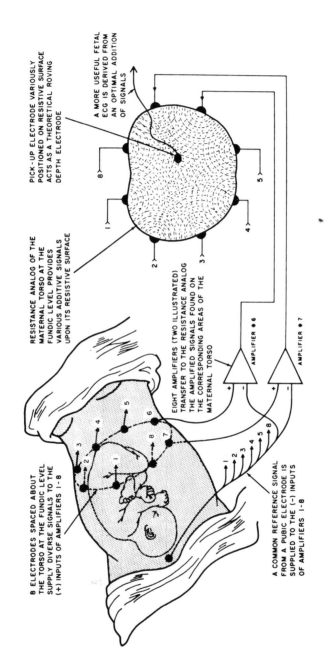

PICK-UP ELECTRODE VARIOUSLY
POSITIONED ON RESISTIVE SURFACE
ACTS AS A THEORETICAL ROVING
DEPTH ELECTRODE

A MORE USEFUL FETAL
ECG IS DERIVED FROM
AN OPTIMAL ADDITION
OF SIGNALS

RESISTANCE ANALOG OF THE
MATERNAL TORSO AT THE
FUNDIC LEVEL PROVIDES
VARIOUS ADDITIVE SIGNALS
UPON ITS RESISTIVE SURFACE

8 ELECTRODES SPACED ABOUT
THE TORSO AT THE FUNDIC LEVEL
SUPPLY DIVERSE SIGNALS TO THE
(+) INPUTS OF AMPLIFIERS 1-8

EIGHT AMPLIFIERS (TWO ILLUSTRATED)
TRANSFER TO THE RESISTANCE ANALOG
THE AMPLIFIED SIGNALS FOUND ON
THE CORRESPONDING AREAS OF THE
MATERNAL TORSO

AMPLIFIER ● 6

AMPLIFIER ● 7

A COMMON REFERENCE SIGNAL
FROM A PUBIC ELECTRODE IS
SUPPLIED TO THE (-) INPUTS
OF AMPLIFIERS 1-8

FIG. 10   Method of selective signal addition of Schuler, Puddicombe and Park. (Reproduced from *Am. J. Obstet. Gynecol.* **101**, 1120, 1968, with permission).

tion of the first, makes contact with the mother via the electrolytic content of the genital tract secretions. An artefact-free record can be obtained throughout most of labour and delivery and it is usually possible to identify all the components of the fetal ECG waveform. Most direct electrodes are now made of stainless steel and though more robust are electrically noisier than the original silver–silver chloride clip electrodes (Strong, 1970) and therefore less suitable for examining the waveform. Narrow band pass filtering with automatic gain control is used for heart rate triggering.

It is clinically important to know that low amplitude maternal signals are usually present on the direct ECG and with an intrauterine fetal death the automatic gain control may amplify these enough to trigger the rate meter.

3.2.3 Phonocardiography

The fetal heart can be detected aurally from about 24 weeks gestation. Phonocardiography (see Vol. 1, Chapter 10) is the detection of the noise of closing of the two sets of heart valves by a microphone strapped to the abdominal wall.

Most of the energy developed by the fetal phonocardiogram lies in the frequency range between 60 and 120 Hz (Shelley, 1967). This unfortunately overlaps with many other noises within the maternal abdomen and the signal is also easily lost or distorted during fetal or maternal movement. There is also the problem of at least two events (1st sound, 2nd sound) each cardiac cycle. The most successful phonocardiographic technique was developed by Hammacher (Hammacher, 1962; Gentner and Hammacher, 1967) using special logic circuits which examined the relationship between the first and second heart sounds to reject artefacts after initial filtering removed extraneous noise. This has formed the basis of the system marketed by the firm of Hewlett-Packard.

Although the earliest method of continuously recording the FHR (Hellman et al., 1958), fetal phonocardiography has been superseded by Doppler ultrasound antenatally and the direct fetal ECG in labour. Its biggest advantage is that it is both physically and electronically safe with a non-invasive transducer and no energy output.

3.2.4 Continuous Wave Ultrasound

The principal ultrasonic method of recording the FHR is by use of the Doppler shift produced when ultrasound is reflected from a moving surface (see Chapter 9). The moving surface alters the frequency of the reflected beam in relation to its velocity and the difference in frequency between transmitted and reflected beams is detected. The small abdominal transducer which contains both a transmitting and receiving crystal generates a continuous ultrasound beam at a frequency of about 2 MHz. When directed towards the heart Doppler frequency shifts may occur from opening and closing of valves, blood flow or movement of the cardiac walls or surrounding structures. The signal output is thus as complex as the phonocardiogram but the signal-to-noise ratio is better as it is less susceptible to fetal movement. It is improved further by selective filtering and variation in beam width using multicrystal arrays. The output signal is not electronically "tidy" and accurate counting is not easy. Beat-to-beat intervals are therefore usually averaged over several beats. Reference has already been made to the possibilities of double or half counting if the actual FHR is outside the algorithm limits of the monitor.

There seem to be no harmful effects of the ultrasound energy used to investigate the fetus at the intensity employed. It has, however, been suggested (David et al., 1975) that ultrasound may stimulate fetal activity though this is disputed (Hertz et al., 1979).

3.2.5 Other Physical Methods

Hukkinen et al. (1976) have described instantaneous FHR monitoring by electromagnetocardiography by detecting the equivalent of fetal QRS complexes by a superconducting magnetometer above the maternal abdomen. It is, however, essential to have an environment of low magnetic noise—a sauna-like cabin in a forest I understand! True beat-to-beat measurements are apparently possible and, unlike the fetal abdominal ECG, there is no signal loss later in pregnancy.

Curran (1975) reports that attempts have been made to obtain indirect fetal cardiographs by ballistocardiography and impedance plethysmography.

3.3 Heart Rate and Heart Rate Variation Analysis

3.3.1 Observation

Subjective visual analysis of FHR recordings have proved useful in assessing fetal health both intrapartum (Hon, 1963; Caldeyro-Barcia et al., 1966; Beard et al., 1971) and antepartum (Visser and Huisjes, 1977; Flynn and Kelly, 1977), but in widespread clinical practice the benefits of continuous heart rate monitoring by this means are questionable (Haverkamp et al., 1979; Beard, 1977; Brown et al., 1982). Gross normality and gross abnormality are usually obvious but the complexity of control of

TABLE I

Recommended classification of heart rate by observation

---

1. Baseline FHR

Baseline FHR is described in beats per minute (beats min^{-1}) and in terms of its variability. It is the rate at which the heart is set for the majority of a 10-min period.

(a) Rate:

Marked bradycardia	99 beats min^{-1} or less
Mild bradycardia	100–119 beats min^{-1}
Normal	120–160 beats min^{-1}
Mild tachycardia	161–180 beats min^{-1}
Marked tachycardia	181 beats min^{-1} or greater

A change in rate should only be described as a new baseline if present for 10 min or more.

(b) Variability:

Short term—Otherwise known as beat-to-beat change, the short-term variability is due to the varying duration between the same cardiac events, e.g. the R–R interval of the ECG.

Long term—Changes in baseline rate which have a frequency of 2–6 cycles min^{-1} and a normal amplitude of change of 6–10 beats min^{-1}.

2. Periodic changes

(a) Accelerations:

These are transitory increases in baseline rate, lasting less than 10 min.

(b) Decelerations:

I. Uniform patterns:	i. EARLY DECELERATIONS The deceleration has an onset, fall and recovery which reflects the onset, peak and fall of a contraction.
	ii. LATE DECELERATIONS The deceleration has an onset, fall and recovery, all of which are delayed in relation to the onset, peak and fall of a contraction.
II. Variable patterns:	Variable in onset and tend to be non-repetitive.
III. Mixed patterns:	Such patterns are difficult to define, containing mixtures of the above.

From *American College of Obstetricians and Gynecologists Technical Bulletin* (1975).

heart rate make judgements of other changes largely a matter of opinion based on clinical experience.

The accepted classification of heart rate by observation is summarized in Table I and is based on the original work of Hon (1963) and Caldeyro-Barcia *et al.* (1966). It has been suggested that the characterization of the type of deceleration is not as helpful as a consideration of the lost beats or dip area of any decelerations. It has been shown that dip area, in the hour prior to delivery, correlates with umbilical cord pHs at delivery (Tipton and Shelley, 1971), but this was not so high as to be predictive and this approach has not gained acceptance.

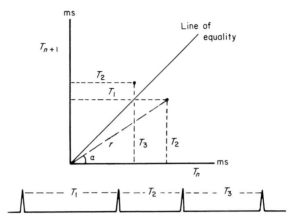

FIG. 11   Illustrating plotting of successive R-R intervals (T_1, T_2, T_3 . . .) on the abscissa and ordinate and definition of the argument (α) and the modulus (r) as proposed by de Haan *et al.* (1971). The line of equality (fixed heart rate) is shown.

3.3.2 *Computer Analysis*

Computer analysis is the only practical means of heart rate quantification. The first step (Hon and Yeh, 1969) was to build up histograms of R–R intervals and beat-to-beat differences over periods of time so that simple statistics (mean, standard deviation and coefficient of variation) could be calculated. Although this summarizes the information without data loss the major problem is the loss of time relationship between events. Intervals, gradually shortening or lengthening, give the same interval distribution as intervals cyclically varying around a mean interval though the pattern over time is quite different. To overcome this a two-dimensional distribution of heart intervals was proposed (de Haan *et al.*, 1971) in which consecutive heart intervals were plotted against each other, the first along the abscissa and the second along the ordinate (i.e. T_1/T_2; T_2/T_3, etc) (Fig. 11). In this form beat-to-beat variation is expressed by the scatter of the points to each side of the line of equality (changes in argument) whilst long-term variation produces scatter along the line (changes in modulus). Quantification of these changes is possible by constructing histograms of the argument and modulus.

Dalton *et al.* (1977), in a thorough study of heart rate and heart rate variation in the fetal lamb, used the simpler concept of quantification by

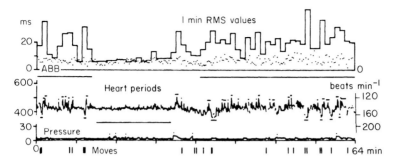

FIG. 12   To show method of data presentation of Dawes *et al.* (Reproduced from *Am. J. Obstet. Gynecol.* **141**, 43, 1981, with permission.) From above downward the histogram of 1-min RMS values, 0·25 min mean beat-to-beat variation (ABB, as dots), original 1/16 min mean heart periods (values > 40 ms from the filtered mode identified by a short line above for decelerations and below for accelerations), pressure from an abdominal transducer, time in 8-min intervals, and fetal movements identified by nurse and patient. The horizontal lines above and below the heart period trace show high and low variation episodes, respectively. Gaps indicate missing data points.

mean heart rate, mean interval differences and the standard deviation of heart periods over specified time epochs (from 15 s to 12 h). Wheeler *et al.* (1979) and Greene *et al.* (1980) have continued this approach in investigations of the human fetus from the ECG.

More recently Dawes *et al.* (1981, 1982a, b) have concentrated on the problem of data reduction and quantification of the large number of antenatal heart rate traces made by ultrasound, to enable objective analysis and display (Fig. 12). Ultrasound pulse intervals (heart periods) were averaged over 1/16 min after validation by an error rejection algorithm. Where no valid average was available a value was interpolated linearly to obtain the smoothed frequency distribution of mean pulse intervals for each record. From this the modal value was used to fit a baseline to the series of validated mean pulse intervals with a digital filter centred around $0·1 \text{ min}^{-1}$. Brief or more sustained accelerations or decelerations in FHR are identified as deviations from this baseline. Episodes of low or high heart-rate variation (discussed later) are also identified. Fetal movements are recorded by hand-held buttons.

A linear relationship exists between the SD of R–R intervals and the logarithm of the observational epoch (Dalton *et al.*, 1977) which is explained by the presence of multiple slow rhythms in heart rate. Indeed the heart rate trace we see can be considered as the summation of many recurrent cycles of differing wavelength superimposed on one another. An analysis of the component rhythms would be very useful but unfor-

tunately biological signals, unlike physical ones, are always a little inconsistent and conventional time series analysis has not so far proved helpful.
Nevertheless this remains an interesting and important concept for if
fundamental rhythms can be identified their presence or absence may be of
considerable clinical importance.

3.4 Observational Findings

3.4.1 *Normal patterns and responses*

Figure 13 illustrates a normal continuous FHR recording obtained at
term. There are obvious periodic changes in both heart rate and heart rate
variation. The striking "rhythm" on such long recordings is episodes of
increased variation and increased heart rate alternating with episodes of
decreased variation and decreased heart rate; often referred to as rest–
activity cycles (Wheeler and Murrills, 1978; Timor-Tritsch *et al.*, 1978;
Junge, 1979). But real-time scanning has shown that though most fetal
activity (body movement and fetal breathing) occurs during the episodes
of increased variation, movements also occur during the episodes of
decreased variation (Greene *et al.*, 1980; Visser *et al.*, 1982) and these
cycles continue through unsedated labour when most fetal activity has
ceased (Greene *et al.*, 1980). The term reactive–non-reactive is more
appropriate, particularly since the modulation of heart rate in association
with fetal movement is different in these episodes of high and low variation
(Fig. 13).

   Cycles in fetal movement have been observed from 24 weeks' gestation
(Sterman and Hoppenbrouwers, 1971) but significant clustering of fetal
movement within episodes of high heart rate variation only occurs from
28 weeks onwards (Visser *et al.*, 1981; Dawes *et al.*, 1982b). Movements
are only significantly associated with bursts of acceleration from 34 weeks
(Dawes *et al.*, 1982b) which then produce a significant difference in mean
heart rate between episodes of high and low variation (Fig. 14). From
this time the length of low variation episodes increases, reaching up to
40 min by term. Thus, the relationship of changes in heart rate and
variation with fetal movement seem to reflect a pattern of neurological
development and integration. The timing of this is similar to that of the
ontogeny of sleep states in newborn preterm infants (Dreyfus-Brisac,
1970; Parmelee and Stern, 1972). Prior to 36 weeks there is no synchronization of cycles of the EEG, rapid eye movements, respiration and tonic
activity (Prechtl *et al.*, 1979) and it is only after this age that periods of
quiet and rapid eye movement (REM) sleep can be consistently identified.
It has been postulated on the basis of visual analysis, cycle length and
movement incidence that episodes of low heart rate variation represent

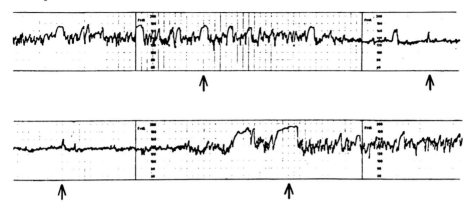

FIG. 13   A continuous fetal heart rate recording (each strip represents a consecutive 40 min) to illustrate periodic changes in both rate and variation. The transient accelerations are all associated with fetal movements and the arrows on each strip are only to show the different magnitude of acceleration during episodes of high variation compared with low.

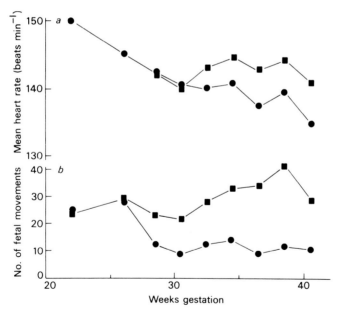

FIG. 14   Mean values for heart rate (a: beats min⁻¹) and the number of fetal movements (b: standardized to 32 min for each type of episode) in low (●) or high (■) heart rate variation. Note the change for fetal movements from 28 weeks and for heart rate from 34 weeks gestation. (Reproduced from Dawes, Houghton, Redman, Visser, *Br. J. Obstet. Gynaecol.* **89**, 276, 1982, with permission.)

quiet sleep in the fetus (Timor-Tritsch *et al.*, 1978; Junge, 1979) but there is as yet no direct evidence for this.

As well as accelerations, short decelerations have been noted in many normal FHR recordings (Ruttgers *et al.*, 1972; Wheeler and Murrills, 1978). Towards the end of gestation these are usually associated with an immediately preceding acceleration in episodes of high heart period variation (Dawes *et al.*, 1982b). Whilst there is a large increase in the number and size of accelerations which dominate the FHR trace towards the end of gestation there is little change in the number or area of decelerations with increasing gestation. Decelerations which are either excessively large, or displaced from periods of high variation or occur following Braxton Hicks contractions cannot be considered normal.

There is a well-known fall in heart rate with advancing gestation and an increase in heart period variation. This increase is mainly due to the increase in size and number of accelerations and small decelerations during the reactive episodes. The variation in episodes of low heart period variation does not increase after 30 weeks and perhaps even falls near term (Visser *et al.*, 1981). As in the fetal lamb (Dalton *et al.*, 1977), there are quite large diurnal changes in both heart rate and heart rate variation in the human fetus and these rhythms do not coincide (Visser *et al.*, 1982). It is therefore not surprising that simple, equal weighting score systems (Pearson and Weaver, 1978) do not detect clinical abnormality with any sensitivity (Brown *et al.*, 1982), for they take no account of age or periodicity.

3.4.2. *Abnormal FHR Changes*

(a) *Baseline heart rate and variation.* We have discussed the normal pattern of FHR and its variation both antepartum and intrapartum and this will correctly predict a good outcome in the fetus in more than 99% of cases (Schifrin and Dame, 1972). The absence of accelerations, though unusual antepartum, is not associated with an abnormal outcome provided that cycles of increased and decreased heart rate variation remain. Loss of episodes of high variation, which is synonymous with a trace with continuously little variation is suspicious (in the absence of sedation), particularly if coupled with decelerations both antepartum (Visser and Huisjes, 1977; Flynn and Kelly, 1977; Visser *et al.*, 1980) and intrapartum (Beard *et al.*, 1971).

The presumed progression of change to abnormal over a period of time antenatally is diagrammatically shown in Fig. 15. This progression frequently seems to occur in association with a reported decrease in fetal movements (Sadovsky and Yaffe, 1973; Pearson and Weaver, 1976).

Fetal animal studies and human observations suggest that the initial basal heart rate response to hypoxaemia is a tachycardia as a result of catecholamine release, with a bradycardia occurring only terminally, unless

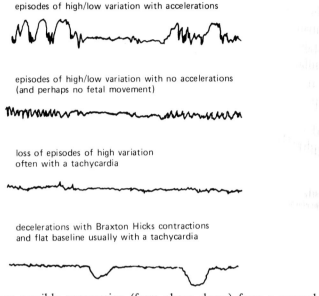

episodes of high/low variation with accelerations

episodes of high/low variation with no accelerations
(and perhaps no fetal movement)

loss of episodes of high variation
often with a tachycardia

decelerations with Braxton Hicks contractions
and flat baseline usually with a tachycardia

FIG. 15   To show possible progression (from above down) from a normal to abnormal fetal heart rate trace antepartum.

the insult is particularly severe. Heart rate variation increases with acute hypoxaemia in the fetal lamb (Dalton *et al.*, 1977) and probably also in the human fetus (Miller and Paul, 1978) but, as already discussed, a more prolonged insult reduces the variation to produce a flat trace possibly by suppression of the episodes of high variation.

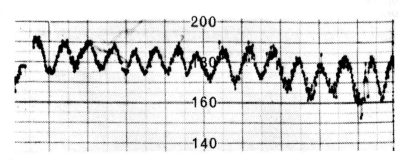

FIG. 16   A sinusoidal fetal heart rate pattern.

From time to time an obvious sinusoidal pattern is seen in the heart rate (Fig. 16) with a frequency of 1–5 cycles min^{-1}. This may be short-lived and of low amplitude (5–10 beats min^{-1}) or persistent with sometimes twice that amplitude.

The pattern has been reported in a number of different clinical situations but it is most frequently associated in the literature with fetal anaemia either resulting from an acute bleed (Yambao *et al.*, 1982; Modanlou *et al.*, 1977) or rhesus isoimmunization (Verma *et al.*, 1980). The antecedent history will often help in deciding management—though transient patterns may often be associated with a normal outcome the persistent patterns are best regarded as ominous. It is of interest that similar looking waveforms in heart rate and blood pressure (Meyer waves) were described by the old physiologists in adult animal exsanguination experiments.

(b) *Decelerations* (see Fig. 17). Late decelerations occur with direct asphyxial depression of the myocardium and are either associated with

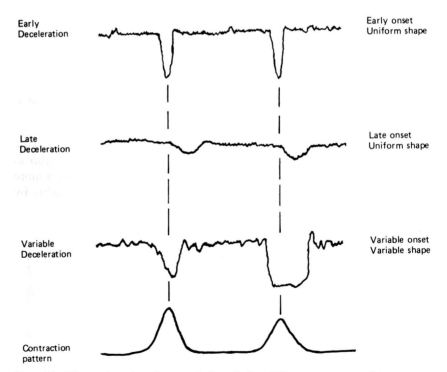

FIG. 17   Illustrating the characteristics of the different patterns of heart rate decelerations in relation to uterine contractions.

pre-existing hypoxaemia or prolonged contractions which reduce oxygenation by restricting placental intervillous perfusion (Myers, 1972). All the features of an abnormal trace are illustrated by Fig. 18 obtained before the onset of labour.

Once labour is established, often with membranes ruptured, a variety of other heart rate changes occur as a result of more frequent contractions, umbilical cord compression, maternal hypotension and pressure on the head as it descends in the pelvis. Between the extremes of normality of Fig. 13 and definite abnormality of Fig. 18 these changes are much less readily interpreted as normal or abnormal. Indeed without some other arbiter of normality, such as fetal scalp pH measurement, they give rise to much unnecessary intervention (Haverkamp *et al.*, 1979).

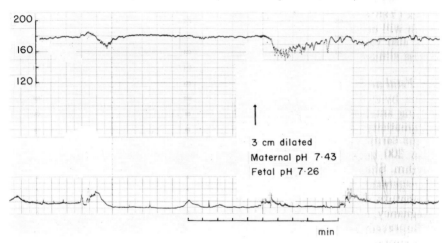

FIG. 18 An abnormal fetal heart rate trace obtained shortly after induction of labour. There is a tachycardia with late decelerations. Heart rate variation is generally decreased but increases during a late deceleration. The discrepancy between fetal and maternal PH (normally about 0·1 pH unit) increased over the next hour and the patient was delivered by caesarean section.

Early decelerations have a symmetrical waveform with the onset and recovery coinciding with the beginning and end of the contraction. They characteristically occur when the cervix is about 6–7 cm dilated or a high head descends into the pelvis and are probably more common when the membranes are ruptured (Steer *et al.*, 1976). They are vagally mediated, being completely abolished by atropine, though the exact pathway is uncertain. It may be by relative ischaemia of the medullary centres as a result of decreased blood flow stimulating chemoreceptors (Paul *et al.*, 1964), by raised intracranial pressure directly activating the cardioinhibitory centre of the brainstem (Moscary *et al.*, 1970) or by a variant of the oculocardiac reflex. They are considered benign unless combined with some other heart rate abnormality.

Variable decelerations are both variable in shape, amplitude and time relationship to the uterine contraction. Though traditionally thought of as being due to umbilical cord compression (Bauer, 1937; Hon, 1963) they also occur with maternal hypotension associated with the supine position and epidural top-ups. The initial onset is probably related to a rise in fetal blood pressure activating the baroreceptors to give a vagal reflex bradycardia which is sometimes sufficient to cause a conduction block (Yeh *et al.*, 1975). If sustained, asphyxial depression of the myocardium will occur to prolong the dip. The relationship to asphyxia is therefore also variable and these are the most difficult heart rate patterns to assess clinically.

(c) *Fetal arrhythmias.* From time to time a cardiac arrhythmia is detected either by auscultation or more frequently by continuous heart rate monitoring antepartum or during labour. Arrhythmias result from abnormal automaticity, abnormal conduction or a combination of both in some part of the cardiac conduction system. They present with a heart rate of more than 200 beats min^{-1}, less than 100 beats min^{-1} or an irregularity of rhythm. Shenker (1979) has recently reviewed the literature, diagnosis and management. An irregularity of rhythm is usully due to atrial or ventricular premature systoles (ectopic foci) which are common, especially in early pregnancy, and are generally benign.

Supraventricular tachyarrhythmias, such as atrial flutter and fibrillation and supraventricular tachycardia, have a 5–10% incidence of congenital heart disease and are prone to fetal and neonatal cardiac failure with ascites and hydrops (Silber and Durnin, 1969; Van der Horst, 1970).

Persistent bradycardias are usually due to congenital heart block with the rate commonly between 50–70 beats min^{-1}. Congenital heart block has an association with maternal auto-immune connective tissue disease (McCue *et al.*, 1977; Berube *et al.*, 1978), and there is a 40% incidence of congenital heart disease in the fetus (Michaelsson and Engle, 1972). Again heart failure is not uncommon (Alterburger *et al.*, 1977).

As already noted, these abnormal rates may confuse the logic of heart rate monitors. The definitive diagnosis of an arrhythmia is from the electrocardiogram. The abdominal fetomaternal ECG is usually unhelpful as the P waves cannot be identified, but a direct recording in labour will allow identification, though it is unlikely to alter management. Observation of the relative rates and synchrony of atrial and ventricular contraction by ultrasound may also determine the nature of the arrhythmia.

More important to the management of arrhythmias is the identification of associated congenital heart disease and cardiac failure antenatally by the use of M-mode and two-dimensional echocardiography (Kleinman *et al.*, 1980; Allan *et al.*, 1981; Crawford, 1982). This will enable appropriate decisions on the timing, mode and place of delivery to be made and may also monitor intrauterine treatment of congestive failure by digitalis (Crawford, 1982) until maturity is reached. In labour, serial fetal pH values should be measured to exclude asphyxia as fetuses with supraventricular tachyarrhythmias or heart block may not show the usual heart rate responses with hypoxia.

4.0 ECG WAVEFORM

4.1 General

Like the adult ECG the fetal ECG consists of P, QRS and T waves separated by the PR and ST intervals. These represent the summation of electrical events within the heart as seen from the body surface. The P wave occurs with atrial contraction initiated by an impulse from the sino-atrial node, the PR interval is an isoelectric interval the length of which represents the conduction time across the atrium from the sino-atrial to the atrioventricular node. Depolarization with contraction of the ventricles then occurs promptly to produce the QRS complex and the ST waveform represents repolarization of the ventricular myocardium to return it to its iso-electric state.

Since Cremer's demonstration of the fetal ECG in 1906 workers have strived, so far without success, to establish a diagnostic place for changes in the fetal ECG waveform. However, with the clinical realization that continuous heart rate monitoring is an insensitive index of asphyxia (Beard and Rivers, 1979) interest is once again being focused on the ECG waveform.

Many combinations of ECG leads have been used: rectal, vaginal, cervical, abdominal, intrauterine and since the 1960s fetal scalp, but the limitations of methods or equipment have often not been appreciated. Thus Southern (1957), Larks and Anderson (1962) and Kendall *et al.* (1964) claimed to observe various alterations in the ECG waveform from abdominal electrodes, in association with fetal distress. Examining their published illustrations of fetomaternal traces it is difficult to see how they made measurements of the fetal waveform in background noise. Apparent alterations in the P wave, QRS complex and T wave may occur by superimposition of some part of the maternal complex—a point well made by

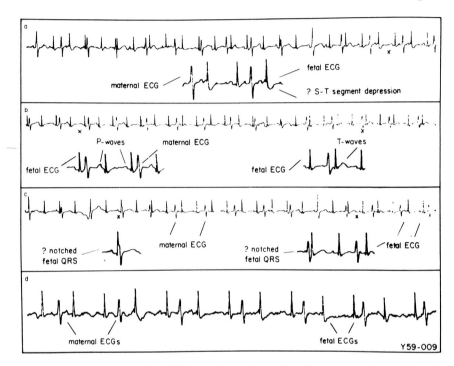

FIG. 19   (a) Apparent fetal ST waveform depression due to superimposition of maternal T wave. (b) Prominent fetal P waves due to superimposition of maternal T wave. (c) Bizarre "fetal" QRS complexes due to coincident maternal and fetal QRS complexes. (d) Fetal P and T waves cannot be identified with certainty because of the masking effect of baseline noise. (Reproduced from Hon and Lee, *Am. J. Obstet. Gynecol.* **87**, 804, 1963, with permission.)

Hon (see Fig. 19). The inaccuracy of their normal values for waveform time intervals is confirmed by later measurements of complexes obtained from direct electrodes (Figueroa-Longo *et al.*, 1966; Symonds, 1971). Unless special averaging techniques are employed (Polvani *et al.*, 1971) the fetomaternal ECG can only give information on rate and occasionally disorders of rhythm (e.g. Smyth, 1953). Though the signal-to-noise ratio is very much better on the recordings from a scalp electrode in labour and there is a constant relationship of the electrode to the fetus, there remains much low frequency noise. This may produce apparent changes in the slower moving portions of the signal in a similar way to the maternal ECG of fetomaternal recordings.

4.2 Objective Measurements

There is large variation between individuals and even by the same individual at a later date in the interpretation of pathological from normal ECGs in the adult (Davis, 1958; Acheson, 1960). Subjective interpretation of the fetal ECG waveform is more difficult because of the higher repetition rate (2–4 complexes s^{-1}) and frequent changes in signal gain and heart rate (Greene and Wickham, submitted). The essential objective measurements of time intervals and amplitudes of various parts of the waveform may need correction for changes in heart rate and signal gain respectively.

Meaningful measurement can only be made on meaningful data and one of the most difficult problems is to improve the signal-to-noise ratio of the fetal ECG without distorting it.

4.3 Signal Processing

4.3.1 *Filtering*

Apart from careful electrode design the simplest technique for noise reduction is analogue filtering and this has been widely used, particularly for eliminating baseline drift. But the frequency components of the ECG waveform are from 0·05 to 80 Hz and high-pass filtering will distort biological changes in the slowest moving component, the ST waveform. Though the effects of filters are well known, and Smyth and Farrow (1958) and Curran (1975) have drawn attention to these in the context of fetal ECG analysis, many authors have neglected this in the past, either failing to record filter characteristics or assuming there were no changes in the fetal ECG waveform whilst using filters which precluded examination of all of it.

Even the classic paper on the dying fetus (Hon and Lee, 1963a) failed to quote the frequency response of the recording system yet concluded that heart rate changes occurred long before late and inconsistent changes in the ECG waveform. The baseline stability of all the illustrated signals (see Fig. 19) is striking, however, and suggests a low frequency cut-off filter was in use.

Recently Marvell *et al.* (1980) have described a system for initially recording a signal between 5 and 250 Hz bandwidth to reduce noise, but later boosting the low frequency components by a software equalization routine. The equalizer was combined with a two-pole high pass digital filter designed by Ackroyd (1973) and optimized to cause no distortion to simulated ECGs of varying waveform (Marvell, 1979). The equalizer restored frequency components above 0·16 Hz and the high pass filter provided additional attenuation of low frequency noise below this.

4.3.2. *Averaging*

The use of signal averaging to improve the signal-to-noise of the fetal ECG was first described by Hon and Lee (1963b) and has subsequently been used by others (Rhyne, 1969; Curran and MacGregor, 1970; Polvani *et al.*, 1971) to extract the fetal ECG from the abdominal fetomaternal ECG. It has been particularly useful for examining ST waveform changes in chronic fetal lamb experiments (Greene *et al.*, 1982) and in human labour (Pardi *et al.*, 1974; Greene and Wickham, submitted). The process can now be performed with considerable sophistication by small, low cost microprocessors (Wickham, 1982).

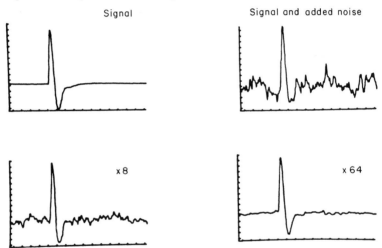

FIG. 20   Averaged plots of the fetal ECG (with number of complexes in each indicated) to show progressive improvement in the signal-to-noise with an averaging technique. Most of this improvement occurs by averaging less than 32 complexes.

Signal averaging recovers a wanted repetitive signal from a background of unwanted noise (Fig. 20) by exploiting the statistical properties of random noise. A trigger pulse is required that has a fixed time relationship to the desired signal and this is used to line up by superimposition in an array each incoming signal. At each pass any component of the input signal which is synchronous with the trigger pulse will add its value into the same locations, whereas components with no relationship to the trigger will add into different locations. So the time-locked signals will add linearly with the number of samples (n) but the background noise will add only as the square root of n. Thus 64 passes should improve the signal-to-noise ratio by a factor of

$$n/\sqrt{n} = \sqrt{n} = 8.$$

This theoretical improvement in signal-to-noise ratio may not be wholly achieved in practice for there are a number of ways in which averaging the fetal ECG can misrepresent biological data (Greene and Wickham, submitted). Transient changes in waveform which may be biologically significant but are not representative of the group are treated as noise and obscured. However, large gain, principally artefactual transients and noise spikes often operate the trigger and may dramatically distort the accumulating average unless some transient rejection routine is used.

The principle of averaging depends on time coherence and time inconsistencies (time jitter) between the trigger pulse (the fetal R wave) and the remainder of the signal will smooth out significant signal. The resultant averaged waveform will be of longer duration and shorter amplitude than if these time inconsistencies did not exist. Noise on the raw R wave signal can vary the actual time of triggering to introduce a "trigger jitter". This may be overcome by producing two parallel signals and filtering one to obtain a clean R wave trigger to average the other raw signal.

A "biological jitter" of the T wave will occur if there are large variations in heart rate during averaging because the ST interval varies with heart rate (Symonds, 1971).

The amount of smoothing which occurs depends on the number of passes as well as heart rate variation. For this reason unnecessary noise needs to be kept to a minimum, to reduce the number of required passes, by good electrode and amplifier design.

4.4 Investigative Findings

4.4.1 *Time Constants*

The normal values at term for the duration of the P wave (50–52 ms), the PR interval (103–109 ms) and the QRS width (50–65 ms excluding Symonds, 1971) are available from manual measurements of the reproduced waveform (Figueroa-Longo *et al.*, 1966; Symonds, 1971; Lee and Blackwell, 1974; Hioki, 1975) and by computer measurement (Polvani *et al.*, 1971; Marvell *et al.*, 1980). These are in reasonable agreement with each other and with measurement of the first day neonatal ECG (Ziegler, 1959). Polvani *et al.* (1971) using averaged fetal ECGs from the maternal abdomen, have shown a progressive increase in these measurements as pregnancy advances. This is supported by comparison of the measurements made directly from the fetus at 10–23 weeks (Stern *et al.*, 1961; Gennser *et al.*, 1968) with those at term.

Measurements of the QT interval are not so consistent—probably because of difficulties in identifying a specific point on the T wave (Polvani et al., 1971; Marvell et al., 1980), differences in the points of measurement (e.g. Marvell et al.—T peak, Figueroa-Longo et al. and Symonds—end of T, and Hioki—unstated), and variation in interval length with heart rate (Symonds, 1971).

The significance of alterations with asphyxia is far less certain. Hioki (1975) and Pardi et al. (1974) have described changes in the P wave and PR interval with heart rate changes suggestive of fetal distress but Marvell et al. (1980) found that changes in P wave shape (and QRS notching) were not uncommon in their study of an unquestionably normal group of 37 fetuses monitored through labour. They found a significant diminution in the P wave and shortening of the PR interval with decreased heart rate as labour progressed. Pardi et al. (1974) suggests increased vagal tone is responsible for some of these changes, shifting pacemaker activity from the SA node to an intermediate position between the SA and AV nodes (biphasic P, shortened PR) or to the AV node itself (absent P wave). Only Symonds (1971) has examined the time constants in relation to maternal and fetal acid–base status. The only significant finding was a prolongation of electrical systole (QT) corrected for heart rate, with an umbilical vein pH of less than 7·20; but corrected QT times were a poor predictor of Apgar score compared with scalp pH. His group (Marvell et al., 1980) now find QT measurement impractical throughout normal labour because of the often poor definition of the T wave.

4.4.2 QRS Complex

As stated earlier the QRS complex represents the electrical events occurring with depolarization of the ventricular muscle mass and this occurs at virtually the same time in both ventricles. In theory, asynchrony between them could produce notching and widening of the complex.

Brambati and Pardi (1980) have shown a progressive increase in the duration of the QRS complex with increasing gestation. Their assumption that this represents increasing ventricular size seems justified by their findings that the QRS interval length was less than normal for gestation in most infants below the fifth centile by weight and more than normal in diabetic large-for-dates babies.

Using fetomaternal ECGs, Larks and Anderson (1962) have suggested that widened or notched QRS complexes occurred with fetal asphyxia but Lee and Hon (1965) found no such evidence on reviewing their scalp electrode recordings, and Marvell et al. (1980) have found QRS notching from time to time in normal labours. In the fetal lamb, we, too, have found intermittent QRS notching unrelated to hypoxaemia. However, two or three fetuses with QRS notching throughout labour in Lee and Hon's (1965) paper probably had congenital heart lesions and in all three with intermittent changes there were proven episodes of cord compression. Brambati and Pardi (1981) have found a definite widening of the QRS in infants dying *in utero* of cardiac failure with rhesus hydrops. And in live-born rhesus affected infants the QRS was found to be in the normal range if the haemoglobin was more than 10 g 100 ml^{-1}, but prolonged if it was less. These observations suggest that ventricular dilatation as a result of cardiac failure can cause QRS widening, whilst intermittent notching may reflect sudden or disparate changes in ventricular dilatation or work load due to profound changes in haemodynamics with cord compression. The effect of QRS widening on systolic time interval measurement is discussed in Sections 5.1 and 5.3.1.

4.4.3 *ST Waveform*

Much of the renewed interest in the ECG waveform has been concerned with the ST waveform. This represents repolarization of the cardiac ventricles. The shape of the waveform is dependent on the sequence of ventricular depolarization and on metabolic events about the myocardial cell influencing the Na^+–K^+ pump (Noble and Cohen, 1978). Vigorous anaerobic work in healthy human adults (Kahn and Simonson, 1957) and ligation of a coronary artery in dogs (Karlsson *et al.*, 1973) produce elevation of the ST segment and T wave. Similar changes have been noted casually in several fetal monkey studies (e.g. Myers, 1972; Morishima *et al.*, 1975) but were considered too variable for practical use. In acute experiments on the exteriorized fetal lamb, however, hypoxia with acidaemia was found to cause a progressive increase in the amplitude of the ST segment and T wave (Pardi *et al.*, 1971; Rosén and Kjellmer, 1975). This finding has been confirmed (Fig. 21) in chronically instrumented fetal lambs (Greene *et al.*, 1982) which have also provided some explanation for the variable nature of this change. The amplitude of the T wave elevation (expressed as the relative amplitude of the T wave to QRS complex), had previously been correlated with myocardial anaerobic metabolism with glycogen utilization (Rosén and Isaksson, 1976; Hökegård *et al.*, 1981) in the exteriorized fetus, and the few chronic fetuses which failed to show a change in the ST waveform during hypoxaemia, also failed to show a rise in plasma lactate. This suggests that they were able to maintain oxygen supply to the myocardium so that metabolism remained aerobic (Fisher *et al.*, 1982) despite peripheral hypoxaemia. Overall there was a strong correlation of the T/QRS ratio to rate of rise of plasma lactate. Thus an

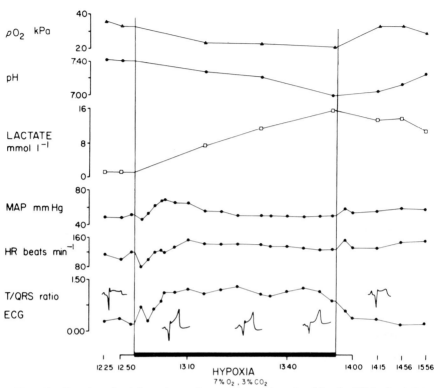

FIG. 21   Results of a 1-hour hypoxia experiment in a fetal lamb. With the fall in
PO_2 there was an increase in the ST waveform expressed by the T/QRS ratio with a
rise in plasma lactate and a progressive fall in pH. Typical ECG complexes are
illustrated at 20-min intervals. MAP, mean arterial pressure; HR, heart rate. (From
Greene, Davis, Lilja and Rosen, 1982.)

elevated ST waveform, which can be expressed quantitatively as a T/QRS
ratio, seems to mark the switch to anaerobic metabolism during hypoxaemia. Further this change in the exteriorized fetus precedes evidence of
failing cardiovascular function and at least initially is β adrenoreceptor
mediated (Rosén et al., 1976; Hökegård et al. 1981).

    Direct ECG studies (with appropriate filter characteristics) of the
human fetus in labour (Polvani et al., 1971; Symonds, 1971; Pardi et al.
1974; Hioki, 1975; Marvell et al., 1980; Lilja et al., submitted) find the
T wave to be normally iso-electric or with a positive deflection no larger
than the P wave. Both Symonds (1971) and Hioki (1975) consider T
wave flattening or inversion to be associated with fetal distress but those
studies specifically examining the ST waveform (Pardi et al., 1974;
Lilja et al., submitted) find ST waveform elevation occurs, as it does in
the fetal lamb. Both the latter studies found a strong but not invariable

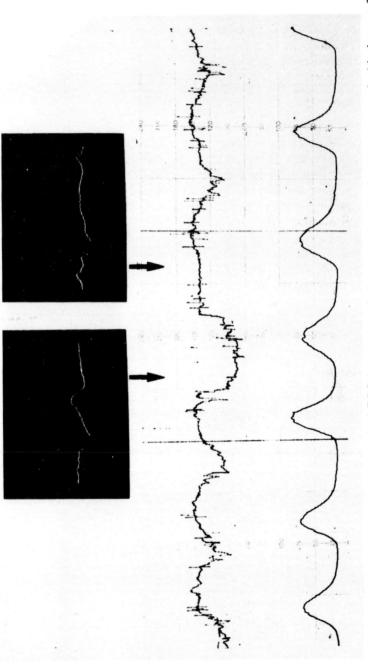

FIG. 22 ST waveform changes in a human fetal ECG intrapartum during severe late decelerations compared with the waveform between decelerations. (Reproduced from Pardi, Tucci, Uderzo, Zanini, *Am. J. Obstet. Gynecol.* **118**, 243, 1974, with permission.)

association with large variable or late decelerations on continuous heart rate traces (Fig. 22). In that of Lilja and colleagues there was a correlation of the T/QRS ratio to umbilical cord lactate values though there was no case with overt birth asphyxia.

4.5 Future

The ECG waveform has the particular advantage that it is obtained from the same signal source in labour as heart rate and heart rate variation, but provides new data. Whereas heart rate and its variation are a complex interplay of neural, humoral and cardiac mechanisms, analysis of the ECG waveform provides direct information on the myocardium and conducting system and may help to interpret heart rate changes more appropriately. There are obvious difficulties in studying and documenting (Sykes *et al.*, 1982) intrapartum asphyxia in man, but there is now a need for a thorough quantitative evaluation of the ECG with other indices of asphyxia to define its role. If it proves to be useful then it is technically feasible to examine it antenatally also.

5.0 SYSTOLIC TIME INTERVALS

5.1 General

The heart's efficiency as a pump is related to the force of contraction of the myocardium. The measurement of systolic time intervals has been pursued as a non-invasive means of assessing this in the adult since the observations of Wiggers (1921) and Katz and Feil (1922). Lewis *et al.* (1977) provide a useful critical review of these measurements in the adult. Ventricular systole begins with depolarization initiating contraction of the ventricular muscle and this is recognized by the onset of the Q wave of the ECG. As the ventricles contract the pressure within them exceeds that in the atria and the atrioventricular valves close. Until the pressure in the ventricles exceeds that in the aorta and pulmonary artery the aortic and pulmonary valves remain closed. So for a short time there is increasing ventricular wall tension without change in ventricular volume (the isovolumic contraction phase). One of the better physiological, but invasive, measurements of ventricular performance is the maximum rate of rise of left ventricular pressure (dP/dt max) (Reeves *et al.*, 1960) which ordinarily occurs during the isovolumic contraction phase, just before aortic valve opening. Metzger *et al.* (1970) have shown an inverse relationship between isovolumic contraction time and maximal dP/dt in the healthy animal and

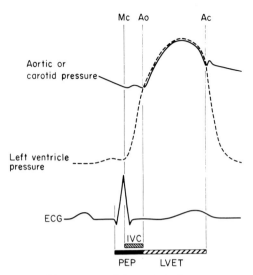

FIG. 23   Aortic or carotid pressure, left ventricular pressure and the ECG to show relationships to pre-ejection period (PEP), isovolumic contraction time (IVC) and left ventricular ejection time (LVET); Mc: mitral valve closure; Ao: aortic valve opening; Ac: aortic valve closure. (Reproduced from Organ, Bernstein, Rowe, Smith, *Am. J. Obstet. Gynecol.* **115**, 369, 1973, with permission.)

this is the rationale for measurement of systolic time intervals as a non-invasive assessment of ventricular performance.

The principal systolic time interval measurements are of pre-ejection period (PEP), and ventricular ejection time (VET) (Fig. 23). PEP is the time interval from the beginning of the QRS complex (ventricular depolarization) to the beginning of ventricular ejection which occurs with aortic valve opening. This includes the intervals from Q wave to mitral valve closure (Q–Mc) and mitral valve closure to aortic valve opening which is the isovolumic contraction time. And since Q–Mc is quite constant in the cardiac cycle PEP varies with the isovolumic contraction time. The relationship of PEP to contractility (measured by maximum dP/dt) is such that increases in contractility produce decreases in PEP and decreases in contractility produce increases in PEP, provided there is no major change in ventricular and diastolic pressure (preload) or aortic diastolic pressure (after-load). Changes in the duration of the QRS complex, e.g. with a bundle branch block, by altering the duration of systole will also alter PEP with no change in contractility (Metzger *et al.*, 1970).

VET is the length of time the aortic valve remains open (A_o–A_c interval). It is a less useful measurement than PEP for assessing contractility for it shortens with both positive and negative inotrophy and with ventricular failure. Both PEP and VET are known to vary inversely with heart rate in the adult and to be properly interpreted correction must be made for variation related to differences in resting heart rate by slightly different equations for males and females (Weissler *et al.*, 1968) and children

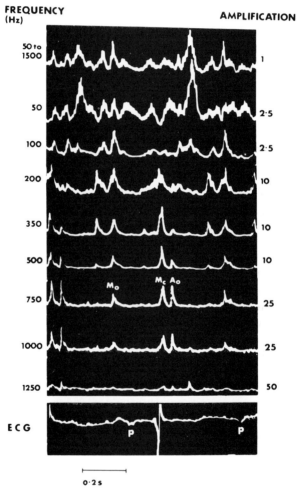

FIG. 24 Oscilloscope photograph of analysis of Doppler ultrasound data at different frequencies. Filtering frequency is indicated to the left and relative amplitude to the right. Three valvular movements are identified in the 750 Hz record: Mo: mitral opening; Mc: mitral closing; Ao: aortic opening. The P wave of the fetal electrocardiogram is shown on the bottom tracing. (Reproduced from Organ, Bernstein, Rowe, Smith, *Am. J. Obstet. Gynecol.* **115**, 369, 1973, with permission.)

(Harris *et al.*, 1964). Systolic time intervals also steadily increase in duration from infancy to puberty (the PEP more than the VET).

The comparison of electrical and mechanical events of the fetal heart is not new. The first simultaneous study of the fetal ECG and phonocardiogram was performed in 1941 (Dressler and Moskowitz, 1941) and fetal systolic time intervals were initially measured between these two signals (e.g. Kelly, 1965; Goodlin *et al.*, 1972). But the advent of Doppler ultrasound (Chapter 9) coupled with the ECG has made measurement of systolic time intervals more practicable both antepartum and intrapartum. Unfiltered Doppler signals are unsuitable for identification of valvular movement but the potential for improvement by high-pass filtering was first demonstrated by Murata and colleagues in 1971.

5.2 Methodology of Measurement

5.2.1 Pre-ejection Period

The methodology for measurement of PEP was refined by Organ *et al.* (1973a) who elegantly demonstrated the improvement, particularly in identifying aortic valve opening (A_o), with selective bandpass filtering from 750 to 1000 Hz (Fig. 24) of the Doppler signal obtained from the heart by a standard 2 MHz transducer. This bandpass filtering removes the lower frequency myocardial and vessel wall components as well as any high frequency electronic noise. Most of the Doppler energy at 750–1000 Hz is probably related to the aortic valve which moves appreciably faster than the pulmonary valve (Hernberg *et al.*, 1970); but since both move simultaneously in the fetus it is impossible to distinguish the reflecting valve and thus more correct to refer to semilunar valve opening.

Murata and Martin (1974) confirmed the accuracy of this method with bandpass filtering of 600–2000 Hz, additionally demonstrating that the signal conditioning did not introduce significant time delay in the Doppler signal. Usable high amplitude signals could only be recorded 10° either side of the direction of valve movement and the variation in time delay of the Doppler signal, from differences in the angle of sound beam and direction of valve motion, was small enough (less than 1 ms) over this sector not to introduce significant measurement error.

These initial studies used averaged Doppler and ECG signals on random samples off line and measured the time intervals manually. Since myocardial function is an integral part of each cardiac cycle measurement and display on a beat-to-beat basis is likely to yield more useful information. Several workers have designed display systems recording successive cardiac cycles on recording film by triggering each sweep by the R wave of the QRS complex (Hon *et al.*, 1974; Organ *et al.*, 1974; Zacutti *et al.*, 1977) and one has described a fetal cardiac interval recorder (Goodlin *et al.*, 1975). More recently processing and measurement has been performed by computer (Bärtling and Klock, 1979; Hawrylyshyn *et al.*, 1980).

To simplify electronic processing of PEP several modified measurements have been proposed for although the onset of the Q wave is easy to visualize electronic detection has proved difficult. As an alternative a "peak" in the QRS complex is identified as the starting point for processing. Wolfson *et al.* (1977) used the Q wave onset of the filtered ECG applying a constant correction of 10 ms to compensate for filtering delay in the processing of the fetomaternal ECG. Bärtling and Klock (1979) and Goodlin (1977) chose the peak of the R wave, the former with a correction factor of 14·8 ms for Q_o–R, and Hon *et al.* (1974) uses the onset of the R wave. Each of these modifications was dismissed as a constant and insignificant error but Hawrylyshyn *et al.* (1980) believe this may introduce an unacceptable error of 10–15%. For the interval Q_o–R not only increases with maturity (as the QRS widens) but varies from 20 to 34 ms depending on whether the QRS complex is biphasic, triphasic or multiphasic.

These workers have developed an initial interactive routine for their computerized analysis which allows characterization of the Q_o–R interval for an individual fetus by obtaining an average QRS complex with user definition of Q_o. Subsequent instantaneous measurements of the R–A_o time are automatically corrected with this. Such automated PEP measurements showed less than 2 ms difference from visual inspection and manual measurement. Whilst this precise interactive definition of the onset of the QRS is important when comparing individual measurements in different fetuses, or at different gestations in the same fetus, it will not alter the assessment of short-term dynamic changes within the same fetus in labour.

There seems to be no difficulty in obtaining instantaneous PEP measurements in labour: Robinson *et al.* (1978) report success in 90% of cases with experience. The ability to do so antepartum is limited by the need to obtain a reasonable abdominal fetomaternal ECG recording.

Wolfson *et al.* (1977) have drawn attention to the analogue delay (of some 10 ms) in processing the fetal QRS complex using an automatic fetomaternal signal processor, which would obviously shorten PEP measurements unless a correction was applied. But even measurements from the raw signal have been smaller antepartum, by some 9·5%, than in labour at similar gestational ages (Murata *et al.*, 1978a). This probably represents inaccuracies of timing of the onset of ventricular depolarization from the abdominal ECG rather than a biological difference. Many of these difficulties could be overcome by averaging the fetal ECG and Doppler signals over short periods. Since the clinical requirements antenatally are rather different than in labour intermittent measurement of averaged PEP may be as useful as instantaneous plots antenatally.

5.2.2 *Ventricular Ejection Time*

The other major systolic interval, ventricular ejection time (VET), is measured as the interval between semilunar valve opening and closing. It, too, may be measured in the fetus from Doppler cardiogram signals. Whilst Organ and associates (1973a) specifically chose their bandpass filtering to identify "aortic valve" opening and measure PEP other investigators have found no difficulty in identifying opening and closing of semilunar and atrioventricular valves by widening the bandpass up to 600–2000 Hz (Murata and Martin, 1974; Wolfson *et al.*, 1977; Robinson *et al.*, 1978). Obviously measurement of VET does not require a precise fetal ECG but this is necessary to identify the valve movements represented by the Doppler signal.

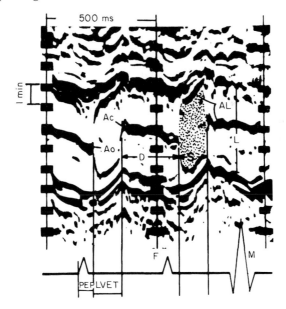

Fig. 25   The M-mode fetal echocardiogram obtained through the aortic valves to show measurement of the PEP from the beginning of the QRS complex to aortic valve opening (Ao) and VET from aortic valve opening to closing (Ac). D is the duration of diastole. S the duration of systole. AL indicates the aortic valve leaflets open and closed. L is the diameter of the aortic lumen. F indicates the fetal and M the maternal QRS complex. The stippled area indicates the opening of the aortic valve during systole. (Reproduced from De Vore, Donnerstein, Kleinman, Hobbins, *Am. J. Obstet. Gynecol.* **141**, 470, 1981, with permission.)

5.2.3 *Further Developments*

DeVore *et al.* (1981) have recently described a system of measuring systolic time intervals by combining real time and M-mode capabilities within the same ultrasound transducer. The semilunar valves are identified by real time and the M mode cursor then aligned across the valves to produce a tracing which indicates the precise moment of valve opening and closing (Fig. 25). A fetomaternal abdominal ECG is obtained simultaneously to allow measurement of PEP and VET. The aortic and pulmonary valves can be visualized separately so it is theoretically feasible to measure and compare right and left ventricular time intervals.

5.3 Investigative Findings

5.3.1 *Pre-ejection Period*

In normal labour in humans, at 38–40 weeks gestation fetal PEP, measured from scalp ECG recordings, has a mean range of 70–73·6 ms (Organ *et al.*, 1973a; Murata and Martin, 1974; Robinson *et al.*, 1978). The mean values from antepartum studies at a similar gestation are more variable, being from 59·7 to 72·9 ms (Wolfson *et al.*, 1977; Murata *et al.*, 1978a; Organ *et al.*, 1980). This is probably related to differences and difficulties in identifying the onset of the Q wave from abdominal recordings.

All published work is agreed that fetal PEP lengthens with advancing gestational age. This has been found in man by comparing fetuses between 36 and 40 weeks gestation both in labour (Murata and Martin, 1974) and antenatally (Wolfson *et al.*, 1977; Organ *et al.*, 1980; Murata *et al.*, 1978a) from as early as 20 weeks to term. Similar results were found in longitudinal studies in rhesus monkeys (Murata *et al.*, 1978b). Indeed, PEP duration continues to lengthen throughout infancy and childhood (Harris *et al.*, 1964; Golde and Burstin, 1970). These findings have prompted the suggestion (Organ *et al.*, 1980) that PEP values should be corrected for gestational age when using them to assess fetal welfare. However, there is no uniformity on the correction factor to be applied (Table II). This variation may be due to different study populations for a difference in fetal weight distribution is likely to alter results since there is a significant correlation of PEP measured in the last week of pregnancy with neonatal body weight (Murata *et al.*, 1979b; Organ *et al.*, 1980). However, the scatter of values was too large for PEP to be used as a predictor of birth weight.

QRS width which is included in the PEP measurement, Q–A_o interval (Fig. 23) also increases with gestation (Organ *et al.*, 1980; Brambati and Pardi, 1980). Further, in fetal monkeys the value PEP length minus QRS

TABLE II

Comparison of regression equations for PEP vs. gestational age on which a correction factor would be based

	Regression equation	Age range (weeks)
Organ et al., 1980	PEP = 1·4 GA + 15	21–43
Murata et al., 1979b	PEP = 1·7 GA − 1	32–42
Murata et al., 1978a	PEP = 1·52 GA + 4	33–42
Wolfson et al., 1977	PEP = 0·85 GA + 24	20–40

GA, gestational age in weeks.

duration does not change over the last trimester (Murata et al. 1978b) and it has been suggested (Organ et al., 1980) that the same is true in humans. If this be the case the lengthening of PEP with gestational age and infant weight reflects an increase in myocardial mass or volume with concomitant broadening of the QRS complex. Volume may be more contributory than mass since the correlation of PEP to heart weight in the fetal lamb was less than expected and less than the correlation with gestational age (Murata et al., 1979a). It also suggests that no correction factor need be applied for gestational age; for the interval PEP minus QRS width could be interpreted without knowledge of this.

There is a shortening of PEP with increasing heart rate in infants, children and adults. Organ and associates have observed a similar relationship in sheep (Organ et al., 1973b) and human (Organ et al., 1974, 1980) fetuses independent of gestational age. But no such relationship has been found by Murata and Martin (1974) and Robinson et al. (1978) in man or monkeys (Murata et al., 1978b). The explanation for this discrepancy is not clear but it may be due to differing selection of heart rate epochs containing transient rather than basal changes in rate. This possiblity is illustrated by the finding that PEP is prolonged during heart rate accelerations associated with fetal movement by increased after-load (Organ et al., 1980) but shortened with the tachycardia produced by adrenaline infusion with increased inotrophy (Murata et al., 1978b). Until the situation is clarified it would seem unwise to correct automatically PEP measurements to a constant heart rate as suggested by Organ et al. (1980).

There is also disagreement in the literature on the changes in PEP with fetal asphyxia. Organ and colleagues have consistently observed shortening of PEP with short lasting hypoxia in the exteriorized fetal lamb (Organ et al., 1973b) whilst Murata and associates have reported a lengthening with longer hypoxia in monkey and lamb fetuses in utero (Murata et al., 1978b, 1979a). These findings are not necessarily contradictory but probably represent differences in the severity of hypoxaemia and the ability of the fetus to compensate. For whilst Organ found an increase in systolic and diastolic pressures with widened pulse pressure Murata reports a fall in both pressures during hypoxia, suggesting an inability to compensate.

In a healthy fetus subject to hypoxia an initial increase in myocardial contractility with shortening of PEP is to be expected with increased catecholamine output (Comline et al., 1965) and utilization of myocardial glycogen by anaerobic metabolism (Dawes et al., 1959). With continuing hypoxia and acidosis the myocardium will fail with lengthening of PEP and a fall in blood pressure. It is interesting to speculate whether these changes might parallel those in the ST waveform (see p. 37). Difficulties are obviously encountered in objectively measuring asphyxia in the human fetus and authors have tended to use observations to fit notions already held. Presently the data are not sufficiently well documented to draw firm conclusions.

Interpretation of PEP is further complicated in human fetal monitoring since asphyxia not only occurs with decreased placental perfusion but may also occur with cord occlusion. This not only reduces oxygenation but produces profound changes in cardiovascular haemodynamics which affect PEP. Occlusion of the umbilical cord or vein alone in the fetal lamb produces a lengthened PEP with a shortening immediately after its release (Organ et al., 1973b). Murata et al. (1978b) have confirmed an inverse relationship of PEP to left ventricular end diastolic pressure (preload) in the fetal monkey as in adult man and cord occlusion lengthens PEP by decreasing ventricular filling and reducing ventricular preload. Within the limits of the Frank–Starling law which operates in the fetal heart (Kirkpatrick et al., 1976) increased preload evokes increasing force of contraction with increased dP/dt max and a decreased PEP and vice versa. Cardiac arrhythmias may also affect PEP by this mechanism: abnormally short R–R intervals prolonging PEP and prolonged R–R intervals shortening it.

Aortic diastolic pressure (after-load) also affects PEP if myocardial contractility remains the same, for the greater the after-load the longer it takes to increase intraventricular pressure to exceed intra-aortic pressure and effect opening of the aortic valves. Aortic diastolic pressure or after-load is principally a reflection of increased peripheral resistance, so it, too, will increase with cord occlusion if the umbilical arteries are compressed. The relative contribution of compression of the umbilical arteries (increased after-load) or vein (decreased preload) to PEP prolongation, during cord occlusion, can be assessed by examining the separate $Q-M_c$ or M_c-A_o components of PEP (Fig. 26), if mitral valve closure (M_c) can be identified on the Doppler cardiogram. After-load also seems to be increased by an increase in peripheral resistance during the fetal move-

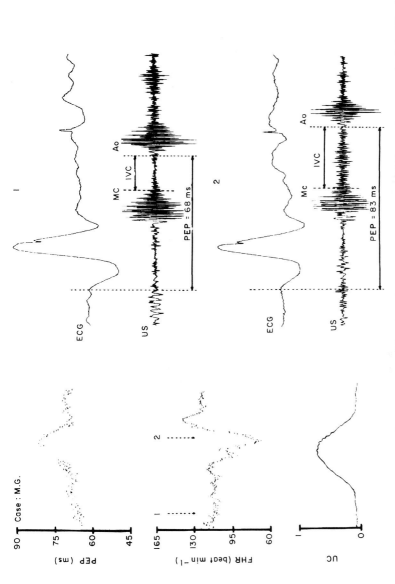

FIG. 26 Continuous recordings of PEP, fetal heart rate (FHR) and uterine contractions (UC) are shown on the left. The ECG and Doppler cardiogram (US) are shown on the right before (1) and during (2) cord occlusion. With cord occlusion there is a delay in closure of the mitral valve (Mc) due to decreased venous return (occlusion of vein) followed by a further delay in aortic valve opening (Ao) due to increased peripheral resistance (occlusion of arteries). PEP is thus increased by a prolongation of both the interval (Qo–MC) and the isovolumic contraction time, IVC (MC–Ao). (Reproduced from Hawrylyshyn, Organ, Bernstein, *Am. J. Obstet. Gynecol.* **137**, 801, 1980, with permission.)

ment associated with heart rate accelerations when PEP is prolonged (Hawrylyshyn *et al.*, 1980).

It is apparent that the interpretation of PEP is complex and that continuous fetal PEP measurements are essential if they are to contribute anything to monitoring cardiac function in labour. In analysis it is important to appreciate the influence of loading of the heart (preload and after-load) and myocardial contractility (inotropic state) which contribute to the dynamics of the mechanical activity of the heart and the pre-ejection period.

5.3.2 *Ventricular Ejection Time*

Ventricular ejection time (VET) in the fetus has not received as much attention as PEP. It is measured between aortic valve opening and its closure and is a function of stroke volume and peripheral resistance. In the human fetus, as in the adult, it varies inversely with rate but only between the range of 115–180 beats min^{-1} (Murata and Martin, 1974; Robinson *et al.*, 1978). Below a heart rate of 115 beats min^{-1} there is no further prolongation. This suggests that rate-related changes in cardiac output are reduced by an increase in stroke volume over the physiological range of heart rate (see Kirkpatrick *et al.*, 1976). Below 115 beats min^{-1}, however, no further compensation occurs and cardiac output will fall.

Murata *et al.* (1980) have studied VET further in the monkey fetus and unlike PEP found no correlation with gestational age. VET, corrected for heart rate, was shortened in the presence of combined hypoxaemia and acidaemia but not when either hypoxaemia or acidaemia was present alone.

6.0 CONCLUSION

The better understanding of fetal cardiovascular physiology coupled with improved technology for non-invasive study of the fetus now enable much more detailed assessment of fetal cardiac status than by heart rate alone. Even the latter, relatively simple, measurement contains much more information than was previously realized.

It is also increasingly clear that no single measurement will provide the answer to all clinical dilemmas either on cardiac function or the welfare of the fetus as a whole. There are obvious clinical advantages in measuring several variables from one signal and the measurement of heart rate, heart rate variation and waveform from the ECG in labour is a potentially useful combination. Systolic time intervals or flow measurements could easily be added or used separately by combining real-time and Doppler ultrasound probes.

But it is no longer helpful to just record a fetal variable; it must be measured objectively and compared with equally objective data on fetal outcome to establish its clinical significance. We also need to define the most appropriate measurement or measurements in given clinical circumstances both before and during labour. Fetal animal studies in which there is simultaneous evaluation of several variables with other indices of cardiac function and asphyxia will be important in this evaluation.

If the obstetric clinician feels this is all too academic and a caesarean section resolves most of his dilemmas he needs to realize that many similar measurements may need to be done on the prematurely delivered sick neonate. He may be encouraged, as should all workers in this field, by the, now possible, logical steps in assessment of the infant with a cardiac arrhythmia.

From detection by auscultation, confirmation by a continuous Doppler FHR, or even diagnosis by an ECG, the fetus can be assessed by real-time ultrasound to exclude a congenital heart lesion and cardiac failure. If present the infant may be treated by digitalis and transferred *in utero* for delivery and assessment in a regional cardiac centre. It is the definition of such a logical sequence from low- to high-level investigation that we must seek to clarify in the care of the fetus in the future.

References

Acheson, R. M. (1960). Observer error in ECG assessment. *Br. J. Prev. Social Med.* **14**, 99.

Ackroyd, M. H. (1973). Digital filters. *In* "Computers in Medicine Series" (Ed. D. W. Hill), p. 64, Butterworths, London.

Allan, L. D., Tynan, M. J., Campbell, S., Wilkinson, J. L. and Anderson, R. H. (1980). Echocardiographic and anatomical correlates in the fetus. *Br. Heart J.* **44**, 444.

Allan, L. D., Tynan, M. J., Campbell, S. and Anderson, R. H. (1981). Normal fetal cardiac anatomy. A basis for the echocardiographic detection of abnormalities. *Prenatal Diag.* **1**, 131.

Alterburger, K. M., Jedzliniak, H., Roper, W. C. and Hernandez, J. (1977). Congenital complete heart block associated with hydrops fetalis. *J. Pediatr.* **9**, 618.

Assali, N. S., Brinkman III, C. R. and Nuwayhid, B. (1974). Comparison of maternal and fetal cardiovascular functions in acute and chronic experiments in the sheep. *Am. J. Obstet. Gynecol.* **120**, 411.

Barclay, A. E., Franklin, K. J. and Prichard, M. M. L. (1944). "The Fetal Circulation and Cardiovascular System and Changes That They Undergo at Birth," Blackwell Scientific Publications, Oxford.

Barcroft, J. (1946). "Researches on Prenatal Life," Blackwell Scientific Publications, Oxford.

Bärtling, Th. and Klock, F. K. (1979). Die pre-ejection-period des menschlichen fetalen Herzens: Bedeutung der Base-Line-Schwankung in der perinatal periode. *Z. Geburtshilfe Perinatol.* **183**, 202.

Bauer, D. J. (1937). The slowing of the heart rate produced by clamping the umbilical cord in the fetal sheep. *J. Physiol.* **90**, 25.

Beard, R. W. (1977). Is intrapartum monitoring worthwhile? *In* "The Current Status of Fetal Heart Monitoring and Ultrasound in Obstetrics" (Eds R. W. Beard and S. Campbell), R.C.O.G. Publication, London.

Beard, R. W. and Rivers, R. P. A. (1979). Fetal asphyxia in labour. *Lancet* **ii**, 1117.

Beard, R. W., Filshie, G. M., Knight, C. A. and Roberts, G. M. (1971). Intensive care of the high-risk fetus in labour. *J. Obstet. Gynaecol. Br. Commonw.* **78**, 865.

Bemmel, J. H., van (1968). Detection of weak fetal electrocardiograms by auto-correlation and cross-correlation of envelopes. *IEEE Trans. Biomed. Engin.* BMF **15**, 17.

Berube, S., Lister, G. Jr, Toews, W. H., Creasy, R. K. and Heymann, M. A. (1978). Congenital heart block and maternal systemic lupus erythematosus. *Am. J. Obstet. Gynecol.* **130**, 595.

Bolte, A. (1961). The derivation and evaluation of fetal electrocardiogram in pregnant women. *Arch. Gynaek.* **194**, 594.

Born, G. V. R., Dawes, G. S., Mott, J. C. and Widdicombe, J. G. (1954). Changes in the heart and lung at birth. *Cold Spring Harb. Quant. Biol.* **19**, 102.

Brambati, B. and Pardi, G. (1980). The intraventricular conduction time of fetal heart in uncomplicated pregnancies. *Br. J. Obstet. Gynaecol.* **87**, 941.

Brambati, B. and Pardi, G. (1981). The intraventricular conduction time of fetal heart in pregnancies complicated by rhesus haemolytic disease. *Br. J. Obstet. Gynaecol.* **88**, 1233.

Brown, V. A., Sawyers, R. S., Parsons, R. J., Duncan, S. L. B. and Cooke, I. D. (1982). The value of antenatal cardiotocography in the management of high-risk pregnancy: A randomised trial. *Br. J. Obstet. Gynaecol.* **89**, 716.

Caldeyro-Barcia, R., Mendez-Bauer, C., Poseiro, J. J. *et al.* (1966). Control of human fetal heart rate during labour. *In* "The Heart and Circulation in the Newborn and Infant" (Ed. D. E. Cassels), pp. 7–36, Grune and Stratton, New York and London.

Caughey, A. F., Jr and Krohn, L. H. (1963). Variation in the fetal electro-cardiogram with period of gestation. *Am. J. Obstet. Gynecol.* **87**, 525.

Comline, R. S., Silver, I. A. and Silver, M. (1965). Factors responsible for the stimulation of the adrenal medulla during asphyxia in the fetal lamb. *J. Physiol.* **178**, 211.

Crawford, C. S. (1982). Antenatal diagnosis of fetal cardiac abnormalities. *Ann. Clin. Lab. Sci.* **12**, 99.

Cremer, M. (1906). Uber die direkte Ableitung der Aktionsstrome des menschlichen Herzens vom Oesophagus und uber das Elektrokardiogramm des Fotus. *Munch. Med. Wschr.* **53**, 811.

Curran, J. T. (1975). "Fetal Heart Monitoring", Butterworths, London and Boston.

Curran, J. T. and MacGregor, J. (1969). A practical system of foetal electro-cardiogram analysis of computer. *J. Physiol., Lond.* **203**, 65.

Dalton, K. J., Dawes, G. S. and Patrick, J. E. (1977). Diurnal, respiratory and other rhythms of fetal heart rate in lambs. *Am. J. Obstet. Gynecol.* **127**, 414.

David, H., Weaver, J. B. and Pearson, J. F. (1975). Doppler ultrasound and fetal activity. *Br. Med. J.* **2**, 62.

Davis, L. G. (1958). Observer variation in reports on electrocardiograms. *Br. Heart J.* **20**, 153.

Dawes, G. S. (1968). "Foetal and Neonatal Physiology: A comparative study of the changes at birth", Year Book Medical Publishers, Chicago.

Dawes, G. S., Mott, J. C. and Shelley, H. J. (1959). The importance of cardiac glycogen for the maintenance of life in foetal lambs and newborn animals during anoxia. *J. Physiol.* **146**, 516.

Dawes, G. S., Johnston, B. M. and Walker, D. W. (1980). Relationship of arterial pressure and heart rate in fetal, newborn and adult sheep. *J. Physiol.* **309**, 405.

Dawes, G. S., Visser, G. H. A., Goodman, J. D. S. and Redman, C. W. G. (1981). Numerical analysis of the human fetal heart rate: The quality of ultrasound records. *Am. J. Obstet. Gynecol.* **141**, 43.

Dawes, G. S., Houghton, C. R. S. and Redman, C. W. G. (1982a). Baseline in human fetal heart-rate records. *Br. J. Obstet. Gynaecol.* **89**, 270.

Dawes, G. S., Houghton, C. R. S., Redman, C. W. G. and Visser, G. H. A. (1982b). Pattern of the normal human fetal heart rate. *Br. J. Obstet. Gynaecol.* **89**, 276.

DeVore, G. R., Donnerstein, R. L., Kleinman, C. S. and Hobbins, J. C. (1981). Real-time directed M-mode echocardiography: A new technique for accurate and rapid quantitation of the fetal pre-ejection period and ventricular ejection time of the right and left ventricles. *Am. J. Obstet. Gynaecol.* **141**, 470.

Drefus-Brisac, C. (1970). Ontogenesis of sleep in human prematures after 32 weeks of conceptional age. *Develop. Psychobiol.* **3**, 91.

Dressler, M. and Moskowitz, S. N. (1941). Fetal electrocardiography and stethography; combined study. *Am. J. Obstet. Gynecol.* **41**, 775.

Eik-Nes, S. H., Brubakk, A. O. and Ulstein, M. K. (1980). Measurement of human fetal blood flow. *Br. Med. J.* **280**, 283.

Eik-Nes, S. H., Maršál, K. and Kristoffersen, K. (1981). Transcutaneous measurement of human fetal blood flow. Methodological studies. *Proceedings 8th International Conference on fetal breathing and other measurements*, Maastricht, p. 10.

Figueroa-Longo, J. G., Poseiro, J. J., Alvarez, L. O. and Caldeyro-Barcia, R. (1966). Fetal electrocardiogram at term labour obtained with subcutaneous fetal electrodes. *Am. J. Obstet. Gynecol.* **96**, 556.

Fisher, D. J., Heymann, M. A. and Rudolph, A. M. (1982). Fetal myocardial oxygen and carbohydrate consumption during acutely induced hypoxemia. *Am. J. Physiol.* **242**, 657.

Fitzgerald, D. E. and Drumm, J. E. (1977). Non-invasive measurement of human fetal circulation using ultrasound: A new method. *Br. Med. J.* **2**, 1450.

Flynn, A. M. and Kelly, J. (1977). Evaluation of fetal well-being by antepartum fetal heart rate monitoring. *Br. Med. J.* **1**, 936.

Gentner, O. and Hammacher, K. (1967). An improved method for the determination of the instantaneous fetal heart frequency from the fetal phonocardiogram. "Digest of 7th International Conference on Medical and Biological Engineering" (Ed. B. Jacobson), p. 140, Almquist and Wiksell, Stockholm.

Gennser, G., Johansson, B. W. and Kullander, S. (1968). Electrocardiographic and tissue lactate changes in the hypoxic human fetus in mid-pregnancy. *J. Obstet. Gynaecol. Br. Commonw.* **75**, 941.

Gill, R. W. and Kossoff, G. (1979). Pulsed Doppler combined with B-mode imaging for blood flow measurement. *Contr. Gynec. Obstet.* **6**, 139.

Golde, D. and Burstin, L. (1970). Systolic phases of the cardiac cycle in children. *Circulation* **42**, 1029.

Goodlin, R. C. (1977). Fetal cardiovascular response to distress. *Obstet. Gynecol.* **49**, 371.

Goodlin, R. C., Girard, J. and Hollmen, A. (1972). Systolic time intervals in the fetus and neonate. *Obstet. Gynecol.* **39**, 295.

Goodlin, R. C., Haesslein, H. C., Crocker, K. and Carlson, R. G. (1975). Fetal cardiac interval recorder. *Obstet. Gynecol.* **46**, 69.

Goodwin, J. W. (1976). The fetal circulation. *In* "Perinatal Medicine; the Basic Science Underlying Clinical Practice" (Eds Goodwin, Godden and Chance) Williams and Wilkins, Baltimore.

Greene, K. R., Natale, R. and Harrison, C. Y. (1980). Heart period variation and gross body and breathing movements after amniotomy in the human fetus. *In* "Foetal and Neonatal Physiological Measurement", p. 250, Pitman Medical, Bath.

Greene, K. R. and Wickham, P. J. D. (1983). Quantitative analysis of the ST waveform of the human fetal electrocardiogram. Submitted for publication.

Greene, K. R., Dawes, G. S., Lilja, H. and Rosén, K. G. (1982). Changes in the ST waveform of the fetal lamb electrocardiogram with hypoxemia. *Am. J. Obstet. Gynecol.* **144**, 950.

Haan, J., de, Bemmel, J. H., van, Versteeg, B., Veth, A. F. L., Stolte, L. A. M., Janssens, J. and Eskes, T. K. A. (1971). Quantitative evaluation of fetal heart rate patterns. 1. Processing methods. *Eur. J. Obstet. Gynecol.* **3**, 95.

Hammacher, K. (1962). Neue methode zur selektiven registrierung der fetalen herzschlagfrequenz. *Geburtshilfe Frauenheilkd.* **22**, 1542.

Harris, L. C., Weissler, A. M., Manske, A. O., Bamford, B. H., White, G. D. and Hammill, W. A. (1964). Duration of the phases of mechanical systole in infants and children. *Am. J. Cardiol.* **14**, 448.

Haverkamp, A. D., Orleans, M., Langendoerfer, S., McFee, J., Murphy, J. and Thompson, H. E. (1979). A controlled trial of the differential effects of intrapartum monitoring. *Am. J. Obstet. Gynecol.* **134**, 399.

Hawrylyshyn, P. A., Organ, L. W. and Bernstein, A. (1980). A new computer technique for continuous measurement of the pre-ejection period in the human fetus: Physiologic significance of the pre-ejection period patterns. *Am. J. Obstet. Gynecol.* **137**, 801.

Hellman, L. M., Schiffer, M. A., Kohl, S. G. and Tolles, L. G. (1958). Studies in fetal wellbeing: Variations in fetal heart rate. *Am. J. Obstet. Gynecol.* **76**, 998.

Hernberg, J., Weiss, B. and Keegan, A. (1970). The ultrasonic recording of aortic valve motion. *Radiology* **94**, 361.

Hertz, R. H., Timor-Tritsch, I., Dierker, L. J. *et al.* (1979). Continuous ultrasound and fetal movement. *Am. J. Obstet. Gynecol.* **135**, 152.

Heymann, M. A. and Rudolph, A. M. (1967). Effect of the exteriorisation of the sheep fetus on its cardiovascular function. *Circ. Res.* **21**, 741.

Hioki, T. (1975). Averaged fetal electrocardiogram obtained by direct lead in fetal distress diagnosed by fetal heart rate pattern. *Acta Obstet. Gynec. Jap.* **22**, 162.

Hökegård, K. H., Eriksson, I., Kjellmer, I., Magno, R. and Rosén, K. G. (1981). Alterations in myocardial metabolism in relation to electrocardiographic changes and cardiovascular functions during graded hypoxia in the fetal lamb. *Acta Physiol. Scand.* **113**, 67.

Hon, E. H. (1963). The classification of fetal heart rate. 1. A working classification. *Obstet. Gynecol.* **22**, 137.

Hon, E. H. and Hess, O. W. (1957). Instrumentation of fetal electrocardiography. *Science* **125**, 553.

Hon, E. H. and Lee, S. T. (1963a). The fetal electrocardiogram. I. The electro-cardiogram of the dying fetus. *Am. J. Obstet. Gynecol.* **87**, 804.

Hon, E. H. and Lee, S. T. (1963b). Noise reduction in fetal electrocardiogram. II. Averaging techniques. *Am. J. Obstet. Gynecol.* **87**, 1086.

Hon, E. H. and Yeh, S. Y. (1969). The fetal arrhythmia index. *Med. Res. Eng.* **8**, 14.

Hon, E. H., Murata, Y., Zanini, B., Martin, C. B. and Lewis, D. E. (1974). Continuous microfilm display of the electromechanical intervals of the cardiac cycle. *Obstet. Gynecol.* **43**, 722.

Hukkinen, K., Kariniemi, V., Katila, T. E. *et al.* (1976). Instantaneous fetal heart rate monitoring by electromagnetic methods. *Am. J. Obstet. Gynecol.* **125**, 1115.

Jones, C. T. and Ritchie, J. W. K. (1978). The cardiovascular effects of circulating catecholamines in the foetal sheep. *J. Physiol.* **285**, 381.

Junge, H. D. (1979). Behavioural states and state related heart rate and motor activity patterns in the newborn infant and the fetus antepartum. *J. Perinat. Med.* **7**, 85.

Kahn, K. A. and Simonson, E. (1957). Changes of mean spatial QRS and T vectors and of conventional electrocardiographic items in hard anaerobic work. *Circ. Res.* **5**, 629.

Karlsson, J., Templeton, G. H. and Willerson, J. T. (1973). Relationship between epicardial ST segment changes and myocardial metabolism during acute coronary insufficiency. *Circ. Res.* **32**, 725.

Katz, L. N. and Feil, H. S. (1922). Dynamics of auricular fibrillation and ventri-cular systole. *Arch. Intern. Med.* **32**, 672.

Kegaradec, de (1822). "Memoire sur l'Auscultation Appliques a l'Etude de la Grossesse", Mequignon-Marvis, Paris.

Kelly, J. V. (1965). The fetal heart: Comparison of its electrical and mechanical events. *Am. J. Obstet. Gynecol.* **91**, 1133.

Kendall, B., Farell, D. P. A., Kane, H. A. and van Ostrand, J. R. (1964). Detection of fetal distress by the abnormal fetal radio electrocardiogram. *Am. J. Obstet. Gynecol.* **90**, 340.

Kennedy, E. (1833). "Observations on Obstetric Auscultation", Hodges and Smith, Dublin.

Kirkpatrick, S. E., Covell, J. W. and Friedman, W. F. (1973). A new technique for the continuous assessment of the fetal and neonatal cardiac performance. *Am. J. Obstet. Gynecol.* **116**, 963.

Kirkpatrick, S. E., Pitlick, P. T., Naliboff, J. and Friedman, W. F. (1976). Frank–Starling relationship as an important determinant of fetal cardiac output. *Am. J. Physiol.* **231**, 495.

Kleinman, C. S., Hobbins, J. C., Jaffe, C. C., Lynch, D. C. and Talner, N. S. (1980). Echocardiographic studies of the human fetus. Prenatal diagnosis of congenital heart disease and cardiac dysrhythmias. *Pediatrics* **65**, 1059.

Larks, S. D. and Anderson, G. V. (1962). The abnormal fetal electrocardiogram. *Am. J. Obstet. Gynecol.* **84**, 1893.

Lee, K. H. and Blackwell, R. (1974). Observations on the configuration of the fetal electrocardiogram before and during labour. *J. Obstet. Gynaecol. Br. Commonw.* **81**, 61.

Lee, S. T. and Hon, E. H. (1965). The fetal electrocardiogram. IV. Unusual variations in the QRS complex during labour. *Am. J. Obstet. Gynecol.* **92**, 1140.

Le Goust, P. quoted by Fasbender, H. (1906) *In* "Geschicte der Geburtshilfe", Gustav Fischer, Jena.

Lewis, R. P., Rittgers, S. E., Forester, W. F. and Boudoulas, H. (1977). A critical review of the systolic time intervals. *Circulation* **56**, 146.

Lilja, H., Greene, K. R., Karlsson, K. and Rosén, K. G. (1983). ST waveform changes of the fetal electrocardiogram during labour. A clinical study. Submitted for publication.

McCallum, W. D., Williams, C. S., Napel, S. and Daigle, R. E. (1978). Fetal blood velocity waveforms. *Am. J. Obstet. Gynecol.* **132**, 425.

McCue, C. M., Mantakus, M. E., Tingelstad, J. B. and Ruddy, S. (1977). Congenital heart block in newborns of mothers with connective tissue disease. *Circulation* **56**, 82.

Marvell, C. J. (1979). Ph.D. Thesis, University of Nottingham.

Marvell, C. J., Kirk, D. L., Jenkins, H. M. L. and Symonds, E. M. (1980). The normal condition of the fetal electrocardiogram during labour. *Br. J. Obstet. Gynaecol.* **87**, 786.

Metzger, C. C., Chough, C. B., Kroetz, F. W. and Leonard, J. L. (1970). True isometric contraction time. Its correlation with two external indexes of ventricular performance. *Am. J. Cardiol.* **25**, 434.

Michaelsson, M. and Engle, M. A. (1972). Congenital complete heart block: An international study of the natural history. *Cardiovasc. Clin.* **4**, 85.

Miller, F. C. and Paul, R. H. (1978). Intrapartum fetal heart rate monitoring. *Clin. Obstet. Gynecol.* **21**, 561.

Modanlou, H. D., Freeman, R. K., Ortiz, O., Hinkes, P. and Pillsbury, G., Jr (1977). Sinusoidal fetal heart rate pattern and severe fetal anaemia. *Obstet. Gynecol.* **49**, 537.

Morishima, H. O., Daniel, S. S., Richards, R. T. and James, I. S. (1975). The effect of increased maternal P_aO_2 upon the fetus during labour. *Am. J. Obstet. Gynecol.* **123**, 257.

Moscary, P., Gaal, J., Komaromy, B. *et al.* (1970). Relationship between fetal intracranial pressure and FHR during labour. *Am. J. Obstet. Gynecol.* **106**, 407.

Murata, Y. and Martin, C. B. (1974). Systolic time intervals of the cardiac cycle. *Obstet. Gynecol.* **44**, 224.

Murata, Y., Takemura, H. and Kurachi, K. (1971). Observation of fetal cardiac motion by M-mode ultrasonic cardiography. *Am. J. Obstet. Gynecol.* **111**, 287.

Murata, Y., Martin, C. B., Ikenoue, T. and Lu, P. S. (1978a). Antepartum evaluation of the pre-ejection period of the fetal cardiac cycle. *Am. J. Obstet. Gynecol.* **132**, 278.

Murata, Y., Martin, C. B., Ikenoue, T. and Petrie, R. H. (1978b). Cardiac systolic time intervals in fetal monkeys: Pre-ejection period. *Am. J. Obstet. Gynecol.* **132**, 285.

Murata, Y., Miyake, K. and Quilligan, E. J. (1979a). Pre-ejection period of cardiac cycles in fetal lamb. *Am. J. Obstet. Gynecol.* **133**, 509.

Murata, Y., Pijls, N., Miyake, K., Schmidt, P., Martin, C. B. and Singer, J. (1979b). Antepartum determination of pre-ejection period of the fetal cardiac cycle: Its relation to newborn body weight. *Am. J. Obstet. Gynecol.* **133**, 515.

Murata, Y., Martin, C. B., Ikenoue, T. and Petrie, R. H. (1980). Cardiac systolic time intervals in fetal monkeys: Ventricular ejection time. *Am. J. Obstet. Gynecol.* **136**, 603.

Myers, R. E. (1972). Two patterns of perinatal brain damage and their conditions of occurrence. *Am. J. Obstet. Gynecol.* **112**, 246.

Noble, D. and Cohen, I. (1978). The interpretation of the T wave of the electro-cardiogram. *Cardiovasc. Res.* **12**, 13.

Offner, F. and Moisand, B. (1966). A Coincidence technique for fetal electro-cardiography. *Am. J. Obstet. Gynecol.* **95**, 676.

Organ, L. W., Bernstein, A., Rowe, I. H. and Smith, K. C. (1973a). The pre-ejection period of the fetal heart: Detection during labour with Doppler ultra-sound. *Am. J. Obstet. Gynecol.* **115**, 369.

Organ, L. W., Milligan, J. E., Goodwin, J. W. and Bain, M. (1973b). The pre-ejection period of the fetal heart: Response to stress in the term lamb. *Am. J. Obstet. Gynecol.* **115**, 377.

Organ, L. W., Bernstein, A., Smith, K. D. and Rowe, I. H. (1974). The pre-ejection period of the fetal heart. Patterns of change during labor. *Am. J. Obstet. Gynecol.* **120**, 49.

Organ, L. W., Bernstein, A. and Hawrylyshyn, P. A. (1980). The pre-ejection period as an antepartum indicator of fetal well-being. *Am. J. Obstet. Gynecol.* **137**, 810.

Pardi, G., Uderzo, A., Tucci, E. and Arata, G. D. (1971). Electrocardiographic patterns and cardiovascular performance of the sheep fetus during hypoxia. *In* "Fetal Evaluation during Pregnancy and Labour" (Eds P. G. Crosignani and G. Pardi), p. 157, Academic Press, New York and London.

Pardi, G., Tucci, E., Uderzo, A. and Zanini, D. (1974). Fetal electrocardiogram changes in relation to fetal heart rate patterns during labour. *Am. J. Obstet. Gynecol.* **118**, 243.

Parmelee, A. H. and Stern, E. (1972). Development of states in infants. *In* "Sleep and the Maturing Nervous System" (Eds C. D. Clements, D. P. Purpura and F. E. Mayer), p. 199, Academic Press, New York and London.

Paul, W. M., Quilligan, E. J. and Maclachlan, T. (1964). Cardiovascular pheno-menon associated with fetal head compression. *Am. J. Obstet. Gynecol.* **92**, 824.

Pearson, J. F. and Weaver, J. B. (1976). Fetal activity and fetal well-being. An evaluation. *Br. Med. J.* **1**, 1305.

Pearson, J. F. and Weaver, J. B. (1978). A six-point scoring system for antenatal cardiotocographs. *Br. J. Obstet. Gynaecol.* **85**, 323.

Polvani, F., Brambati, B. and Pardi, G. (1971). Analysis of the fetal electro-cardiogram during pregnancy and labour. *In* "Fetal Evaluation during Preg-nancy and Labour" (Eds P. G. Crosignani and G. Pardi), p. 221, Academic Press, New York and London.

Prechtl, H. F. R., Fargel, J. W., Weinmann, H. M. and Bakker, H. H. (1979). Postures, motility and respiration in the low-risk preterm infant. *Dev. Med. Child Neurol.* **21**, 3.

Reeves, T. J., Hefner, L. L., Jones, W. B. *et al.* (1960). The hemodynamic deter-minations of the rate of change in pressure in the left ventricle during isometric contraction. *Am. Heart. J.* **60**, 745.

Rhyne, V. T. (1969). A digital system for enhancing the fetal electrocardiogram. *IEEE Trans. Biomed. Engin.* **16**, 80.

Robinson, H. P., Adam, A. H., Fleming, J. E. E., Houston, A. and Clarke, D. M. (1978). Fetal electromechanical intervals in labour. *Br. J. Obstet. Gynaecol.* **85**, 172.

Roche, J. B. and Hon, E. H. (1965). The fetal electrocardiogram. V. Comparison of lead systems. *Am. J. Obstet. Gynecol.* **92**, 1149.

Rosén, K. G. and Isaksson, O. (1976). Alterations in fetal heart rate and ECG correlated to glycogen, creatinine phosphate and ATP levels during graded hypoxia. *Biol. Neonate.* **30**, 17.

Rosén, K. G. and Kjellmer, I. (1975). Changes in the fetal heart rate and ECG during hypoxia. *Acta Physiol. Scand.* **93**, 59.

Rosén, K. G., Hökegård, K. H. and Kjellmer, I. (1976). A study of the relationship between the electrocardiogram and hemodynamics in the fetal lamb during asphyxia. *Acta Physiol. Scand.* **98**, 275.

Rudolph, A. M. and Heymann, M. A. (1973). Control of the foetal circulation. *In* "Foetal and Neonatal Physiology", Proceedings of Barcroft Centenary Symposium, Cambridge University Press.

Rudolph, A. M., Itskovitz, J., Iwamoto, H., Reuss, M. L. and Heymann, M. A. (1981). Fetal cardiovascular responses to stress. *Sem. Perinatol.* **5**, 109.

Ruttgers, H., Kubli, F., Haller, U., Bachmann, M. and Grunder, E. (1972). Die antepartale fetale Herzfrequenz 1. *Z. Geburtschilfe Perinatol.* **176**, 294.

Sahn, D. J., Lange, L., Allen, H. D., Goldberg, S. J., Anderson, C., Giles, H. and Haser, K. (1980). Quantitative real-time cross-sectional echocardiography in the developing normal human fetus and newborn. *Circulation* **62**, 588.

Schifrin, B. S. and Dame, L. (1972). Fetal heart rate patterns—prediction of Apgar scores. *J. Am. Med. Assoc.* **219**, 1322.

Schuler, G., Puddicombe, P. and Park, G. (1968). Fetal ECG improvement due to selective signal addition. *Am. J. Obstet. Gynecol.* **101**, 1120.

Shelley, T. (1967). Fetal heart rate measurement in pregnancy and labour. Ph.D. Thesis, University of Sheffield.

Shenker, L. (1979). Fetal cardiac arrhythmias. *Obstet. Gynecol. Surv.* **34**, 561.

Silber, D. L. and Durnin, R. E. (1969). Intrauterine atrial tachycardia (associated with massive edema in a newborn.). *Am. J. Dis. Child.* **117**, 722.

Smyth, C. N. (1953). Experimental electrocardiography of the foetus. *Lancet* i, 1124.

Smyth, C. N. and Farrow, J. L. (1958). Electronic evaluation of the foetus. *Br. Med. J.* **2**, 1005.

Southern, E. M. (1957). Fetal anoxia and its possible relation to changes in the prenatal fetal electrocardiogram. *Am. J. Obstet. Gynecol.* **73**, 233.

Steer, P. J., Little, D. J., Lewis, N. L., Kelly, M. C. M. E. and Beard, R. W. (1976). The effect of membrane rupture on fetal heart rate in induced labour. *Br. J. Obstet. Gynaecol.* **83**, 454.

Sterman, M. B. and Hoppenbrouwers, T. (1971). The development of sleep-waking and rest-activity patterns from fetus to adult in men. *In* "Brain Development and Behaviour" (Eds M. B. Sterman, D. J. McGinty and A. M. Adinolfi), p. 203, Academic Press, New York and London.

Stern, L., Lind, J. and Kaplan, B. (1961). Direct human foetal electrocardiography (with studies of the effects of adrenalin, atropine, clamping of the umbilical cord, and placental separation on the foetal ECG). *Biol. Neonate* **3**, 49.

Strong, P. (1970). *In* "Biophysical Measurements: Measurement Concept Series", Tektronix, Beaverton, Oregon.

Stuart, B., Drumm, J., Fitzgerald, D. E. and Duignan, N. M. (1980). Fetal blood velocity waveforms in normal pregnancy. *Br. J. Obstet. Gynaecol.* **87**, 780.

Stuart, B., Drumm, J., Fitzgerald, D. E. and Duignan, N. M. (1981). Fetal blood velocity waveforms in uncomplicated labour. *Br. J. Obstet. Gynaecol.* **88**, 865.

Sureau, C. and Trocellier, R. (1961). Un probleme technique d' electrocardiographie foetale. Note sur l'elimination de l'electrocardiogramme maternel. *Gynec. Obstet.* **60**, 43.

Su, J. Y. and Friedman, W. F. (1973). Comparison of the responses of the fetal and adult cardiac muscle to hypoxia. *Am. J. Physiol.* **224**, 1249.

Sykes, G. S., Molloy, P. M., Johnson, P., Gu, W., Stirrat, G. M. and Turnbull, A. C. (1982). Do Apgar scores indicate asphyxia? *Lancet* **i**, 494.

Symonds, E. M. (1971). Configuration of the fetal electrocardiogram in relation to fetal acid–base balance and plasma electrolytes. *J. Obstet. Gynaecol. Br. Commonw.* **78**, 957.

Timor-Tritsch, I. E., Dierker, L. J. and Hertz, R. H. *et al.* (1978). Studies of antepartum behavioural state in the human fetus at term. *Am. J. Obstet. Gynecol.* **132**, 524.

Tipton, R. H. and Shelley, T. (1971). Dip area. A quantitative measure of fetal heart rate patterns. *J. Obstet. Gynaecol. Br. Commonw.* **78**, 694.

Van Der Horst, R. L. (1970). Congenital atrial flutter and cardiac failure presenting as hydrops at birth. *S. Afr. Med. J.* **44**, 1037.

Verma, U., Tejani, N., Weiss, R. R., Chatterjee, S. and Halitsky, V. (1980). Sinusoidal fetal heart rate patterns in severe Rh disease. *Obstet. Gynecol.* **55**, 666.

Visser, G. H. A. and Huisjes, H. J. (1977). Diagnostic value of the unstressed antepartum cardiotocogram. *Br. J. Obstet. Gynaecol.* **84**, 321.

Visser, G. H. A., Redman, C. W. G., Huisjes, H. J. and Turnbull, A. C. (1980). Nonstressed antepartum heart rate monitoring: Implications of decelerations after spontaneous contractions. *Am. J. Obstet. Gynecol.* **138**, 429.

Visser, G. H. A., Dawes, G. S. and Redman, C. W. G. (1981). Numerical analysis of the normal human antenatal fetal heart rate. *Br. J. Obstet. Gynaecol.* **88**, 792.

Visser, G. H. A., Goodman, J. D. S., Levine, D. H. and Dawes, G. S. (1982). Diurnal and other cyclic variations in human fetal heart rate near term. *Am. J. Obstet. Gynecol.* **142**, 535.

Weissler, A. M., Harris, W. S. and Schoenfeld, C. D. (1968). Systolic time intervals in heart failure in man. *Circulation* **37**, 149.

Wheeler, T. and Murrills, A. (1978). Patterns of fetal heart rate during normal pregnancy. *Br. J. Obstet. Gynaecol.* **85**, 18.

Wheeler, T., Murrills, A. and Shelley, T. (1978). Measurement of the fetal heart rate during pregnancy by a new electrocardiographic technique. *Br. J. Obstet. Gynaecol.* **85**, 12.

Wheeler, T., Cooke, E. and Murrills, A. (1979). Computer analysis of fetal heart rate variation during normal pregnancy. *Br. J. Obstet. Gynaecol.* **86**, 186.

Wickham, P. J. D. (1982). Microprocessor-based signal averager for analysis of the foetal ECG. *Med. Bio. Eng. and Comp.* **3**, 253.

Wiggers, C. J. (1921). Studies on the consecutive phases of the cardiac cycle. The duration of the consecutive phases of the cardiac cycle and the criteria for their precise determination. *Am. J. Physiol.* **56**, 415.

Wladimiroff, J. W. and McGhie, J. (1981). Ultrasonic assessment of cardiovascular geometry and function in the human fetus. *Br. J. Obstet. Gynaecol.* **88**, 870.

Wolfson, R. N., Zador, I. E., Pillay, S. K., Timor-Tritsch, I. F. and Hertz, R. H. (1977). Antenatal investigation of human fetal systolic time intervals. *Am. J. Obstet. Gynecol.* **129**, 203.

Yambao, T. J., Clark, D., Ashby, W. and Abdul-Karim, R. (1982). Sinusoidal fetal heart rate patterns: Case reports and management. *Br. J. Obstet. Gynaecol.* **89**, 765.

Yeh, M. N., Morishima, H. O., Niemann, W. H. and James, L. S. (1975). Myocardial conduction defects in association with compression of the umbilical cord. *Am. J. Obstet. Gynecol.* **121**, 951.

Zacutti, A. (1977). Detection of systolic and diastolic time intervals of the fetal heart. *Contrib. Gynecol. Obstet.* **3**, 31.

Ziegler, R. F. (1959). *In* "Handbook of Circulation" (Eds D. S. Dittmer and R. M. Grebe), W. B. Saunders, Philadelphia.

2. FETAL BREATHING AND MOVEMENT

K. Lindström and K. Maršál

Department of Biomedical Engineering and Department of Obstetrics and Gynecology, University Hospital, Malmö, Sweden

1.0 INTRODUCTION

Fetal limb and body movements, which are an expression of fetal intra-uterine activity, are usually perceived by the mother from the middle of pregnancy until delivery. On relatively rare occasions, pregnant women describe regular rhythmic movements which are discordant with both their own pulse and the pulse of the fetus. These movements are caused by fetal breathing. For a long time, fetal movements reported by the mother were used to confirm the diagnosis of pregnancy and provide a sign of fetal well-being (Ahlfeld, 1869).

Objective investigation of fetal motor activity largely depends on the availability of non-invasive, safe and reliable measurement techniques. Until recently such methods were not available, and therefore most of our knowledge concerning fetal physiology had to be derived from experiments on animals. Introduction of modern techniques, particularly those based on ultrasound, into perinatal research opened up new possibilities of objectively measuring fetal motor function in humans. The development of the ultrasound real-time B-mode technique rapidly attracted the interest of physiologists and clinicians in this field of fetal medicine.

NON-INVASIVE MEASUREMENTS: 2   *Copyright©1983 by Academic Press Inc. (London) Ltd.*
ISBN 0 12 593402 5   *All rights of reproduction in any form reserved*

2.0 FETAL BREATHING MOVEMENTS (FBM)

2.1 Historical Observations

For many years the existence of movements of the fetal chest and abdomen produced by breathing, as a normal phenomenon during gestation, was doubted. Several observations were reported from animal experiments in various species, such as dogs (Winslowius, 1787), guinea-pigs (Preyer, 1882) and sheep (Barcroft and Barron, 1937). A multitude of animal experiments was performed to decide whether amniotic fluid normally enters the airway of the fetus during breathing movements (Geyl, 1880; Snyder, 1941). Also, in experiments on exteriorized fetuses, i.e. fetuses delivered into a warm saline bath with maintained placental circulation, FBM have been shown to occur in cats (Windle et al., 1938) and in sheep (Dawes et al., 1970). Many of the historical reports were criticized because the observations were made on asphyxiated fetuses. Conclusive proof of the existence of FBM in utero was given by Dawes et al. (1970) and Merlet et al. (1970) in experiments on sheep with chronic indwelling catheters in the fetal trachea and blood vessels. This technique made possible continuous measurements of the fetal tracheal pressure for several weeks after surgery. It was shown beyond doubt that episodic breathing movements appeared in all healthy lamb fetuses (Dawes et al., 1972). By a similar technique, FBM have been detected in the Rhesus monkey (Martin et al., 1974) and in the goat (Towell and Salvador, 1974).

Several aspects of FBM in the lamb were elucidated in the chronic experiments. A diurnal variation in the incidence of FBM was found with a maximum in the late evening (Boddy et al., 1973); a reduction in the amplitude and frequency of FBM was seen in hypoglycaemic fetuses. Hypercapnia was found to increase the incidence and amplitude of FBM, whereas hypoxaemia promptly abolished them. Abnormal patterns of FBM in lambs were found to occur with fetal asphyxia and preceding intrauterine death (Chernick and Bahoric, 1974; Patrick et al., 1976). General anaesthetics and barbiturates administered to the mother depressed FBM, whereas catecholamines (Boddy and Dawes, 1975) and doxapram (Hogg et al., 1977) had a stimulating effect. The results of the observations on FBM in animals have recently been summarized in more detail in several reviews (see Duenhoelter and Pritchard, 1977; Marsál, 1977; Wilds, 1978).

In the human, FBM were described for the first time by Ahlfeld (1888) and Weber (1888). They observed shallow periodic movements of the maternal abdomen where the chest of the fetus was located near the uterine wall. During the last hundred years, much evidence has been presented both for and against the existence of FBM. Jaeger (1919) reported a case of a newborn with multiple exostoses which were in contact between adjacent ribs. Pseudo-articular changes on the surface of the exostoses were interpreted as being a result of intrauterine breathing movements. Dyroff (1927) failed to see FBM during fetoscopy, but he did not state what kind of anaesthesia he used. An amniographic technique showed radiopaque material in the fetal lungs early in pregnancy (Erhardt, 1939) and also in pregnancies near term (Davis and Potter, 1946). However, more recent investigations have not confirmed these results (de Blasio *et al.*, 1960). Several radioactive tracers have been used for intra-amniotic application and for measurements of radioactive particles within the fetal lungs (Abramovich, 1970; Duenhoelter and Pritchard, 1973).

Since Dawes (1973) suggested that the recording of FBM in fetal lambs might give valuable information about fetal well-being, interest has grown in the investigation of the clinical potential of FBM in the human. For this, a non-invasive recording method was required. Such a method, using ultrasonic techniques, was described by Boddy and Robinson in 1971. The authors modified a one-dimensional A-mode system so that movements of an echocomplex, reflected from the fetal chest wall, were recorded on a polygraph. In their study, Boddy and Robinson (1971) presented convincing evidence of the existence of FBM in man. Some years later the application of the real-time B-mode ultrasound method proved the existence of FBM in man and made it possible to study these movements in detail (Gennser and Maršál, 1975).

2.2 Recording Technique

2.2.1. *Mechanical Methods*

Weber (1888) was the first to record FBM objectively in humans. He used a glass funnel connected to a kymograph. The funnel was placed on the maternal abdomen where rhythmic movements transmitted from the fetal chest were observed. The movements were recorded as pressure changes in a closed measuring system. The recorded movements had a mean frequency of 61 min^{-1}. Similar results using the same measuring system were reported by Reifferscheid (1911) and Dietel (1955). Reifferscheid (1911) recorded simultaneously the pulse and respiratory movements of the mother as a confirmation that the recorded movements were not artefacts of maternal origin. Various types of strain-gauges may be used for recording FBM from the maternal abdomen (Boddy and Mantell, 1972; Timor-Tritsch *et al.*, 1976). However, kymographic methods can be used only when FBM are detectable on the maternal abdomen. The

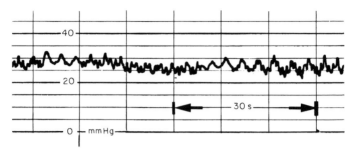

FIG. 1 Tocographic recording of fetal breathing movements (FBM). FBM (rate 62 min^{-1}) are superimposed on maternal breathing movements (rate 20 min^{-1}).

percentage of successful detections of FBM by kymographic methods varies from 34 (Boddy and Mantell, 1972) to 62% (Timor-Tritsch et al., 1976). Sometimes, even conventional external tocographs might be used for this purpose (Fig. 1).

Timor-Tritsch et al. (1979) described the use of two tocodynamometers applied to the maternal abdomen for monitoring human FBM. The signals produced by the tocodynamometers were augmented by an amplifier with a variable gain control. In this way, the authors were able to record FBM in 14 of 16 pregnancies, and they advocated the use of conventional tocodynamometers for studies of FBM. According to our experience and to the above literature, the success rate of FBM recording by mechanical methods is low, and this limits the more general use of those methods.

2.2.2. Ultrasonic Pulse Echo Methods

Physical conditions for the detection of FBM by ultrasound are favourable in late pregnancy. The uterus with its contents lies close to the abdominal wall of the mother and offers good conditions for examination by ultrasound: the fetus is immersed in amniotic fluid, which readily transmits ultrasound waves, and the lungs of the fetus are non-aerated so that echoes from both walls of the fetal chest can be obtained.

(i) *One-dimensional A-mode technique* The one-dimensional A-mode method for recording FBM, as originally described by Boddy and Robinson (1971), employed a modified commercial echoscope (Ekoline 20, Smith Kline Instrument Co.). The transducer was applied to the maternal abdomen and secured by a belt incorporating a universal joint which allowed the transducer to be turned. The emitted narrow beam of pulsed ultrasound (ultrasound frequency 2·25 MHz, pulse repetition frequency 1000 Hz) was directed to the fetal heart. In the A-mode display, the fetal heart echoes, because of their characteristic movements, were easily identified. The heart echoes were kept on the oscilloscope screen for the whole recording period thus ensuring that the ultrasound beam was passing through the fetal thorax. The echoes reflected from the proximal thoracic wall were identified by their appearance and location. The movements of the thoracic echoes were then detected and measured by a time-to-voltage converter. The output analogue signal was recorded graphically (Fig. 2). By this method, it was difficult to identify properly the echo-giving structures crossed by the ultrasound beam. The method was subject to the artefacts caused by the non-specific movements transmitted, for example, from the mother (Maršál *et al.*, 1978a). Only in the periods when the fetus presented regular breathing movements and no general movements could the operator be sure that the recorded signals were true FBM signals.

maternal breathing

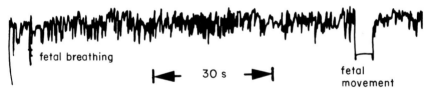

fetal breathing

|◄─ 30 s ─►|

fetal movement

FIG. 2 Fetal breathing movements recorded by A-mode ultrasound. Maternal breathing is recorded simultaneously by a nasal thermistor to facilitate identification of artefactual signals in the FBM record.

Several attempts were made to diminish the influence of artefactual signals. Maternal breathing was recorded simultaneously to subtract these signals from FBM records (Gennser and Maršál, 1975). Two-dimensional B-mode scanning or the use of a Doppler-instrument was recommended for the exact localization of the fetal heart before the FBM recording (Meire *et al.*, 1975). Farman *et al.* (1975) used an improved swept gain system and Meire *et al.* (1975) advocated the use of a large transducer and gating of the posterior chest wall echoes.

In an attempt to obtain true signals of FBM, simultaneous recording of the movements in the proximal and distal chest walls of the fetus was performed. Mantell (1976) used a single ultrasound beam with two gates tracking both chest walls' echoes. Time–motion displays of all echoes received by the transducer made it easier to identify movement in the proximal and distal chest wall, but recordings had to be evaluated further manually (de Wolf, 1976). To increase the reliability of the A-mode system, Maršál *et al.* (1976b) developed a differential echofetoscope (DEFS) which measured on-line the changes in the fetal chest diameter during FBM. The system used two separate transducers emitting ultrasound pulses for recording movements in each of the fetal chest walls by one of the echoscopes (Fig. 3).

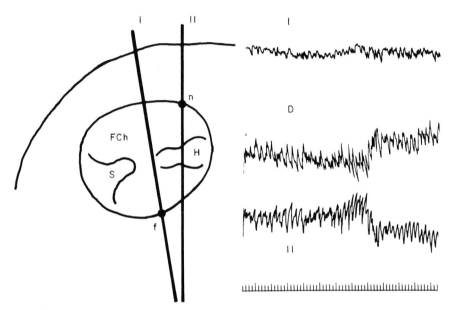

Fig. 3   Recording of fetal breathing movements by the differential echofetoscope (DEFS). In the left part schematically transverse section of the fetal chest (FCh); S = fetal spine; H = fetal heart. Fetal thorax is insonated by two synchronized A-mode transducers (I, II); FBM are recorded in reflecting points n and f on the opposite walls of the fetal chest. In the right part of the figure a chart record of the signals from the transducers (I, II) and of the differential signal (D) corresponding to the changes of the chest diameter $(n-f)$. Time in s.

Despite the improvements, all these methods were still basically one-dimensional and the identification of the echo-giving structures remained difficult.

(ii) *Real-time B-mode technique*   FBM has occasionally been observed during obstetric examinations by mechanical real-time scanners in the early 1970s. To our knowledge, the first demonstration of the normal appearance of FBM in the human fetus in a moving two-dimensional real-time image was presented in 1975 (Gennser and Maršál, 1975). The development of modern, linear array, real-time scanners during recent years prompted the investigations of the FBM. With the real-time technique a detailed moving two-dimensional image of the fetal section is easily obtained. No special training is necessary for the recognition of typical breathing movements of the fetal chest and abdominal wall. Therefore the most widely used method for the quantification of FBM has been simple observation of the screen of the real-time B-mode scanner, each FBM-cycle being marked with an event marker. Several types of real-time scanners have been used for FBM monitoring: mechanical scanners (Maršál et al., 1976b), linear array systems (Hohler and Fox, 1976) and phased array systems (Stephens and Birnholz, 1978).

The real-time ultrasound technique enables FBM to be visualized easily but it offers no possibility to quantify directly the FBM. Marking each FBM cycle by the operator is a relatively rough method including subjective influences, e.g. irresponsiveness of the observer. None the less, this method showed good reproducibility and sufficient accuracy in clinical studies in which only the incidence of FBM was being evaluated (Patrick et al., 1978; Lewis et al., 1978). The accuracy of the method can be improved by slow-motion display of the videorecording of the real-time images. Time resolution of 0·1 s can then be achieved (Patrick et al., 1978). However, such analysis is time consuming.

More detailed analysis of the FBM cycle, particularly the study of breath-to-breath intervals, calls for an objective method with less risk of artefacts. This led us to develop a technical solution combining a real-time B-mode system with a system for the semi-automatic measurement of echo movements along one of the lines in the real-time image.

Time distance recording (TD-recorder; Lindström et al., 1978). The system provides an on-line measurement of the changes in the fetal chest diameter as an expression for FBM. One of the lines in the real-time image is chosen for the measurements and the two echoes whose movements are to be recorded are selected by two electronic markers. The first echo after the proximal marker starts a digital clock which is stopped by the echo next to the distal marker. The echoes from the proximal and distal wall of the fetal chest are chosen, the distance between the two echoes is digitally measured, and a digital-to-analogue converter produces an on-line signal corresponding to the instantaneous fetal chest diameter. The absolute value or the relative changes in the diameter representing FBM can be recorded. The measured distance is visualized on the oscilloscope of the scanner as an intensified line between the two selected echoes. Figure 4 illustrates the principle of the system.

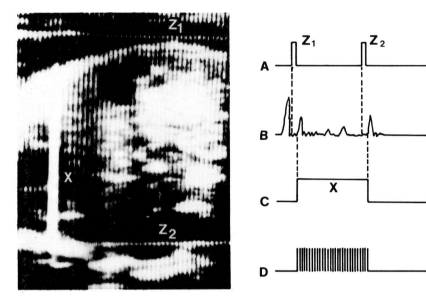

FIG. 4 Function of the time-distance recorder (TD-recorder). Longitudinal section of the fetal body in real-time B-mode image (left) and a sequence function diagram (right). One of the lines in the real-time image is selected for measurement A: Two markers (Z_1, Z_2) are used for choosing the echoes reflected from the near and the far chest wall. B: videosignal of the selected line in A-mode. C: gate signal corresponding to the distance between two selected echoes (X). D: 4 MHz pulses transmitted through the gate representing a digital measure of the length of the signal X. The measured momentary chest diameter (X) is visualized in the real-time image as an intensified part of the selected line.

The whole system including the TD-recorder and the real-time scanner, was thoroughly tested both *in vitro* and *in vivo* (Maršál *et al.*, 1978b). A new filter (slew-rate filter), eliminating the rapid noise spikes but not affecting the true FBM signals, was constructed and incorporated into the system. The system produces an analogue FBM signal, which is recorded on a polygraph (Fig. 5) or stored on a tape-recorder. The real-time images of the FBM movements are stored on a videotape when desired. Since 1976, the TD-recorder has been used at our laboratory for routine examinations of FBM in the last trimester of pregnancy in more than 1200 cases and has proved to be a reliable and easily handled device making possible a quantitative evaluation of FBM.

Korba *et al.* (1979) adopted the principle of the TD recording for measurement of FBM. In their system, they included a differential tracking loop in conjunction with a real-time facility. A line in the real-time image is selected and then the repetition rate of the corresponding trans-

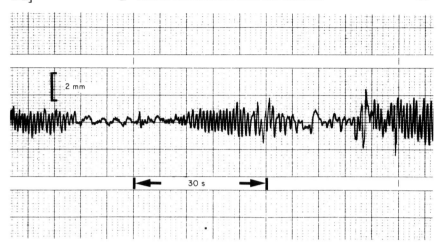

F<small>IG</small>. 5   Episodic fetal breathing movements recorded by the TD-recorder as changes in the fetal chest diameter.

ducer crystal group is increased. Thus the resolution of the system is increased and movements of any portion of the echo complex can be selected for tracking. By a feedback loop, a window pulse is adjusted to a selected zero-crossing of the ultrasonic echocomplex. This eliminates the influence of changing echo-amplitude on the triggering of the system. It also reduces possible artefacts, depending on the changes in the inclination of the ultrasonic beam, and the resolution might be improved beyond the axial resolution of the scanner. The system was tested both electronically and with an *in vitro* set-up. The maximum velocities to be detected by the system were 13·2 and 9·0 cm s^{-1} from 8 to 12 cm depths, respectively. The tracking resolution was found to be 10 μm.

Another solution to the problem of extracting information on FBM from the real-time B-mode image are the systems presented by Bots *et al.* (1976) and Wladimiroff *et al.* (1976). In both systems, TM-mode display of the echo signals from one of the transducers in a multitransducer scanner is recorded. The TM-mode record has to be further analysed manually to yield quantitative data on FBM.

2.2.3 *Doppler Ultrasound Technique*

(i) *Continuous wave Doppler ultrasound*   In 1976, Boyce *et al.* used possible to detect and measure motion of a specific tissue structure or blood They found typical frequency shifts presented as audible rushing noises with a smooth rise and fall. These sounds were probably identical to the unidentified Doppler sounds previously described by Bishop (1966). The rhythmic Doppler shift frequencies were not synchronous with fetal heart action nor with the maternal pulse or breathing. By alternate use of A-mode and continuous Doppler system, Boyce et al. (1976) demonstrated that the shifts were synchronous with FBM. Further investigations of the method in exteriorized lamb fetuses showed that the Doppler signals emanated from blood flow in the inferior vena cava in the region of the diaphragm (Gough and Poore, 1977).

In their more recent paper, Gough and Poore (1979) reported a success rate of 80% in recording FBM after the 30th week of gestation. The Doppler signals of FBM in the human can provide valuable information on the rate and variability of breathing movements. The method using continuous Doppler ultrasound for recording FBM is obviously superior to the original A-mode method, but the optimum transducer orientation still greatly depends on the skill of the operator. Goodman and Mantell (1978) showed that two groups of investigators using the same technique can obtain qualitatively different types of Doppler shift sounds, obviously originating in movements of different structures. The full potential of the continuous Doppler method for clinical recordings of FBM during longer periods of time is yet to be evaluated. One of the obvious drawbacks of the method is the difficulty to identify positively fetal apnoea, i.e. periods of time when the fetus is not performing breathing movements.

(ii) *Pulsed Doppler systems*   By using pulsed Doppler instruments, it is possible to detect and measure motion of a specific tissue structure or blood flow in a vessel at a known depth. Two pulsed Doppler systems for recording FBM have been designed. Tremewan et al. (1976) incorporated a pulsed Doppler system into a compound B-scanner and simultaneously recorded TM-mode traces representing displacement of the fetal chest and traces of the velocity of the movements by detecting Doppler frequency from the selected range. McHugh et al. (1978) developed a system including a real-time B-scanner with a rotating probe, A-mode recording facility, and pulsed Doppler instrument. This arrangement made possible an easy orientation and identification of fetal structures in the real-time image, and recording of Doppler frequencies from the range gate within the fetal chest or abdomen. The authors demonstrated good correlation between A-mode and pulsed Doppler records. The rotating transducer scanner uses two independent ultrasound crystals at 2·5 MHz frequency. The scanner gives a moving two-dimensional image with 25 frames s^{-1} with 120 lines in each image. The fetal thorax is visualized in the real-time B-mode image and FBM can then be recorded either in a TM-trace, when distinct interfaces are demonstrated, or as Doppler shifts caused by motion of weak reflecting tissue structures within the fetal body. For

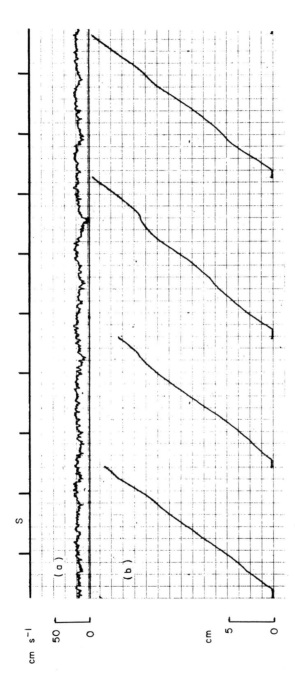

FIG. 6 Pulsed Doppler blood flow measurement in the intra-abdominal part of the umbilical vein during fetal apnoea (left) and during fetal breathing movements (right). (a) analogue signal of the mean blood flow velocity estimated by the Doppler velocitymeter (PEDOF, Vingmed, Oslo, Norway). (b) integral under the velocity curve representing blood flow. Modulations of the blood flow caused by FBM are observed in the right part of the traces.

pulsed Doppler mode, one of the ultrasound crystals is used at a repetition rate of 3 kHz. As the authors stated, further studies are necessary to explore the possibility of measuring quantitatively the velocity of FBM by this method (McHugh *et al.*, 1978).

Recently, the pulsed Doppler technique was combined with real-time B-mode ultrasonography for transcutaneous measurement of fetal blood flow *in utero* (Eik-Nes *et al.*, 1980a). The fetal vessel, e.g. aorta or umbilical vein, is identified in the real-time image and the real-time transducer positioned parallel to the vessel. The 2 MHz pulsed Doppler transducer insonates the vessel at a known angle since the transducers are firmly attached to each other at an angle of 52°. Thus, by this method Doppler shift frequencies are recorded, and these are related to blood flow in an identified vessel at a selected depth. The instrument processes the Doppler shift signals and estimates the mean and maximum blood velocity. For calculation of volume blood flow, fetal vessel diameter is measured in the real-time image. Blood flow in the intra-abdominal part of the umbilical vein showed a typical modulation during FBM (Eik-Nes *et al.*, 1980b) (Fig. 6). Sound characteristics of the Doppler shifts from the umbilical vein were very similar to those reported by Boyce *et al.* (1976). Comparison of the volume blood flow, calculated for periods of fetal apnoea and for periods of high amplitude FBM in the same fetus showed an increase averaging 20% during FBM (Maršál and Eik-Nes, 1980). A similar increase was observed also for the blood flow in the descending part of the fetal thoracic aorta. These preliminary observations suggest that FBM might influence fetal blood circulation, but more information is needed before any conclusions can be reached regarding possible physiological significance of this influence.

2.2.4 *Artefacts in Ultrasonic Measurements of FBM*

Recordings of FBM performed by the original A-mode method often included artefactual signals which could be misinterpreted. This was recognized early by investigators and various artefacts were thoroughly described by Farman *et al.* (1975) and Maršál *et al.* (1978). The use of the two-dimensional real-time ultrasonic technique increased the reliability of the FBM recordings and removed most of the artefactual signals. However, artefacts may still occur if the quantified FBM recording is performed with a system measuring along one selected line in the ultrasonic image.

Artefactual signals may be classified into 3 groups: (a) artefacts depending on biological conditions, e.g. on maternal respiratory movements, movements transmitted from the maternal aorta, or from the fetal heart, limbs, and fetal body; (b) artefacts caused by the physical limitations of the ultrasound method, e.g. by a limited axial or lateral resolution; (c) incorrect use of the recording technique.

Maternal breathing was the major cause of artefacts in the A-mode FBM recordings, and in the real-time recordings, it might well cause difficulties in the interpretation (Maršál *et al.*, 1978). In some cases, signals are obtained with a frequency exactly twice that of maternal breathing. These signals can be due to rhythmical changes in the angle of insonation caused by dislocations of the transducer by maternal breathing, or, when the transducer is stable, by the changes in the inclination of the target in an oscillatory manner. The shape of the fetal chest, with its convex and concave interfaces, may simulate changing inclination when the fetus is moved perpendicularly to the ultrasound beam by the movements of the maternal diaphragm (Fig. 7). Such a situation can, however, be recognized easily in the two-dimensional image and the recording system can be re-adjusted. Therefore careful supervision of the recording by the operator is necessary even when semi-automatic systems coupled to a real-time scanner are used.

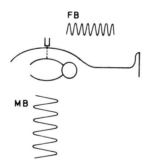

FIG. 7   Origin of false fetal breathing signals (FB) with frequency double that of maternal breathing (MB). The fetus is pushed along its longitudinal axis by maternal breathing, the ultrasonic beam sweeps over the convex outline of the fetal trunk and artefactual FBM signals arise.

When the observer simply views the screen of the real-time scanner and marks each fetal breath, the accuracy of the resulting record much depends on the concentration and responsiveness of the observer. The pattern recognition in the ultrasound image might sometimes cause problems, as demonstrated by Lewis *et al.* (1978). They found a coefficient of variation for six different FBM readings from a videotape to be 6·2%; when using visual analysis, the corresponding value was 0·2%.

TABLE I
**Rate of fetal breathing movements measured in periods of
continuous FBM**

Author	Rate (breaths min^{-1})		Recording technique
	Mean	Range	
Ahlfeld (1905)	61	38– 76	Kymography
Boddy and Dawes (1975)	—	30– 70	A-mode ultrasound
Boyce *et al.* (1976)	—	30– 90	Doppler ultrasound
Wladimiroff *et al.* (1977)	—	20–300	Real-time ultrasound + TM-mode
Maršál (1978)	52	35– 70	Real-time ultrasound + TD-recorder
Patrick *et al.* (1978)	49	10–200	Real-time ultrasound + event marker

2.2.5 *Physiological Characteristics of FBM*

The real-time ultrasonic technique makes possible a detailed study of the *configurative changes* of the fetal trunk during FBM. The main event in the FBM cycle seems to be a contraction of the diaphragm. This hypothesis was supported by the findings from experiments in fetal lambs (Maloney *et al.*, 1975), where EMG activity of the fetal diaphragm was present synchronously with rhythmic changes of tracheal pressure. In the human fetus, the high resolution of real-time scanners allows the visualization of the diaphragm and its movements. Because of the contraction of the diaphragm during "inspiration", the intrathoracic pressure decreases; the fluid-filled lungs cannot expand, and consequently the elastic thoracic cage retracts. The retraction of the chest cage is most prominent in the caudal part of the sternum which is drawn in. At the same time the kyphosis of the thoracic spine is lightly increased and the abdominal wall of the fetus is pushed outwards. During relaxation of the diaphragm in the "expiratory" phase, the fetal chest and abdomen resume their inital shape. During periods of no movements—fetal apnoea—the abdomen and chest wall of the fetus remain in the expiratory position. In a longitudinal section of the body, the fetal chest and abdominal outlines move during FBM in opposite directions, giving a typical appearance of see-saw breathing (Fig. 8). Obviously, the recorded amplitude of the dislocation of the observed structure depends on the point of the measurement. The TD-recorder delivers a calibrated signal of the chest or abdominal movement. However, because it is difficult to standardize the projection for measurement of FBM, the amplitude has not been used as a parameter for characterization of FBM.

TABLE II

Incidence of fetal breathing movements in normal pregnancies

Author	FBM incidence (% of time)	Gestational age (weeks)	Recording time	Recording technique
Boddy and Dawes (1975)	55–90	not stated	not stated	A-mode ultrasound
Wladimiroff et al. (1977)	67	36–41	30 min	Real-time + TM-mode
Maršál (1978)	58	30–40 (mean 32·5)	30 min	Real-time + TD record
Patrick et al. (1978)	32	34–35	24 h	Real-time + event marker
Timor-Tritsch et al. (1979)	28·5	39–40	2 h	Tocodynamometry

The *rate of FBM* in normal pregnancies after the 32nd week of gestation was $52 \pm 6{\cdot}5$ breaths min^{-1} (mean$\pm$SD) in the study by Maršál (1978). This agrees well with frequencies reported from investigation by kymographic methods and by various ultrasonic techniques (Table I).

The *incidence of FBM* in normal pregnancies was reported to be as high as 70% of the time in the A-mode records (Boddy and Dawes, 1975). The recent studies using real-time ultrasound found a significantly lower

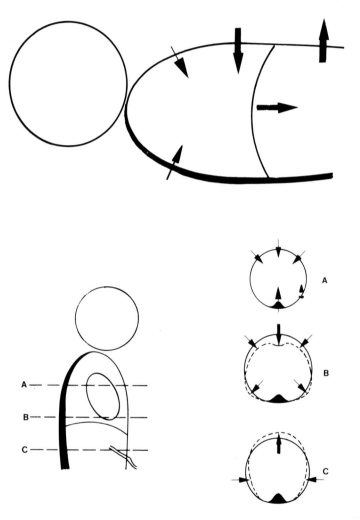

FIG. 8. Schematic presentation of the configurative changes during the "inspiratory" phase of fetal breathing movements. Longitudinal section (upper part) and three transverse sections (A, B, C) of the fetal body are demonstrated.

time incidence of FBM (Table II). The difference is obviously due to the artefactual FBM signals included in the A-mode records. In the studies by real-time B-mode technique, the range of the FBM incidence reported was wide. Several factors which influence the observed incidence of FBM were described. Boddy and Dawes (1975) found considerable diurnal variation of the FBM incidence. This finding was confirmed later by real-time B-mode systems (Roberts et al., 1978 a, b; Patrick et al., 1978). In 24-h studies, Patrick et al. (1978) revealed three distinct patterns of FBM incidence: (a) increase in the incidence 2–3 h after maternal meals; (b) increase in the fetal breathing activity between 0100 and 0700 hours; (c) periodically appearing increases in the percentage of FBM every 1–1·5 h. The periods of increased activity lasted for 20–60 min. These findings seem to be an expression of the sleep state changes in the human fetus in agreement with the reports by Dawes et al. (1972) on the association of FBM and REM sleep in fetal lambs.

The observed increase in the incidence of FBM after maternal meals agrees with experimental studies demonstrating an increase in FBM incidence after intravenous (Boddy and Dawes, 1975) or oral administration of glucose to fasting mothers (Fox et al., 1977; Lewis et al., 1978; Natale et al., 1978). The incidence of FBM 90 min after the glucose load, was twice as high as during the control period. Lewis et al. (1978) suggested a change in the state of the fetus to be a primary effect of the increased maternal glycaemia; Natale et al. (1978) considered the possibility that FBM might be stimulated by local accumulation of carbon dioxide in fetal brain as a consequence of an increased oxidation of glucose. Both groups of investigators suggested an oral glucose loading (50 g) 1 h before the clinical examination of FBM as a way of standardizing the procedure.

When the glucose load is not used, relatively long recording times will be necessary to decide properly on the pattern of FBM incidence. To analyse this problem further, Maršál and Gennser (1980) examined ten uncomplicated pregnancies in the 32nd week of gestation. The recordings lasted for 60 min, starting at 1300 hours. The incidence of FBM for the whole group was $27 \pm 20\cdot5\%$ (mean $\pm$ SD). A moving average method revealed a cycle of activity of 30–40 min duration. Thus, 70–80 min recording was necessary to cover one whole cycle of the FBM incidence and to allow the detection of a particular pattern. An autocorrelation analysis confirmed this conclusion (Fig. 9). Similar findings were reported by Campbell et al. (1980) who analysed 24-h recordings of FBM in normal pregnancies by the Box-Jenkins method.

As mentioned previously, human FBM are episodic and irregular in time. In uncomplicated pregnancies, even long periods of apnoea (up to 108 min) were found to be a normal feature (Patrick et al., 1978). Fetal

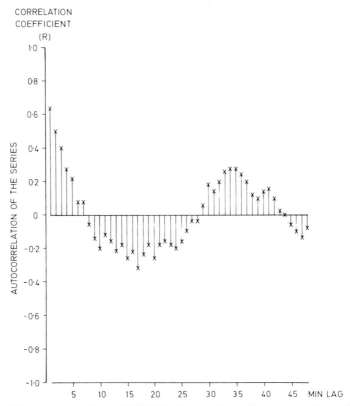

FIG. 9   Time-series analysis of FBM incidence in a 60-min recording of a 32-week-old fetus. FBM incidence was estimated for each 1-min epoch and correlated to the following epochs at increasing time lag. The changes in the correlation coefficients suggest a cyclic pattern in the FBM activity.

apnoea was defined as a period of 6 s or more without FBM (Gennser *et al.*, 1975). This definition was utilized by analogy with apnoea in newborns, and its validity was confirmed statistically in recent work by Patrick *et al.* (1980).

FBM show short-term variability even during periods of seemingly regular activity. Fluctuations in the breath-to-breath intervals, with the oscillations varying between 5 and 10 min^{-1}, were found (Gennser and Maršál, 1979). These oscillations of instantaneous breathing frequency were similar to the variability in the breathing rate in infants (Hathorn, 1975).

The *time course* of an individual FBM cycle was analysed by Gennser (1979), who found that the duration of the active phase occupied most of the breathing cycle and that there was a linear relation between the inspiratory period and the total active breathing phase. This finding suggests that the duration of expiration is determined by the preceding inspiratory time as it is in neonates (Olinsky et al., 1974).

See-saw movements of the fetal chest and abdomen were seen most often during FBM in normal fetuses. Sometimes *other types of FBM* were observed: cycles with prolonged inspiratory phase and lower frequency; and high amplitude breathing movements with a short interruption in the course of inspiration and with expiration following as a continuous movement (similar to augmented breaths in newborns) (Maršál, 1978). On some occasions, Stephens and Birnholz (1978) observed bursts of rapid FBM with a frequency of 200 min^{-1}. Several authors (Lewis et al., 1977; Maršál, 1978; Bots et al., 1978) reported fetal hiccups which were qualitatively similar to FBM but had a low frequency, high amplitude and short duration of the movement cycle. The pathological significance of the two latter types of FBM is obscure. Chapman et al. (1978) described episodes of rapid FBM and gasping movements before intrauterine death in fetal lambs. In the human fetus no conclusive proof of the existence of gasps as a pathological appearance has been presented.

The possible relation between *FBM and the fetal circulation* was discussed by Walz (1922), who postulated that FBM facilitate the return of blood to the fetal heart. Respiratory sinus arrhythmia was observed in the human fetus (Timor-Tritsch et al., 1977); also an increased variability of the fetal heart rate during periods of FBM compared with periods of apnoea was reported (Wheeler et al., 1980). This has been interpreted as being due to both central and peripheral control mechanisms. Quantitative changes in the blood flow in umbilical vein and fetal aorta have been described during periods of high amplitude breathing (Maršál and Eik-Nes, 1980). However, at present, the possible role of FBM in the fetal circulation has not been elucidated.

Similarly, the *control mechanisms of FBM* are not yet sufficiently known. Association of FBM with certain activity states of the central nervous system (CNS) have already been mentioned above. Carbon dioxide is considered to be one of the most potent stimulators of ventilation and was therefore tested also in the fetus *in utero*. In normal pregnancy, the exposure of the mother to 5% CO_2 induced a significant increase in the breathing activity of the fetus (Ritchie and Lackarney, 1979). In growth retarded fetuses, a different response to CO_2 was found to that present in normal fetuses (Wladimiroff et al., 1978). Under hypoxic conditions, caused by placental insufficiency, the mother's inhalation of 10% CO_2 did not stimulate FBM. Probably, sufficient oxygenation of the fetus is necessary for normal response of the respiratory centre to the increased CO_2 level.

2.2.6 *Clinical Application of FBM Examinations*

We still have a great deal to learn about the physiology of FBM and their control mechanisms. Therefore the results of the clinical studies published to date should be considered as being preliminary. For practical reasons, most of the studies used 30-min duration recordings. However, as explained above, 30 min might not be sufficient for valid conclusions. Maršál and Gennser (1980) recorded FBM in six pregnant women four times each day on five consecutive days. The pregnancies were uncomplicated, and the mean gestational age of the group was 35 weeks (range 33–36 weeks). Each recording lasted for 30 min, and the results were analysed regarding incidence of FBM. The mean incidence was highest at 1300 hours on most days. Hierarchic variance analysis revealed a significant dependence on the time of day, but no dependence on the day of the week. The residual was 51% which suggested that the clinical usefulness of quantitative FBM examinations based on 30-min recordings of FBM incidence would be limited as the residual might include uncontrollable factors.

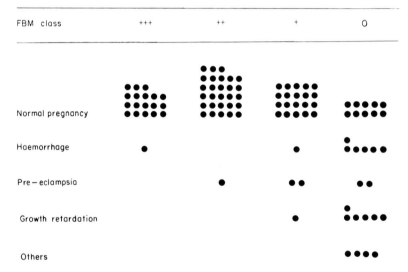

FIG. 10   Distribution of FBM incidence in 100 consecutively examined pregnancies according to diagnosis at the time of examination. FBM incidence in 30-min recordings was classified semiquantitatively as absent or less than 17% (0), 18–34% (+), 35–77% (+ +), and 78–100% (+ + +). Each circle represents one pregnancy.

One hundred consecutive FBM recordings in late pregnancy were analysed for possible relationships between the FBM incidence and the clinical course of pregnancy (Maršál, 1978). Seventy-six of the women examined had uncomplicated pregnancies at the time of recording, the other 24 had various disorders of pregnancy potentially compromising the fetus. The group of pathological pregnancies had a significantly higher proportion of fetuses with low incidence of FBM or with apnoea (Fig. 10). In 92% of the observations with FBM incidence more than 17% of the time, this finding indicated an uncomplicated pregnancy; in 64% of the FBM recordings with FBM incidence below 17% of time, this was indicative of a complication of pregnancy (Fig. 11). The results suggest that the presence of FBM might be a sign of fetal well-being. On the other hand, finding of fetal apnoea seems to have less clinical significance.

In another study, Person and Maršál (1978) found significantly lower mean FBM incidence in the group of 30 pregnancies with suspected intrauterine growth retardation than in the control group (Table III). However, in individual cases, no relation was found to the subsequent

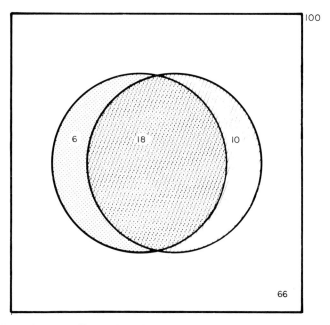

FIG. 11　Venn-diagram illustrating the predictive value of FBM examinations in 100 consecutive pregnancies (assuming that the presence of FBM indicates good fetal health). Twenty-four of the pregnant women had various disorders of pregnancy (stippled circle), the remaining 76 had uncomplicated pregnancies. The hatched circle ($n = 28$) demonstrates the set of women with FBM absent (class 0 in Fig. 10). The predictive value of a negative test (FBM present) was 92% and of a positive test (FBM absent) 64%.

TABLE III
Incidence of fetal breathing movements in fetuses with retarded BPD growth and in normal fetuses

	n	FBM incidence (% of time)		Significance of difference
		Mean	SD	
Growth-retarded group	30	42	$\pm 22 \cdot 0$	$P < 0 \cdot 01$
Control group	53	57	$\pm 23 \cdot 8$	
Growth-retarded group				
Low urinary oestriol	14	34	$\pm 23 \cdot 5$	$P < 0 \cdot 05$
Normal urinary oestriol	16	51	$\pm 18 \cdot 5$	

clinical course of pregnancy. In this study, all except one of the fetuses were symmetrically retarded.

Roberts *et al.* (1978a, b) described a significant decrease in FBM in asymmetrically growth-retarded fetuses, whereas FBM were normal in symmetrically growth-retarded fetuses. This suggests that the hypoxic situation, more likely to occur in asymmetrically growth-retarded fetuses, might be the cause of FBM reduction. Trudinger *et al.* (1979a) also examined growth retarded fetuses and found two patterns of FBM: a decrease in FBM incidence or continuous breathing with a very high percentage of FBM. This, of course, makes the interpretation of clinical results still more difficult.

Platt *et al.* (1978) examined 124 high risk pregnancies within 2 weeks of delivery. They found a significant relationship between the FBM (classified as present or absent) and the outcome of pregnancy judged by Apgar score and birth weight. The same group of authors (Manning and Platt, 1979) found the FBM results useful when interpreting abnormal or equivocal contraction stress tests of fetal heart rate. In a comparative analysis of the prognostic value of the FBM incidence and other routine methods for fetal monitoring, Trudinger *et al.* (1979b) found the FBM examinations had a sensitivity exceeding that of the biochemical methods.

The results of clinical studies suggest that the presence of FBM might be a sign of fetal well-being and the absence of FBM might indicate the fetus at risk. However, caution is necessary when interpreting FBM results. Probably, in future, the prognostic value of the method might improve by evaluating qualities of FBM other than the incidence, using various load tests or including other fetal parameters (e.g. fetal movements, cardiotocogram (CTG)).

3.0 FETAL MOVEMENTS

3.1 Historical Observations

Fetal movements perceived by the mother have always been considered a reassuring sign of fetal life. Reference to Rebecca's description in the Bible of fetal movements is often used as a support for this postulation. In modern times, the time of quickening is used by clinicians as an additional parameter for dating the pregnancy. The motor activity of the fetus, however, is present long before the point of quickening, as has been shown by observations on aborted living fetuses with intact gestational sac (Preyer, 1885) and more recently by real-time ultrasound techniques (Jouppila, 1976). Minkowski (1938) systematically examined the development of the motor behaviour of the fetus. He found various reflex responses to external stimuli and related them to the developing fetal CNS. Mori (1956) observed fetal activity in early pregnancy through a hysteroscope; Hooker (1952) and Humphrey (1970) studied exteriorized human fetuses. The application of the ultrasound technique in obstetrics made it possible to study human FM *in utero*. Reinold (1976) thoroughly analysed various types of FM in early pregnancy both under normal and pathological conditions. He suggested the assessment of the type and frequency of FM, visualized by real-time scanners, as a possible clinical test for evaluation of fetal well-being in early pregnancy.

The disappearance of FM in the second half of pregnancy, when previously felt by the mother, is considered suggestive of fetal death. More than 100 years ago, Ahlfeld (1869) described FM as an indicator of fetal well-being, but it was not until the last decade that this concept was accepted as clinically relevant. Recently, several reports showed that the changes in the incidence of FM, as counted by the mother, might be the first sign of deteriorating fetal health (Sadovsky and Yaffe, 1973; Mathews, 1975; Pearson and Weaver, 1976). However, other authors have shown that a subjective estimate of the daily count of FM is impaired by considerable error (Edwards and Edwards, 1970; Wenderlein, 1975); they discouraged the use of subjective FM counts as a method for assessing the fetal state. The need for an objective method to detect and record FM *in utero* is thus obvious.

3.2 Methods for Objective Recording of FM

3.2.1 *Passive Detection of FM*

The movement of the fetus inside the uterine cavity generates local pressure waves. The waves propagate in the amniotic fluid and tissue with a velocity of approximately 1500 ms^{-1} to the abdominal wall of the mother. These waves might be detected in the simplest way by the *observer's hand* placed on the patient's abdomen (Wood *et al.*, 1977). The observer marks the perceived FM on a chart recorder. However, only FM of high amplitude can be detected in this way and the method also depends on the observer's perceptiveness and reaction time.

The mechanical energy of the FM pressure waves might be detected electronically by placing a suitable receiver or transducer on the maternal abdomen. Possible converters of the mechanical into electrical energy for the detection of FM are the conventional pressure transducers incorporated in cardiotocographic systems used currently. However, the sensitivity of these transducers does not allow the detection of small amplitude FM. Timor-Tritsch *et al.* (1976) used a *tocodynamometer* equipped with custom-built amplifiers. Their method showed a considerable sensitivity by detecting even FBM. They used two transducers placed on the maternal abdomen and were able to recognize four different types of FM (see Section 3.3).

Wood *et al.* (1977) used four *strain gauges* applied to the abdominal wall of the mother. The output signals of the four transducers were recorded graphically on separate channels. The authors found a significant correlation between the number of FM recorded by the instrument and felt by the mother ($r = 0.84$; $P < 0.01$). In all cases, the number of FM recorded by the patient was less than that recorded by the instrument.

In 1977, Sadovsky *et al.* developed a new device for objective recording of FM employing two *piezo-electric crystals*. The crystals converted the sound-pressure wave into an electric signal. The transducers were highly sensitive to rapid movements and less sensitive to slow changes of fetal position. In 20 patients, the authors performed recordings and found that the mothers felt 70% of all FM recorded by the device. In all, the device recorded 90.4% and the women 79.7% of all the FM observed. Later, the authors applied the device on eight women in clinical situations (Sadovsky *et al.* 1979) and could differentiate between weak and strong movements by considering the amplitude of the signals. Furthermore, rolling movements of the fetal body gave biphasic "N" or "W" shaped deflections of the recorded trace with durations of 3–10 s.

In 1978 (c), Maršál *et al.* described a device, called an *FM-detector*, also utilizing the principle of piezo-electric crystals. Preliminary tests with one crystal showed that the recorded signals varied depending on the position of the fetus in the uterine cavity. As the fetus moved around, the spatial distance from the moving fetal parts to the transducer changed, and the recorded pressure waves were differently attenuated. The quality of the recorded signals could be restored by repositioning the transducer, but

(a)

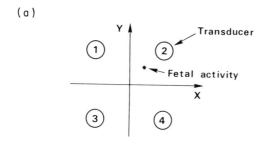

(b)

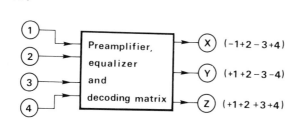

(c)

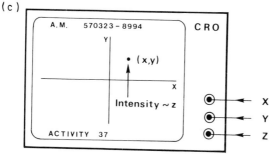

FIG. 12   Principle of the fetal movement detector (FM-detector). (a) Localization of the four transducers on the maternal abdomen. (b) Processing of the primary signals (1, 2, 3, 4). (c) Display of the output vector signals (**X**,**Y**) on an oscilloscope.

this would not be acceptable for clinical use. To reduce the spatial dependence, four separate transducers were used in a fixed-pattern localization on the maternal abdomen. In this way a standardized and consistent total signal of FM could be continuously recorded. This method can recover additional information relating to the spatial distribution of FM (Lindström *et al.*, 1980). The amplitudes of the output signals from the four transducers vary depending on the localization of the transducers in relation to the centre of the momentary maximum of fetal activity. By simple processing of the primary transducer signals, vectors (**X**, **Y**) indicating

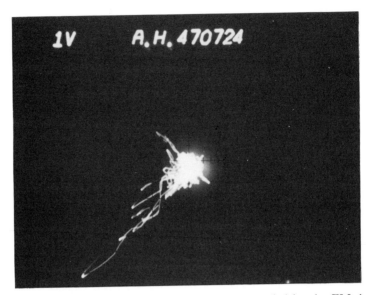

Fɪɢ. 13   Oscilloscope display of fetal movements recorded by the FM-detector during a 5-min period. The fetus is situated with its head downwards and its spine to the right. Movements mainly of the upper extremities were present.

the momentary centre of the fetal activity are obtained (Fig. 12). The sum of the four signals is also calculated (vector **Z**); this represents an arbitrary measure of the total momentary activity.

A data-collecting system which can record and store the signals **X**, **Y** and **Z** is needed. In the present set-up, the signals are displayed on an oscilloscope (Tektronix model 7603) equipped with a read-out unit (7M13) for alphanumeric presentation of results. The position of the oscilloscope beam is at any moment given by the signals **X** and **Y**, and the beam is intensity-modulated by the amplitude of the signal **Z**. Signals displayed on the oscilloscope are photographically integrated during suitable intervals (e.g. 5 min). The resulting photographs (Fig. 13) give an over-all picture of the localization and strength of the total fetal activity during the measurement. Such a graphic presentation has a clinically acceptable accuracy. The amplitude of the total signal (**Z**) is recorded simultaneously on a chart recorder and also electronically integrated to obtain a quantitative measure of the fetal activity.

The signals received by the four transducers differ in two ways: in the amplitude, due to absorption, and in the arrival time, due to the limited velocity of sound in tissue. The original FM-detector displayed the signals directly after full-wave rectification. This caused a display of FM activity signals on the oscilloscope in rather broad loops, similar to

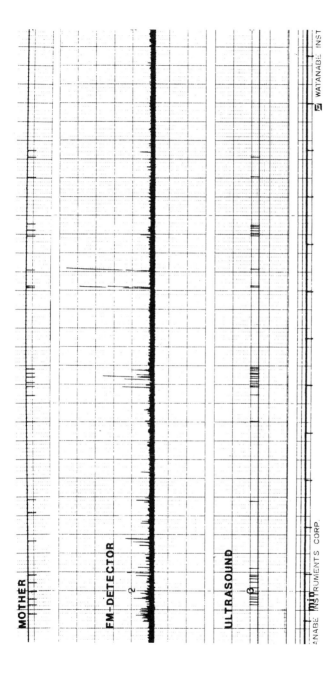

FIG. 14 Polygraphic record of fetal movements (FM) as subjectively felt by the mother, recorded by the FM-detector and observed in a real-time ultrasound image. A good agreement between the three methods is demonstrated.

Lissajou's figures, due to the slightly different arrival times. Recently, a new FM-detector using a sample-and-hold technique in combination with an electronic peak-follower has been designed (Fig. 14) to simplify the

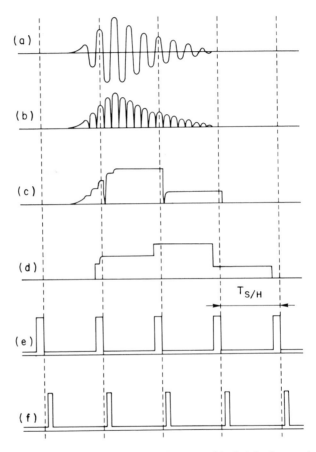

Fig. 15   Signal processing in the FM-detector. (a) Original transducer signal. (b) Absolute value of the signal (a). (c) Positive peak detection of the signal (b) during a predetermined time period ($T_{S/H}$). (d) Sampling and storage of the peak signals (c). Sample period: 1 ms. (e) Pulse signals controlling the sample/hold circuit. (f) Reset signals of the positive peak detector. Duration of the interval $T_{S/H}$ might be predetermined within the range 7–150 ms, thus making the detection of even very fast movements possible.

interpretation of the fetal activity pictures. A future instrument might also utilize the differences in arrival times for an accurate determination of the centre of fetal activity in a way similar to techniques used in seismology.

The FM-detector was clinically tested on 35 pregnant women. The gestational age range was 32–40 weeks, mean 36 weeks. During the recordings, which lasted for 30 min, the patients rested in semirecumbent position. The transducers were attached to the abdomen by double-sided adhesive tape in the manner described above. The output signal (**Z**) was recorded on a polygraph. The fetuses were simultaneously observed with a two-dimensional real-time B-mode scanner; the observer marked each FM on a separate channel of the chart recorder. The women marked the movements they perceived during recording on the third channel (Fig. 15). Photographs of the **X** and **Y** signals as displayed on the oscilloscope were taken for periods of 5 min. The system was arbitrarily calibrated before the start of the recording by dropping a light weight plastic ball on the maternal abdomen from a height of 10 cm. The graphic record of FM was then evaluated manually concerning the incidence and number of FM recorded. The mean number of FM recorded by the FM-detector was 32 per 30 min and the mean incidence 14% of the recording time. Considerable intra- and interindividual variations in the number of movements were present (range 6–177 per 30 min). The FM-detector and the ultrasound method agreed in 84% of the detected FM. The mothers recorded subjectively only 41% of the objectively detected FM; 4% of the FM reported by the mother was not registered by the two-objective methods. The results are shown schematically in Fig. 16.

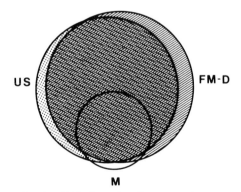

FIG. 16   Agreement in the FM detection between three methods used simultaneously on 35 pregnant women. US: real-time ultrasound (stippled circle); FM-D: FM-detector (hatched circle); M: subjective maternal detection (open circle). The circles are proportional to the detected number of FM.

The FM-detector was found to be very sensitive and easily handled; it produced reliable results. Work is in progress to develop a system for further analysis of the photographs of the two-dimensional display of

FM signals. The FM-detector is now included as one of the clinical tests for surveillance of at-risk pregnancies and for examinations of pregnant women who noted a decrease in subjectively perceived FM. Besides its clinical application, the method can also be used for studies of the basic physiological characteristics of fetal motor activity and for studies of fetal responses to various challenges. Moreover, the FM-detector can be used simultaneously with other methods of fetal monitoring, e.g. fetal heart rate recording or FBM recording. Recently, the transducer array used with the FM-detector has been applied in the analysis of repetitive biological sounds in adults; experiences from the measurements of heart beats and lung sounds are very promising.

3.2.2 Active Detection of FM

The following methods actively introduce a low intensity energy, e.g. alternating current or ultrasound, into the examined object when measuring FM.

Sadovsky et al. (1973) used a device recording changes caused by FM in an *electromagnetic field*. The device could measure the displacements directionally within the magnetic field. In 20 late pregnancies, the electromagnetic device proved more sensitive than the maternal perception of FM. The women recorded 87 % of the FM detected by the instrument.

The principle of *impedance plethysmography* was used by Ehrström (1979). Surface electrodes were applied to the maternal abdomen over the uterus; 30 kHz alternating current was then passed through the tissue. Changes in the electrical impedance of the tissue were recorded on a chart recorder. The results of recordings in 21 pregnancies in the last trimester of gestation were comparable to those reported by Sadovsky et al. (1973) for the electromagnetic device. In the study by Ehrström (1979) the women recorded 90·3 % of the movements detected by the impedance plethysmograph (range 71–100 %).

Various *ultrasound techniques* have been used for the detection and recording of FM. In early pregnancy, deflections of the TM-mode tracing of fetal echoes were interpreted as an indication of FM (Schillinger, 1976). Repeated B-mode compound scanning was used both in early and in late pregnancy (Higginbottom et al. 1976). In late pregnancy, continuous Doppler ultrasound was also applied for FM detection (Resch and Herczog, 1971). All these methods, however, were relatively unreliable and the identification of the moving target unsatisfactory.

Recently, *B-mode real-time scanners* have become generally available in prenatal care. They offer moving, two-dimensional images of the fetus. In early pregnancy, the real-time method is suitable for FM observations, as a sectional image of the whole uterus housing the fetus can be obtained. Jouppila (1976) and Reinold (1976) made extensive examinations of various types of FM before the 20th week of pregnancy. Jouppila (1976) found the real-time technique to be 100% reliable for the detection of FM after the 8th gestational week.

In late pregnancy, usually only a part of the fetal body and/or fetal limbs can be visualized by the real-time scanner. None the less, proper orientation of the real-time scanner can achieve a projection in which at least a part of each moving fetal structure is visualized on most of the FM occasions. Both linear array (Patrick *et al.*, 1978; Roberts *et al.*, 1979) and phased array systems (Stephens and Birnholz, 1978) have been used for FM detection. The ultrasonic images were evaluated visually by the

TABLE IV

Classification of fetal movement types recorded by objective methods

Author	Recording technique	FM types
Lindström and Maršál (present study)	Real-time ultrasound	Rolling movements Stretching movements Head bobbing Isolated limb movements Spasmodic movements (startle-like) Breathing movements Hiccups
Timor-Tritsch *et al.* (1976)	Tocodynamometry	Rolling movement Simple movement High-frequency movement (1) isolated (2) repetitive Respiratory movement
Sadovsky *et al.* (1979)	Piezo-electric crystals	Weak movements Strong movements Rolling movements
Birnholz *et al.* (1978)	Real-time ultrasound	Twitchy movement Independent limb movement Isolated head movement Combined repetitive movement Quasi-startle movement Limb-joint movement Head-face movement Diaphragm movement Respiratory movement

operator, and the observed FM were marked either on a chart or by punching a tape. Usually, FM incidence and the number of FM episodes per time unit were assessed quantitatively. The operator observing the screen of the scanner might be a possible source of error. Roberts *et al.* (1979) investigated the reproducibility of FM records by repeatedly examining videotape recordings of real-time images. They found the mean standard deviation to be 1·98% for different observers and 1·25% for the same observer when analysing the percentage incidence of FM. The reproducibility for the number of FM was somewhat less but still sufficient.

Steps have been taken to analyse objectively other qualities of FM in the real-time display, e.g. the acceleration and velocity of the movements (Henner *et al.*, 1974). In this method videorecordings of real-time images were analysed by following a chosen moving echo manually on the video-screen with an electronic graticule, and each trace was recorded on an **XY** plotter. Obviously, the method was very laborious and did not find any further application.

At present, real-time ultrasound and the FM-detector are the two methods in practical use for the objective recording of FM. Both methods have their advantages and disadvantages. The real-time ultrasound system visualizes even very small movements provided that they pass the plane of the projection. The number and quality of FM have to be evaluated subjectively. The FM-detector functions in a semi-automatic way, but a certain minimum intensity of FM is needed to give a signal. Non-specific movements, e.g. maternal pulse, maternal movements or uterine contractions, may give rise to artefactual signals. In most cases, however, such artefacts are easily recognized in the chart record and disregarded. The FM-detector offers information not only on the number and incidence of FM, but also on the strength and localization of FM.

3.3 Physiological and Clinical Studies of FM

The experience of subjective recordings of FM by mothers showed that there is a considerable interindividual and intra-individual variation in the number of recorded movements (Wood *et al.*, 1979; Valentin *et al.*, 1980). Diurnal variation in FM was observed with a maximum in the evening (Minors and Waterhouse, 1979; Goodlin and Lowe, 1974). Minors and Waterhouse (1979) found a relation between the maternal posture and the number of movements, most movements being detected with mothers in a lying position. They did not find that maternal meal-times affected FM. Several authors reported a decrease in the number of

FM toward term (Pearson and Weaver, 1976; Spellacy et al. 1977), but this was not confirmed by others (Wood et al., 1977; Valentin et al., 1980). To date, a number of papers on the subjective recording of FM have been published, but it is difficult to compare the studies, as they use different recording times and different methods for evaluation of FM.

Subjective assessment of the daily movement count has been suggested as useful in the evaluation of fetal well-being (Mathews, 1975; Sadovsky and Yaffe, 1973; Pearson and Weaver, 1976; Spellacy et al., 1977). A reduction of FM usually precedes intrauterine fetal death and is considered an alarm signal (Sadovsky and Yaffe, 1973). Gettinger et al. (1978) demonstrated the relatively good reliability of subjective counting of FM by comparing it with real-time ultrasound recording. In pregnancies with low FM counts, however, this reliability was impaired. Therefore it is desirable to confirm the decrease of FM by objective methods. Valentin et al. (1980) developed a method for assessment of individual low limits in subjective recording of FM. This method is now used for screening the whole pregnant population in the Malmö district, and fetuses with a decrease in FM count below their own low limit are subjected to further evaluation by the FM-detector described above.

By using objective methods for the detection of FM, several types of movements have been described (Table IV). The mean FM incidence in uncomplicated pregnancy during the last trimester of gestation varied from 8 to 18% of recording time (Patrick et al., 1978; Roberts et al., 1979). Birnholz et al. (1978) distinguished nine types of FM and localized in time their first occurrence during pregnancy. The latter study demonstrated the suitability of the real-time ultrasonic technique for studies of fetal motor behaviour through the pregnancy. In their 24-h study, Patrick et al. (1978) observed distinct periods of increased activity interspersed by periods of fetal inactivity. However, no pronounced periodicity similar to that of FBM could be discovered (Patrick et al., 1978; Maršál and Gennser, 1980). Roberts et al. (1979) calculated a percentage incidence of total fetal activity (TFA) which they defined as a percentage of the total observation time spent by the fetus either performing FBM or FM. The TFA incidence in normal pregnancies was 48%. The normal TFA rarely falls below 10% in any half-hour period during the day. In a group of fetuses with intrauterine growth retardation, the same authors found a decreased incidence of FM and TFA and suggested that this method was a clinically useful test for evaluating risk pregnancies. Trudinger et al. (1978) observed FM and FBM in 50 high-risk pregnancies and found an incidence lower than that in the control group. However, the FM incidence was found to be a less predictive parameter than FBM incidence, probably due to the wide distribution of FM in normal fetuses.

The potential of objective recording of FM in late pregnancy is still relatively unexplored. Normal fetal activity may be reassuring in terms of fetal well-being. A change in the type and number of FM is indicative of impaired fetal health and should initiate an intensified surveillance of the fetus.

4.0 SUMMARY

Fetal breathing movements (FBM) have been observed previously in various animal species. The existence of FBM in the human has been proven conclusively by real-time B-mode ultrasound. FBM are of episodic character and appear normally in uncomplicated pregnancies. Configurative changes of the fetal chest and abdomen during FBM usually have a see-saw character with retraction of the fetal chest and expansion of the fetal abdominal wall in the inspiratory phase with a return to the initial shape during the expiratory phase. FBM are mainly diaphragmatic and appear on average for 30% of the observation time. FBM show a great interindividual and intra-individual variability; prolonged periods of apnoea are observed in normal pregnancies. FBM incidence is subject to diurnal variation and is correlated with changes in maternal blood glucose levels.

FBM may be recorded by classical kymographic methods as pressure changes at the surface of maternal abdomen, but these methods have a low success rate. Therefore ultrasonic techniques have become the method of choice for recording FBM. The original A-mode method was discontinued because of the presence of artefacts. The Doppler shift frequencies caused by FBM can be recorded both using continuous and pulsed wave Doppler ultrasound. The most detailed information on FBM is obtained by using the real-time B-mode ultrasound technique. FBM are readily visualized in the real-time image and the incidence and rate of FBM can be evaluated by marking each observed FBM cycle. This simple method of evaluating FBM is widely used. An objective method is necessary for measurements of breath-to-breath intervals and of the time-course of a single FBM cycle. For this purpose, time-distance recording and TM-mode recording have been applied. Both methods process the echo signals from a selected line within the real-time image and provide a chart record of FBM signals for further quantitative evaluation.

Attempts were made to evaluate the clinical usefulness of the FBM investigations. The presence of FBM seems to be a reassuring sign of fetal health. The absence of FBM is less predictive in individual cases.

Preloading with 50 g glucose *per os* was recommended for optimization and standardization of the FBM examination. The recording time should be sufficiently long: at least 1 h. More knowledge is needed about the physiology of FBM and about their control mechanisms before large-scale clinical application of FBM recordings.

Fetal limb and body movements (FM) can be perceived subjectively by the mother in late pregnancy and can also be recorded objectively by several methods. Movements of various fetal parts can be visualized, counted and assessed quantitatively in real-time ultrasound images. However, FM outside the plane of the sectional image might be missed.

Mechanical energy transmitted to the maternal abdominal wall as a pressure wave from moving fetal parts can be detected and recorded. A new device, the FM-detector, has been designed which uses four piezo-electric crystals as transducers attached to the maternal abdomen. FM are recorded as analogue signals. The intensity, number, and incidence of movements are evaluated by a computer program and their localization is recorded in a two-dimensional **XY**-mode on an oscilloscope screen. Both the FM-detector and the real-time ultrasound method showed high reliability and reproducibility in the detection of FM and were superior to subjective counting. The two methods seem to complement each other and can be used in combination.

The results of preliminary clinical studies suggest that the presence of FBM and a normal pattern of FM might be a sign of fetal well-being and that the absence of FBM and a changed pattern of FM indicate a fetus at risk. The fetal activity examination seems to possess valuable potential as an additional clinical test to be used in the assessment of the fetal condition.

Acknowledgements

The studies by the authors, referred to in this review, were supported in part by grants from the Faculty of Medicine, University of Lund, the "Expressen" Prenatal Research Foundation, and the Swedish Medical Research Council (Grant B 81-17X-05980-01).

References

Abramovich, D. R. (1970). Fetal factors influencing the volume and composition of liquor amnii. *J. Obstet. Gynaecol. Br. Commonw.* 77, 865–877.

Ahlfeld, F. (1869). Ueber die Dauer der Schwangerschaft. *Monatsschrift für Geburtskunde.* 34, 180.

Ahlfeld, F. (1888). Ueber bisher noch nicht beschriebene intrauterine Bewegungen des Kindes. *Verhandlungen der Deutschen Gesellschaft für Gynäkologie, Zweiter Kongress*, pp. 203–210, Breitkopf und Härtel, Leipzig.

Barcroft, J. and Barron, D. H. (1937). The genesis of respiratory movements in the foetus of the sheep. *J. Physiol.* **88**, 56–61.

Birnholz, J. C., Stephens, J. C. and Faria, M. (1978). Fetal movement patterns: A possible means of defining neurologic developmental milestones *in utero. Am. J. Roentgenol.* **130**, 537–540.

Bishop, E. H. (1966). Obstetric uses of the ultrasonic motion sensor. *Am. J. Obstet. Gynecol.* **96**, 863–867.

Boddy, K. and Dawes, G. S. (1975). Fetal breathing. *Br. Med. Bull.* **31**, 3–7.

Boddy, K. and Mantell, C. D. (1972). Observations of fetal breathing movements transmitted through maternal abdominal wall. *Lancet* **ii**, 1219–1220.

Boddy, K. and Robinson, J. S. (1971). External method for detection of fetal breathing *in utero. Lancet* **ii**, 1231–1233.

Boddy, K., Dawes, G. S. and Robinson, J. S. (1973). A 24-hour rhythm in the foetus. *In* "Foetal and Neonatal Physiology", Proceedings of the Sir J. Barcroft Centenary Symposium (Eds R. S. Comline, K. W. Cross, G. S. Dawes and P. W. Nathanielsz), pp. 63–66, Cambridge University Press, Cambridge.

Bots, R. S. G. M., Farman, D. J. and Broeders, G. H. B. (1976). Multiscan echofetography: Application to the study of fetal breathing movements. *Eur. J. Obstet. Gynecol. Reprod. Biol.* **6**, 271–275.

Bots, R. S. G. M., Broeders, G. H. B., Farman, D. J., Haverkorn, M. J. and Stolte, L. A. M. (1978). Fetal breathing movements in the normal and growth-retarded human fetus: A multiscan M-mode echofetographic study. *Eur. J. Obstet. Gynecol. Reprod. Biol.* **8**, 21–29.

Boyce, E. S., Dawes, G. S., Gough, J. D. and Poore, E. R. (1976). Doppler ultrasound method for detecting human fetal breathing *in utero. Br. Med. J.* **2**, 17–18.

Campbell, K., MacNeill, J. and Patrick, J. (1980). Time series analysis of human fetal breathing movements at 30–39 weeks gestational age. *J. Biomed. Engng* **2**, 108–112.

Chapman, R. L. K., Dawes, G. S., Rurak, D. W. and Wilds, P. L. (1978). Intermittent breathing before death in fetal lambs. *Am. J. Obstet. Gynecol.* **131**, 894.

Chernick, V. and Bahoric, A. (1974). Output of the fetal respiratory centre in utero. *Paediatr. Res.* **8**, 465.

Davis, M. E. and Potter, E. L. (1946). Intrauterine respiration of the human fetus. *J. Am. Med. Assoc.* **131**, 1194–1201.

Dawes, G. S. (1973). Breathing and rapid-eye-movement sleep before birth. *In* "Foetal and Neonatal Physiology," Proceedings of the Sir. J. Barcroft Centenary Symposium (Eds R. S. Comline, K. W. Cross, G. S. Dawes and P. W. Nathanielsz), pp. 49–62, Cambridge University Press, Cambridge.

Dawes, G. S., Fox, H. E., Leduc, B. M. Liggins, G. C. and Richards, R. T. (1970). Respiratory movements and paradoxical sleep in the foetal lamb. *J. Physiol.* **210**, 47P–48P.

Dawes, G. S., Fox, H. E., Leduc, B. M., Liggins, G. C. and Richards, R. T. (1972). Respiratory movements and rapid eye movement sleep in the foetal lamb. *J. Physiol.* **220**, 119–143.

de Blasio, A., Ambrosio, G. and D'Amora, G. (1960). Sui movimenti respiratori endouterini del feto umano a termine. *Pediatria (Napoli)* **68**, 1124–1141.

de Wolf, F. (1976). Permanent recording of fetal chest-wall movements *in utero. Lancet* **ii**, 914.

Dietel, K. (1955). Die Atemzahlen des Kindes vor und nach der *Geburt. Wschr. Kinderheilk.* **103**, 449–451.

Duenholter, J. H. and Pritchard, J. A. (1973). Human fetal respiration. *Obstet. Gynecol.* **42**, 746–750.

Duenholter, J. H. and Pritchard, J. A. (1977). Fetal respiration. A review. *Am. J. Obstet. Gynecol.* **129**, 326–338.

Dyroff, R. (1927). Gibt es regelmässige intrauterine Atembewegungen? *Zbl. Gynäk.* **51**, 967–970.

Edwards, D. D. and Edwards, J. S. (1970). Fetal movement. Development and time course. *Science* **169**, 95–97.

Ehrström, C. (1979). Fetal movement monitoring in normal and high-risk pregnancy. *Acta Obstet. Gynecol. Scand.* Suppl. 80.

Eik-Nes, S. H., Brubakk, A. O. and Ulstein, M. (1980a). Measurement of human fetal blood flow. *Br. Med. J.* **280**, 283–284.

Eik-Nes, S. H., Maršál, K., Brubakk, A. O. and Ulstein, M. (1980b). Ultrasonic measurements of human fetal blood flow in aorta and umbilical vein: Influence of fetal breathing movements. *In* "Recent Advances in Ultrasound Diagnosis", Vol. 2, (Ed. A. Kurjak), pp. 233–240, Excerpta Medica, Amsterdam.

Erhardt, K. (1939). Atmet das Kind im Mutterleib? *Münch. med. Wschr.* **86**, 915–918.

Farman, D. J., Thomas, G. and Blackwell, R. J. (1975). Errors and artifacts encountered in the monitoring of fetal respiratory movements using ultrasound. *Ultrasound Med. Biol.* **2**, 1–6.

Fox, H. E., Hohler, C. W., Jaeger, H., Steinbrecher, M. and Peco, N. (1977). A preliminary report of an alteration of human fetal breathing associated with the glucose tolerance test and the oxytocin challenge test. *Proceedings 3rd Conference on Fetal Breathing*, Malmö, June 8, 1976 (Eds G. Gennser, K. Maršál and T. Wheeler), pp. 56–62.

Gennser, G. (1979). Spatial and temporal characteristics of fetal breathing movements in man. *In* "Central Nervous Control Mechanisms in Breathing" (Ed. C. v. Euler and H. Lagercrantz), pp. 375–388, Pergamon Press, Oxford.

Gennser, G. and Maršál, K. (1975). Fetal breathing movements in man. Film presented at the 2nd Conference on Fetal Breathing, Nuffield Institute for Medical Research, Oxford, October 3rd, 1975.

Gennser, G. and Maršál, K. (1979). Fetal breathing movements monitored by real-time B-mode ultrasound: Basal appearance and response to challenges. *In* "Contributions to Gynecology and Obstetrics", Vol. 6 (Ed. R. Chef), pp. 66–79, Karger, Basel.

Gennser, G., Maršál, K. and Brantmark, B. (1975). Maternal smoking and fetal breathing movements. *Am. J. Obstet. Gynecol.* **123**, 861–867.

Gettinger, A., Roberts, A. B. and Campbell, S. (1978). Comparison between subjective and ultrasound assessments of fetal movement. *Br. Med. J.* **2**, 88–90.

Geyl, A. (1880). Die Aetiologie der sogennanten "puerperalen Infection" des Fötus und des Neugeborenen. *Arch. Gynäk.* **15**, 384–411.

Goodlin, R. C. and Lowe, E. W. (1974). Multiphasic fetal monitoring *Am. J. Obstet. Gynecol.* **119**, 341–357.

Goodman, I. D. S. and Mantell, C. D. (1980). A second means of identifying fetal breathing movements using Doppler ultrasound. *Am. J. Obstet. Gynecol.* **136**, 73–74.

Gough, J. D. and Poore, E. R. (1977). Directional Doppler measurements of foetal breathing. *J. Physiol.* **272**, 12P–13P.

Gough, J. D. and Poore, E. R. (1979). A continuous wave Doppler ultrasound method of recording fetal breathing in utero. *Ultrasound Med. Biol.* **5**, 249–256.

Hathorn, M. K. S. (1975). Analysis of the rhythm of infantile breathing. *Br. Med. J.* **31**, 8–12.

Henner, H., Haller, U., Wolf-Zimper, O., Lorenz, W. J., Bader, R., Müller, B. and Kubli, F. (1975). Quantification of fetal movement in normal and pathologic pregnancy. *In* "Ultrasonics in Medicine" (Eds E. Kazner, M. deVlieger, H. R. Müller and V. R. McCready), pp. 316–319, Excerpta Medica, Amsterdam.

Higginbottom, J., Bagnall, K. M., Harris, P. F., Slater, J. H. and Porter, G. A. (1976). Ultrasound monitoring of fetal movements. *Lancet* i, 719–721.

Hogg, M. I. J., Golding, R. H. and Rosen, M. (1977). The effect of doxapram on fetal breathing in the sheep. *Br. J. Obstet. Gynaecol.* **84**, 48–50.

Hohler, C. W. and Fox, H. E. (1976). Real-time grayscale B-scan ultrasound recording of human fetal breathing in utero. "Ultrasound in Medicine," Vol. 2 (Eds D. N. White and R. Barnes), pp. 203–206, Plenum Publishing, New York.

Hooker, D. (1952). "The Prenatal Origin of Behaviour", Porter Lectures, Series 18, University of Kansas Press, Lawrence, Kansas.

Humphrey, T. (1970). *In* "The Physiology of the Perinatal Period" (Ed. A. Stove), Appleton-Century-Crofts, New York.

Jaeger, H. (1919). Kongenitale gelenkige Verbindung von Exostosen der Rippen und Ahlfelds Lehre der intrauterinen Atembewegungen. *Correspondenz-Blatt f. Schweizer Ärzte.* **39**, 1461–1464.

Jouppila, P. (1976). Fetal movements diagnosed by ultrasound in early pregnancy. *Acta Obstet. Gynecol. Scand.* **55**, 131–135.

Korba, L. W., Cobbold, R. S. and Cousin, A. J. (1979). Ultrasonic imaging and differential measurement system for the study of fetal respiratory movements. *Ultrasound Med. Biol.* **5**, 139–148.

Lewis, P. J., Trudinger, B. J. and Mangez, J. (1978). Effect of maternal glucose ingestion on fetal breathing and body movements in late pregnancy. *Br. J. Obstet. Gynaecol.* **85**, 86–89.

Lindström, K., Maršál, K., Gennser, G., Bengtsson, L., Benthin, M. and Dahl, P. (1977). Device for measurement of fetal breathing movements. I. The TD-recorder. A new system for recording the distance between two echo-generating structures as a function of time. *Ultrasound Med. Biol.* **3**, 143–151.

Lindström, K., Maršál, K. and Ulmsten, U. (1981). (Ed. P. Rolfe), Fetal movement detector *In* "Foetal and Neonatal Physiological Measurements", pp. 233–243. Pitman Medical Ltd, London.

McHugh, R., McDicken, W. N., Bow, C. R., Anderson, T. and Boddy, K. (1978). An ultrasonic pulsed Doppler instrument for monitoring human fetal breathing *in utero*. *Ultrasound Med. Biol.* **3**, 381.

Maloney, J. E., Adamson, T. M., Brodecky, V., Cranage, S., Lambert, T. F. and Ritchie, B. C. (1975). Diaphragmatic activity and lung liquid flow in the unanesthetized fetal sheep. *J. Appl. Physiol.* **39**, 423–428.

Manning, F. A. and Platt, L. D. (1979). Fetal breathing movements and the abnormal contraction stress test. *Am. J. Obstet. Gynecol.* **133**, 590.

Mantell, C. D. (1976). Breathing movements in the human fetus. *Am. J. Obstet. Gynecol.* **125**, 550–553.

Maršál, K. (1977). Ultrasonic measurements of fetal breathing movements in man. Thesis, Malmö.

Maršál, K. (1978). Fetal breathing movements—characteristics and clinical significance. *Obstet. Gynecol.* **52**, 394–401.

Maršál, K. and Eik-Nes, S. H. (1980). Effects of fetal breathing on blood flow in human fetus. *Proceedings 7th International Workshop on Fetal Breathing*, Oxford 1980, Abstract No. 19.

Maršál, K. and Gennser, G. (1980). Fetal breathing movements examinations: Research tool and/or clinical test? *In* "Recent Advances in Ultrasound Diagnosis" Vol. 2 (Ed. A. Kurjak), pp. 262–274, Excerpta Medica, Amsterdam.

Maršál, K., Gennser, G., Hansson, G.-Å., Lindström, K. and Mauritzsson, L. (1976a). New ultrasonic device for monitoring fetal breathing movements. *Biomed. Engng* **11**, 47–52.

Maršál, K., Gennser, G. and Lindström, K. (1976b). Real-time ultrasonography for quantified analysis of fetal breathing movements in man. *Lancet* i, 718–719.

Maršál, K., Gennser, G., Lindström, K. and Ulmsten, U. (1978a). Errors and pitfalls in ultrasonic measurements of fetal breathing movements. *In* "Recent Advances in Ultrasound Diagnosis" (Ed. A. Kurjak), pp. 200–208, Excerpta Medica, Amsterdam.

Maršál, K., Ulmsten, U. and Lindström, K. (1978b). Device for measurement of fetal breathing movements. II. Accuracy of *in vitro* measurements, filtering of output signals, and clinical application. *Ultrasound Med. Biol.* **4**, 13–26.

Maršál, K., Ulmsten, U. and Lindström, K. (1978c). Detektor för kontinuerlig registrering av fosterrörelser. *Hygiea, Acta Societat. Medic. Suecan.* **87**, 432.

Martin, C. B., Jr, Murata, Y., Petrie, R. H. and Parer, J. T. (1974). Respiratory movements in fetal rhesus monkeys. *Am. J. Obstet. Gynecol.* **119**, 939–948.

Mathews, D. D. (1975). Maternal assessment of fetal activity in small-for-dates infants. *Obstet. Gynecol.* **45**, 488–493.

Meire, H. B., Fish, P. J. and Wheeler, T. (1975). Ultrasound recording of fetal breathing. *Br. J. Radiol.* **48**, 477–480.

Merlet, C., Hoerter, J., Devilleneuve, C. and Tchobroutsky, C. (1970). Mise en évidence de mouvements respiratoires chez le foetus d'agneau *in utero* au cours du dernier mois de la gestation. *C. R. Acad. Sci. Paris* **270**, 2462–2464.

Minkowski, M. (1938). Neurobiologische Studien am menschlichen Fötus. *In* "Handbuch der biologischen Arbeitsmetoden" (Ed. Abderhalden), Abt. 5, Teil 5B, Urban and Schwarzenberg, Berlin.

Minors, D. S. and Waterhouse, J. M. (1979). The effect of maternal posture, meals and time of day on fetal movements. *Br. J. Obstet. Gynaecol.* **86**, 717–723.

Mori, C. (1956). *Jpn. J. Obstet. Gynecol.* **3**, 374.

Natale, R., Patrick, J. and Richardson, B. (1978). Effects of human maternal venous plasma glucose concentrations on fetal breathing movements. *Am. J. Obstet. Gynecol.* **132**, 36–41.

Olinsky, A., Bryan, M. H. and Bryan, A. C. (1974). Influence of lung inflation on respiratory control in neonates. *J. Appl. Physiol.* **36**, 426–429.

Patrick, J. E., Dalton, K. J. and Dawes, G. S. (1976). Breathing patterns before death in fetal lambs. *Am. J. Obstet. Gynecol.* **125**, 73–78.

Patrick, J., Natale, R. and Richardson, B. (1978). Patterns of human fetal breathing activity at 34 to 35 weeks' gestational age. *Am. J. Obstet. Gynecol.* **132**, 507–513.

Patrick, J., Campbell, K., Carmichael, L., Natale, R. and Richardson, B. (1980). A definition of human fetal apnea and the distribution of fetal apneic intervals during the last ten weeks of pregnancy. *Am. J. Obstet. Gynecol.* **136**, 471–477.

Pearson, J. F. and Weaver, J. B. (1976). Fetal activity and fetal wellbeing: an evaluation. *Br. Med. J.* **1**, 1305–1307.

Persson, P.-H. and Maršál, K. (1978). Monitoring of fetuses with retarded BPD growth. *Acta Obstet. Gynecol. Scand.* Suppl. 78, 49–55.

Plätt, L. D., Manning, F. A., Lemay, M. and Sipos, L. (1978). Human fetal breathing: Relationship to fetal condition. *Am. J. Obstet. Gynecol.* **132**, 514–518.

Preyer, W. (1882). Über die erste Athembewegung des Neugeborenen. *Z. Geburtshilfe* **7**, 241–253.

Preyer, W. (1885). "Specielle Physiologie des Embryo", Griebens Verlag, Leipzig.

Reifferscheid, K. (1911). Über intrauterine im Rhythmus der Atmung erfolgende Muskelbewegungen des Fötus. (Intrauterine Atmung.) *Pflügers Arch. ges. Physiol.* **140**, 1–16.

Reinold, E. (1976). Ultrasonics in early pregnancy. "Contributions to Gynecology and Obstetrics", Vol. 1, Karger, Basel.

Resch, B. and Herczog, J. (1971). Klinische Erfahrungen mit dem Ultraschall-"Bewegungsdetektor" Doptone. *Zbl. Gynäk.* **83**, 86–92.

Ritchie, K. J. and Lackarney, K. (1979). Fetal breathing movements in normal antenatal patients in response to 5 percent carbon dioxide inhalation. *Br. J. Obstet. Gynaecol.* **86**, 491.

Roberts, A. B., Little, D. and Campbell, S. (1978a). 24 hours studies of fetal respiratory movements and fetal body movements: Relationship to glucose, catecholamine, oestriol, and cortisol levels. *In* "Recent Advances in Ultrasound Diagnosis" (Ed. A. Kurjak), pp. 189–191, Excerpta Medica, Amsterdam.

Roberts, A. B., Little, D. and Campbell, S. (1978b). 24 hour studies of fetal respiratory movements and fetal body movements in five growth-retarded fetuses. *In* "Recent Advances in Ultrasound Diagnosis" (Ed. A. Kurjak), pp. 192–193, Excerpta Medica, Amsterdam.

Roberts, A. B., Little, D., Cooper, D. and Campbell, S. (1979). Normal patterns of fetal activity in the third trimester. *Br. J. Obstet. Gynaecol.* **86**, 4.

Sadovsky, E. and Yaffe, H. (1973). Daily fetal movement recording and fetal prognosis. *Obstet. Gynecol.* **41**, 845–850.

Sadovsky, E., Polishuk, W. Z., Mahler, Y. and Malkin, A. (1973). Correlation between electromagnetic recording and maternal assessment of fetal movement. *Lancet* **i**, 1141–1143.

Sadovsky, E., Polishuk, W. Z., Yaffe, H., Adler, D., Pachys, F. and Mahler, M. (1977). Fetal movements recorder, use and indications. *Int. J. Gynecol. Obstet.* **15**, 20–24.

Sadovsky, E., Laufer, N. and Allen, J. W. (1979). The incidence of different types of fetal movements during pregnancy. *Br. J. Obstet. Gynaecol.* **86**, 10–14.

Schillinger, H. (1977). Quantitative Untersuchungen zur embryonalen Motorik mit dem Ultraschall Time-motion Verfahren. *Arch. Gynäk.* **222**, 137–147.

Snyder, F. F. (1941). The rate of entrance of amniotic fluid into the pulmonary alveoli during fetal respiration. *Am. J. Obstet. Gynecol.* **41**, 224–230.

Spellacy, W. N., Cruz, A. C., Gelmán, S. R. and Buhi, W. C. (1977). Fetal movements and placental lactogen levels for fetal-placental evaluation. *Obstet. Gynecol.* **49**, 113–115.

Stephens, J. D. and Birnholz, J. C. (1978). Noninvasive verification of fetal respiratory movements in normal pregnancy. *J. Am. Med. Assoc.* **240**, 35–37.

Timor-Tritsch, I., Zador, I., Hertz, R. H. and Rosen, M. G. (1976). Classification of human fetal movement. *Am. J. Obstet. Gynecol.* **126**, 70–77.

Timor-Tritsch, I., Zador, I., Hertz, R. H. and Rosen, M. G. (1977). Human fetal respiratory arrhythmia. *Am. J. Obstet. Gynecol.* **127**, 662–666.

Timor-Tritsch, I., Dierkel, L. J. Jr, Hertz, R. H., Zador, I. and Rosen, M. G. (1979). Human fetal respiratory movements: Techniques for non-invasive monitoring with the use of a tocodynamometer. *Biol. Neonate* **36**, 18–24.

Towell, M. E. and Salvador, H. S. (1974). Intrauterine asphyxia and respiratory movements in the fetal goat. *Am. J. Obstet. Gynecol.* **118**, 1124–1131.

Tremewan, R. N., Aickin, D. R. and Tait, J. J. (1976). Ultrasonic monitoring of fetal respiratory movement. *Br. Med. J.* **1**, 1434–1435.

Trudinger, B. J., Lewis, P. J., Mangez, J. and O'Connor, E. (1978). Fetal breathing movements in high-risk pregnancy. *Br. J. Obstet. Gynaecol.* **85**, 662.

Trudinger, B. J., Lewis, P. J. and Pettit, B. (1979a). Fetal breathing patterns in intrauterine growth retardation. *Br. J. Obstet. Gynaecol.* **86**, 432.

Trudinger, B. J., Gordon, Y. B., Grudzinskas, J. G., Hull, M. G. R., Lewis, P. J. and Lozana Arrans, M. E. (1979b). Fetal breathing movements and other tests of fetal well-being: A comparative evaluation. *Br. Med. J.* **2**, 577–579.

Valentin, L., Löfgren, O. and Maršál, K. (1980). Subjective recording of fetal movements as a screening method for detection of fetal jeopardy. *Acta Obstet. Gynecol. Scand., Suppl.* **93**, 39.

Walz, W. (1922) Über die Bedeutung der intrauterinen Atembewegungen. *Mschr. Geburtshilfe Gynäk.* **60**, 331–341.

Weber, H. (1888). Über physiologische Athmungsbewegungen des Kindes im Uterus. Inaugural Dissertation, Universität Marburg. G. Schirling, Marburg.

Wenderlein, J. M. (1975). Erleben der Fetalbewegungen. *Z. Geburtshilfe Perinatol.* **179**, 377–382.

Wheeler, T., Gennser, G., Lindvall, R. and Murrills, A. J. (1980). Changes in the fetal heart rate associated with fetal breathing and fetal movement. *Br. J. Obstet. Gynaecol.* **87**, 1068–1079.

Wilds, P. L. (1978). Observations of intrauterine fetal breathing movements— A review. *Am. J. Obstet. Gynecol.* **131**, 315–338.

Windle, W. F., Monnier, M. and Steele, A. G. (1938). Fetal respiratory movements in the cat. *Physiol. Zool.* **11**, 425–433.

Winslowius (1787). Quoted by P. Scheel (1798), Dissertatio inauguralis physiologica de liquore amnii asperae arteriae foetuum humanorum. N. Christensen, Hafniae.

Wladimiroff, J. W., Ligtvoet, C. M. and Spermon, J. A. (1976). Combined one- and two-dimensional ultrasound system for monitoring fetal breathing movements. *Br. Med. J.* **2**, 975.

Wladimiroff, J. W., van Weering, H. K. and Roodenburg, P. J. (1978). Fetal respiratory responses to changes in fetal blood gases and maternal drug administration. *In* "Medicina Fetale", Simposio Internazionale, Gorizia 1978, pp. 97–104, Monduzzi Edizioni, Bologna.

Wood, C., Walters, W. A. W. and Trigg, P. (1977). Methods of recording fetal movements, *Br. J. Obstet. Gynaecol.* **84**, 561–567.

Wood, C., Gilbert, M., O'Connor, A. and Walters, W. A. W. (1979). Subjective recording of fetal movement. *Br. J. Obstet. Gynaecol.* **86**, 836–842.

3. THE NON-INVASIVE ASSESSMENT OF UTERINE ACTIVITY

J. Nagel and M. Schaldach

Zentralinstitut für Biomedizinische Technik der Universität Erlangen-Nürnberg, Erlangen, FRG

1.0 INTRODUCTION

The monitoring of uterine activity is of importance in prenatal medicine for the surveillance of the course of the pregnancy, and for assessing the condition of the fetus. The relevant information about uterine activity can be derived from mechanical effects, such as deformation, tension in the abdominal wall, changes in intrauterine pressure, and from electrical phenomena produced by the uterus, i.e. the muscle potentials which trigger the mechanical activity.

Since the uterus is not readily accessible for the direct recording of the mechanical aspects of its activity, the results of such measurements, as performed clinically, provide only little information about the state of the uterus. Normally, the activity is evaluated in terms of the intrauterine pressure, usually in the form of the so-called tocogram. Two methods, one invasive and the other non-invasive, may be employed. For internal (invasive) tocography, the pressure is determined by introducing a catheter-tip pressure-transducer, or a catheter connected to an external transducer, into the uterine cavity. With this method, it is possible to determine the absolute intrauterine pressure. Often, however, internal tocography is not employed because of the practical difficulties involved. Moreover, the insertion of a catheter may stimulate parturition, so that this method may be contra-indicated for possible premature deliveries,

NON-INVASIVE MEASUREMENTS: 2 *Copyright©1983 by Academic Press Inc. (London) Ltd.*
ISBN 0 12 593402 5

where monitoring of the uterine contraction might be of particular clinical value. External (non-invasive) tocography, in which intrauterine pressure changes are inferred from measurements on the abdominal wall, avoids these disadvantages of the invasive measuring techniques. However, it has the drawback of being only a relative measurement, and is also subject to numerous interfering influences which may severely affect the results it provides.

Although invasive and non-invasive tocography have, until fairly recently, been the only methods employed for the assessment of uterine activity, it has been shown that they do not allow detailed information about the mechanical activity of the myometrium to be determined. Simultaneous recordings of intrauterine and intramural pressure reveal local increases in wall tension that are not always mirrored by the intrauterine pressure.

Since the fundamental problems of tocography cannot be eliminated by technical improvements to the measuring equipment, it would seem justifiable to consider assessing uterine activity by processing the uterine electrical action potentials, which are directly related to the activity of the myometrium. As clinical trials have shown, this technique yields much more information, in particular with respect to the excitation and propagation of the contractions.

2.0 TOCOGRAPHY

The procedure of recording uterine activity can be traced back to the period around 1870. At that time, Kehrer (1867) and Schatz (1872), published the first ever intrauterine pressure curves which, in their precision and quality, remain exemplary even today. They measured the intrauterine pressure by means of a liquid-filled balloon catheter, which they had introduced between the uterine wall and the membranes. The disadvantages of invasive pressure recording, in particular the associated danger of infection, precluded its use in the clinical setting for a long time. Since internal (invasive) tocography is still relatively rarely employed, and since this article is limited to non-invasive techniques, this method will not be dealt with further here.

External (non-invasive) tocography, which still remains the most commonly employed method for the monitoring of uterine activity, can also be traced back to the last century. In the obstetrical literature, reference can be found to an initial attempt to carry out external tocography using a large, air-filled metal box made by Schäffer in the year 1896. Subsequently described external devices for the measurement of uterine contractions,

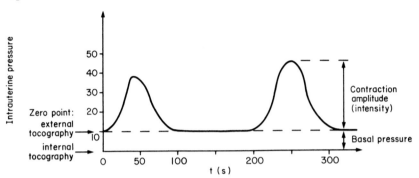

FIG. 1 Schematic representation of the course of uterine contraction. With internal tocography, pressure and intensity are recorded in absolute values, external tocography providing only relative values, without the possibility of determining the basal pressure. In the monitoring of uterine activity, apart from pressure, the form of the contractions also plays an important part.

such as the tocodynamometer (Crodel, 1927) and the external hysterograph (Rübsamen, 1920) had considerable shortcomings, but showed that there was very early interest in obtaining a continuous, external measurement of uterine contractions. Further progress was made by the development of a hysterotonograph (Frey, 1933) and a tocograph (Löwi, 1933), but they required that the entire recording unit be affixed to the abdomen of the

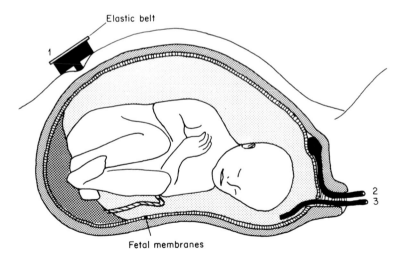

FIG. 2 Common methods of tocometry. (1) External tocometer with "tocometer pin", which is affixed to the abdomen by means of a strap. (2) Intrauterine, extra-amniotic method using a fluid-filled balloon catheter. (3) Trans-cervical intra-amniotic pressure measurement using an open-end catheter.

mother. An electromechanical uterine contraction recording device that is relatively insensitive to interference has been known since the work of Rech (1934). The uterine contraction transducer is attached to the abdomen by means of a rubber strap, and the recording device is set up at a distance from the patient.

The transducers in common use today still employ the principle described by Rech. Contractions of the uterus cause a sensing pin contained within the transducer to be deflected mechanically from its resting position, and the deflection is detected by a strain gauge. A number of strain gauges are connected together to form a measuring bridge (full or half-bridge). The change in resistance of the strain gauge accompanying a mechanical deformation is converted into an electrical signal, which represents a measure of the deformation and thus of the strength of the uterine contraction.

Despite very thorough investigations, agreement has still not been reached as to what, in fact, this transducer is measuring. The "hardness" of the uterine wall is a complex parameter which depends upon the wall tension, radius, wall thickness, internal pressure, transverse elasticity and the hardness of the uterine musculature, and is just as involved in the measurement, as is the deformation of the uterus and its "erection" during contractions. A further component of the measured signal results from the elastic nature of the transducer attachment. It can thus readily be appreciated that external tocography does not permit any statement about the absolute contraction amplitude nor the absolute level of the basal pressure or tone. All that we can obtain is an impression of the relative intensity of uterine contractions and changes in the basal pressure. Even this restricted performance may only be obtained when it is ensured that the transducer remains at its original site throughout the recording period and that the measuring conditions are not changed by any alteration of the patient's position. In very many cases, these requirements cannot be met; indeed, the maintenance of a given position by the patient might even be dangerous for her and the fetus.

In general, external tocography can reliably reproduce contraction rate and approximate form. Movements of the fetus can be recognized as small arrhythmic peaks, while respiratory movements appear as superimposed rhythmic waves. Care must be exercised, however, in the identification of the movements of the fetus since similar peaks in the contraction curve can also be produced by brief contractions of the abdominal wall musculature.

Although tocography has become a routine method for the monitoring of uterine contractions, and, today, almost no pregnancy remains without tocographic monitoring, it is not capable of providing information on the detailed mechanical activity of the myometrium. Cibils and Hendricks (1969) made simultaneous recordings of intrauterine and intramural pressures by means of open-tip catheters inserted into the uterine cavity and the myometrium, respectively. They observed local increases in wall tension, which were not always accompanied by measurable increases in intrauterine pressure. These measurements were made in the post-partum uterus, but this does not invalidate the conclusion that a recording of intrauterine pressure is not a faithful reproduction of detailed activity of the uterine musculature.

3.0 THE ASSESSMENT OF UTERINE ACTIVITY ON THE BASIS OF ITS MYO-ELECTRIC SIGNALS

The myo-electric signal is the electrical manifestation of any contracting muscle. It seems reasonable to assess the possibility of obtaining accurate information about the mechanical activity of a muscle by analysing its myo-electric signals (electromyography, EMG). Many attempts have been made to record and analyse the electrical activity of the pregnant uterus. The lack of suitable measuring and signal-processing methods, and the resulting poor, or even false, results, meant that the possibilities of electromyographic monitoring of uterine activity long remained unrecognized. Only recently has a procedure been described (Nagel and Schaldach, 1980a) which resolves the earlier problems and permits the reliable determination of uterine activity on the basis of its myogram.

3.1 The Electrical Activity of the Uterus

Numerous attempts have been made to record the electrical activity of the uterus. Larks (1960) and Wolfs and van Leeuwen (1979) published a detailed review of the historical development in this area of research. In the majority of cases, the myo-electrical signals were picked up via skin electrodes affixed to the abdomen; in a number of cases, the EMG was recorded invasively using needle or micro-electrodes. A wide variety of different measuring methods and equipment were employed. Thus, it is not surprising that the numerous investigations produced widely varying results. The signals measured were simply attributed to the uterine activity without any attempt to check their actual origin. There are, of course, numerous possible sources of artefacts, such as movement of the patient, including respiratory movements, the electrical activity of the muscles of the abdominal wall, the smooth muscles of the intestine and the bladder, the electrocardiogram of the mother, skin potentials and movement artefacts caused by the contracting uterus, all of which tend to degrade the signal-to-noise ratio.

Bode (1931), Clason (1934) and Mestwerdt (1944) recorded slow, biphasic waves, with faster fluctuations superimposed upon them. They suggested that part of the activity they had recorded was due to the heart, or the mechanical or electrical activity of the respiratory and abdominal wall muscles. Many other investigators, such as Dill and Maiden (1946), Steer and Hertsch (1950) and Halliday and Heins (1950) also recorded very low-frequency electrical signals (0·1–2 Hz), their measurements varying greatly with respect to the shape and occurrence of the signals. Müller and Liechty (1954) discovered more electrical activity between contractions than during contractions. Steer (1954) described two different types of electrical activity during uterine contraction: slow waves having a periodicity of several seconds, upon which faster waves (0·3–2 Hz) were superimposed. Sureau is one of the leading investigators in this field (Sureau, 1955, 1956, 1964; Sureau et al., 1965). During contraction, he recorded sinusoidal waves having a frequency of 0·3–1 Hz. Larks (1956) obtained results similar to those of Steer. The amplitude of the electrical signals measured by different investigators varies considerably, covering a range of between 50 µV and 150 mV. Very extensive investigations were carried out by Wolfs and Leeuwen (1979), and although they did not succeed in obtaining all of the information contained within the uterine EMG, their measurements did reveal a marked correlation between the EMG and intrauterine pressure.

A very relevant question is why so many working groups recording the uterine myopotentials have obtained such a wide variety of different results. The reason for this would appear to be that the investigations have been based on incorrect or incomplete theoretical models of the origin, propagation and measurement of the electrical signals, and on the mode of electromechanical coupling. The result of this was the use of unsuitable measuring equipment for the recording of the signals. Thus, for example, in many cases, DC-coupled or very low-frequency measuring amplifiers were employed because adequate attention had not been paid to the problem of temporarily changing electrode potentials. The frequencies of these changes are in the same frequency range as the signals under investigation and can almost completely mask the useful signal. Thus, the question as to the frequency spectrum of the signals was completely ignored, and as a result, most measurements recorded only the fluctuating resting potentials, but not the action potentials typical for the activity of the musculature. Furthermore the selection of the most suitable combination of electrodes and recording positions and of the origin and composition of the signals, was not subjected to a systematic examination. The resolution of these problems is, however, an essential pre-condition for the careful analysis of the electromyographic signals picked up from the uterus. An understanding of the physiology of labour presupposes a knowledge of a number of basic biological principles. Here, therefore, the fundamental processes in the formation of bio-electrical potentials, and electromechanical coupling, are briefly described.

Every animal cell is bounded by a highly differentiated membrane (cell membrane) which regulates the exchange of substances between the intracellular and extracellular spaces. This membrane has the capability of being selectively permeable and effecting active transport. The intracellular and extracellular spaces differ in their ionic concentrations. Within the cell, the concentration of potassium ions is 40 to 50 times as high as that on the outside. Sodium ions, on the other hand, have an extracellular concentration 3 to 10 times that of the intracellular concentration. Owing to this difference in ion concentrations, an electrical potential difference develops between the inside and outside of the cell. The resting potential of the cell membrane is between -60 and -90 mV. In the resting state of the cell (polarized), the intracellular potential is negative with respect to the extracellular space.

Nerve and muscle cells are characterized by the fact that stimulated or autonomous activation can have an effect on the cell membrane. As a

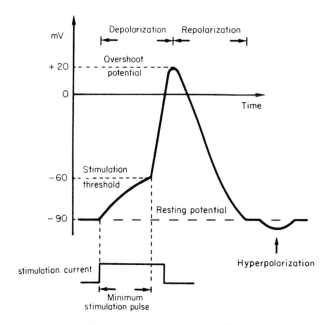

FIG. 3   The action potential of a cell membrane.

result of considerable changes in membrane permeability during activation, the ion concentration gradients undergo a shift, which leads to a change in the trans-membrane potential. The time course of membrane potential change from the start of excitation to the return to the resting state is defined as the action potential of the cell (Fig. 3).

The excitation takes place when the membrane potential is raised beyond a critical value ($\approx 80\%$ of the resting potential) by means of an impressed ion current. When this threshold has been crossed, the permeability of the membrane to sodium ions is markedly increased. As a result, driven by their concentration gradient, sodium ions pass into the cell where, on account of their positive charges, they give rise to a further depolarization which, in turn, leads to a further increase in the permeability of the membrane to sodium ions. This mechanism leads to an extremely rapid, avalanche-like, inflow of sodium ions. Before the membrane potential reaches a sodium-equilibrium potential, however, the permeability of the membrane to sodium ions drops again. The permeability of the membrane for potassium ions also increases during the depolarization procedure, leading to an increased outflow of potassium. When the permeability of the membrane to sodium ions decreases again, the outflow of potassium ions predominates, so that the cell again returns to its resting potential. At this point, the specific permeabilities of the membrane have returned to their original levels.

The depolarization that occurs when the critical threshold is exceeded represents an automatic process, invariable with respect to the course of potential changes, and independent of any further increase in the intensity of the stimulation. This behaviour is described as the "all or nothing" law of excitation. Since the energy required to activate the cell has to be provided by the cell membrane itself, a certain period of time, the refractory period, has to elapse before the cell has returned to its original state, and depolarization can occur again.

Locally occurring currents (flows of ions) accompanying the depolarization of the cell result in the stimulation of neighbouring areas within the same cell, thus propagating the stimulation throughout the cell body. If the coupling impedance to the neighbouring cells is small enough, the excitation spreads beyond the cell boundary.

The physiology of the uterus musculature has been studied in particular by Bozler (1942), Jung (1972) and Wolfs and Leeuwen (1979). The resting potential of the uterine musculature is dependent upon hormonal influences; the action potentials occur in salvos and in series. The musculature of the uterus produces its own excitation, the conduction of the excitation also occurring within the muscle fibres. No nervous pulses are needed to trigger uterine contraction. In principle, autonomic excitation appears possible in all parts of the muscle although various investigations seem to indicate the presence of certain local "excitation centres" or pacemakers, where excitation preferentially originates. The rate of conduction of the excitation is also dependent upon hormonal influences.

In any contracting cell, mechanical shortening is triggered by depolarization. This electromechanical coupling represents a fundamental biochemical process, in which calcium ions have a particular role to play. At excitation, Ca^{2+} ions flow into the intracellular space where they activate the myofibril ATPase; ATP is split and energy liberated for the contraction process. The contraction itself involves an interaction of myosin, actin and ATP. For the interaction of the contractile apparatus with ATP, Ca^{2+} ions are required. By employing calcium inhibitors, an electromechanical decoupling can be brought about, so that, although the bio-electric excitation processes may persist, no further contraction of the muscle fibre occurs.

3.2 Measurement of the Uterine Myo-electric Potential

The action potentials of individual cells can be measured either directly inside or at the cell surface with the aid of micro-electrodes. Here, however, we are concerned with measuring the state of excitation of the entire uterus. A further point is that invasive procedures may not be used for such measurement. This means that the recording of the action potentials cannot be achieved at the cellular level using micro-electrodes, but must be made at a distance, on the surface of the body. The distribution of the flow of ions over the cell membrane gives rise to a characteristic electromagnetic field, which spreads throughout the neighbouring space. By measuring this field, conclusions can be drawn as to the time-space behaviour of the field-producing cell community, here the uterus. The total field resulting from the summation of the action potentials of the individual cells depends upon the form of the action potentials, their time sequence and also on the distribution of the cells in space, and the material properties and form of the medium surrounding them. It can be shown that, instead of measuring the overall electromagnetic field, the measurement of the scalar electric potentials produced at the surface of the body suffices to provide the information needed about the source functions, i.e. the condition of the field-producing musculature (Faust, 1965; Nagel, 1979). However, to permit conclusions to be drawn as to the state of excitation of the muscle, or rather the uterus, from the electrical potentials measured, the *a priori* knowledge of the physiology, geometry and the possible states of the uterus, is a prerequisite. The electrical potential at any given point of the body is a function of the space- and time-dependent ion currents

$\vec{G}\,(\vec{r},\,t)$ and the conductivity of the transmission medium for the electromagnetic field. Assuming, for the sake of simplicity, a homogeneous medium (uniform conductivity), the potential $U\,(\vec{r},\,t)$ at the measuring point $P\,(\vec{r})$ is given by

$$U\,(\,\vec{r},\,t\,) = \frac{1}{4\pi\sigma} \int_{V'} \frac{-\operatorname{div}\vec{G}\,(\vec{r}',\,t)}{|\vec{r}-\vec{r}'|}\,\mathrm{d}V', \qquad (1)$$

where $\vec{r}'$ is the distance vector from the origin of the co-ordinate system to the current source, $\vec{r}$ is the distance vector from the origin to the measuring point, V' is the source volume, and σ is the conductivity.

Equation (1) permits the evaluation of the potential field provided that the source function is known and the conductivity is uniform. When a region contains inhomogeneities, it is usually convenient to take them into account by subdividing the region into a finite number of uniformly conducting subregions. In such a case one can account for the inhomogeneities by determining the secondary sources that necessarily arise at the interface between regions of different conductivity. In order to simplify the understanding of the following explanations, such effects are not included here; the results remain essentially unaffected.

In accordance with Eqn (1) the electrical potential of the uterus musculature as measured at any given point, is given by the volume integral of the ion currents that flow on contraction of the musculature. Conversely, this means that when adequate information is available about the possible physiological states and the resulting potentials, the measurement of the potential distribution at the surface of the body permits us to draw conclusions about the activity of the uterus. In this connection, our major interest is the question as to whether or not it is possible to derive from the electrical signal a measure for the strength of contraction of the uterus. Theoretically, on account of the strict electromechanical coupling, such a possibility should exist, provided this coupling has not been interfered with by the use of calcium inhibitors. In order to represent the relationship between the externally measurable EMG and the strength of uterine contractions in a more readily appreciable manner, Eqn (1) must be put into a somewhat different form.

In order to represent the sequence of excitations of the individual muscle fibres of the uterine musculature in time, the term spike train has been introduced. The spike train $a_n\,(t)$ of the nth muscle fibre is represented with the aid of a delta function $\delta(t)$, and can be expressed as follows:

$$a_n\,(t) = \sum_k \delta(t - t_{nk}). \qquad (2)$$

In this equation, t_{nk} is the time at which the fibre is stimulated for the kth time. Although the electromagnetic fields produced by the action potentials spread throughout the body at the speed of light, the conduction velocity for the propagation of the excitation in the biological tissue is very small (approximately $0.1\text{--}100 \text{ ms}^{-1}$). As a result, the excitation of the individual, spatially distributed muscle fibres is associated with a time delay, which, vis-à-vis the duration of the action potentials, is not negligible and which is of considerable importance for the form of the EMG curve.

If the potential of the electrical field produced at the site of a measuring electrode by a single stimulation of a single muscle fibre n, is given as $h_n(\vec{r}, t)$, then the time course of the potential of this muscle fibre at the measuring point is given by the convolution (*) of h_n and the associated spike train a_n:

$$U_n(\vec{r}, t) = h_n(\vec{r}, t) * a_n(t). \tag{3}$$

In the representation of U_n, use is made of the fact that the course of the action potential is a function which is characteristic for the individual cell involved, and this does not change in time, i.e. it is independent of the number of times the cell has been stimulated. The potential of the muscle as a whole is given from the summation of the individual potentials of all N muscle fibres in the muscle:

$$U(\vec{r}, t) = \sum_{n=1}^{N} h_n(\vec{r}, t) * a_n(t). \tag{4}$$

According to Eqn (1), the potential of the individual excitation of a muscle fibre $h_n(r, t)$ can be expressed, as a function of the ion current, as:

$$h_n(\vec{r}, t) = \frac{1}{4\pi\sigma} \int_{V'} \frac{-\text{div } \vec{G}_n(\vec{r'}, t)}{|\vec{r} - \vec{r'}|}. \tag{5}$$

Integration of the volume of the muscle fibre n must be carried out. Equation (4), then becomes:

$$U(\vec{r}, t) = \sum_{n=1}^{N} \left\{ \frac{1}{4\pi\delta} \int_{V'} \frac{-\text{div } \vec{G}_n(\vec{r'}, t)}{|\vec{r} - \vec{r'}|} \, dV' * a_n(t) \right\}. \tag{6}$$

If the biphasic impulse responses $h(\vec{r}, t)$ are summated at the surface electrode in accordance with Eqn (4), or (6), a complicated interference pattern occurs, which is dependent upon the firing rate and firing pattern of the individual muscle fibres. For the interspike intervals, greatly varying distribution functions are observed. In the literature, they are characterized as having a Poisson or a gamma distribution (Sanderson et al., 1973), a Gaussian distribution (Clamann, 1969) or a Weibull distribution (De Luca and Forrest, 1973). In practice, it is impossible to characterize precisely the EMG, since the probability distributions for the patterns of excitation must be known, and so it is impossible to obtain more than rough estimates. A more suitable measure for the evaluation of the EMG is its modulation envelope $I(\vec{r}, t)$. This is determined by rectification and subsequent low-pass filtering of the EMG signal.

If the impulse response of the filter is designated $m(t)$, then the intensity of the EMG is given by:

$$I(\vec{r}, t) = |U(t)| * m(t) \tag{7}$$

The waveform of the envelope is strongly dependent upon the time constant of the low-pass filter. If small time constants ($\tau < 1$ s) are employed, $I(\vec{r}, t)$ reveals a marked structuring. Greater values of τ result in smoother waveforms which, however, are associated with a loss of information on short-term changes in intensity. To keep these losses to a minimum, great care must be exercised when choosing the cut-off frequency of the low-pass filter to be employed. Below, the question is examined as to whether or not, and under what conditions, $I(\vec{r}, t)$ represents a measure of the force of contraction of the muscle.

3.3 Contraction Intensity and Intrauterine Pressure

If a muscle fibre is stimulated with a single pulse, it responds with a twitch, i.e. a brief contraction that exerts a force of $\vec{g}(t)$. Since the cells can only be either in an active or a passive state, and the active state of a cell is an unchangeable state ("all-or-nothing" law), the force produced by a single twitch is invariable. The gradation of the force of contraction of the muscle as a whole involves two regulating mechanisms. On the one hand, the number of activated muscle fibres can be tailored to the force requirement, on the other, with an increasing requirement of force, the individual fibres depolarize at ever-shortening time intervals. In the presence of a persisting spike train, the sequential contractions sum to form a maximum, resulting in complete tetanic contraction. Here, the stimulation interval must be larger than the refractory period of the muscle fibre membranes and smaller than the decay time of the individual twitch.

The pulse response of the contraction mechanism to stimulation manifests a considerably larger time constant than the action potential of the stimulated cell. On account of the resulting low-pass filtering of the spike train, and the fact that the contraction of the fibres cannot be negative, the force produced by the muscle is continuous, despite the quasi-stochastic distribution of the stimulation pulses, in contrast to the superimposition of the biphasic action potentials.

The time course of the force produced by a muscle fibre is obtained from the convolution of the associated spike train $a_n(t)$ with the force $\vec{g}_n(t)$ produced by an individual twitch:

$$\vec{f}_n(t) = \vec{g}_n(t) * a_n(t). \tag{8}$$

The force developed by the muscle as a whole is obtained from the summation of all N fibres:

$$\vec{f}(t) = \sum_{n=1}^{N} \vec{g}_n(t) * a_n(t). \tag{9}$$

The computation is restricted to a linear superimposition of the individual forces. Non-linearities can, as experimental studies confirm, be neglected. For the further derivation, it is assumed that, to a first approximation, the forces of the individual muscle fibres are directed tangentially to the surface of the uterus.

The variable, which is of diagnostic importance and which may be measured directly, is not the force developed by the uterine musculature, but the increase in intrauterine pressure during the contractions. The relationship between muscle action and pressure increase can be approximated from a simple theoretical model. With the fetal membranes intact, the uterine musculature encloses a fluid-filled space, in which the pressure may be considered to be uniformly effective in all directions. The force acts tangentially to the surface. It is well known in mechanics that with such an arrangement the external force and the internal increase in pressure are proportional, i.e.:

$$\Delta p(t) = k_1 \cdot f(t), \tag{10}$$

Thus, for intrauterine pressure as a function of the force exerted by the uterine musculature, we have the equation

$$p(t) = p_0 + k_1 \cdot f(t), \tag{11}$$

where p_0 represents the basal pressure which is present in the absence of uterine musculature contraction. It is dependent upon a number of physiological factors, such as, for example, the elasticity of the uterus. It is not intended to discuss such details here, since they are of no importance for the measurement of the uterine activity.

3.4 Relationship between Mechanical Activity and Myo-electric Potential

As a result of the electromechanical coupling, there is a strong correlation between the strength of the uterine contraction, or the intrauterine pressure, and the myo-electrical potential. An essential difference is the fact that, in contrast to the intrauterine pressure, the EMG is dependent upon the site of the measuring electrode. With the aid of a number of simplifications the relationship can be made clear and a basis for the measurement of the uterine activity or the pressure changes from the EMG, found. Because of the low-pass filtering of the spike train through the mechanical contractile mechanism, as mentioned above, and the assumed uniform tangential direction of the individual forces contributed by the muscle fibres, Eqn. (9) can be simplified. This is achieved by the approximation of the temporarily accurately defined spike train by a medium stimulation frequency $\omega_n(t)$, which, of course, is dependent upon time. The force $f(t)$ can then be expressed thus

$$f(t) = C_1 \cdot \sum_{n=1}^{N} g_n(t) \cdot \omega_n(t). \tag{12}$$

If it be assumed that the individual muscle fibres are excited synchronously, and, further, that they each develop an identical contraction force, then we obtain, for the force $f(t)$

$$f(t) = C_2 \cdot k(t) \cdot \omega(t), \tag{13}$$

in which $k(t)$ is the number of activated muscle fibres. The question must now be examined as to whether the intensity of the EMG can be expressed in a similar manner, with the aid of the stimulation frequency. For this purpose, Eqn (6) is put into a more easily interpretable form. We shall first consider the case in which the measuring point and the co-ordinate origin coincide—in the centre of the spherical theoretical uterus. Here, $r = 0$. Equation (6) reduces to:

$$U(t) = \sum_{n=1}^{N} \left\{ \frac{1}{4\pi\sigma} \int_{V'} \frac{-\operatorname{div} G_n(\vec{r'}, t)}{r'} \, dV' * a_n(t) \right\}. \tag{14}$$

The point of departure for further simplification of (14) is the vector identity:

$$\nabla \cdot (\vec{G}/r) = (\nabla \cdot \vec{G})/r + \vec{G} \cdot \nabla (1/r). \tag{15}$$

The integration of all three terms in volume V, which contains all sources, provides us, on applying the divergence theorem to the first term, with the equation

$$\int_S \left(\frac{\vec{G}}{r}\right) \cdot dS = \int_V \frac{(\nabla \cdot \vec{G})}{r} dV + \int_V \vec{G} \cdot \nabla\left(\frac{1}{r}\right) dV. \tag{16}$$

Since $G = 0$ on the entire limiting surface area S, it follows from (16) that

$$\int_V \frac{(\nabla \cdot \vec{G})}{r} dV = - \int_V \vec{G} \cdot \nabla\left(\frac{1}{r}\right) dV, \tag{17}$$

So that (14) can be re-formulated as follows:

$$U(t) = \sum_{n=1}^{N} \left\{ \frac{1}{4\pi\sigma} \int_{V'} \vec{G}_n (\vec{r'}, t) \cdot \nabla\left(\frac{1}{r'}\right) dV' * a_n(t) \right\} \tag{18}$$

or

$$U(t) = \sum_{n=1}^{N} \left\{ \frac{1}{4\pi\sigma} \int_V - \vec{G}_n (\vec{r'}, t) \cdot \frac{\vec{r'}}{r'^3} dV' * a_n(t) \right\}. \tag{19}$$

The flow of ions, G, is a source function, which is interpreted as a dipole moment per unit of volume. A contribution to the potential at the measuring point is made only by the radial (G_{nr}), but not by the tangential, components of the dipole vectors $\vec{G}_n$. Accordingly, the following equation

$$U(t) = \sum_{n=1}^{N} \left\{ \frac{1}{4\pi\sigma} \int_{V'} \frac{- G_{nr} (\vec{r'}, t)}{r'^2} dV' * a_n(t) \right\} \tag{20}$$

applies.

With the assumed spherical symmetry and the relatively small thickness of the myometrium, for an estimation of the potential the contribution of the radius vector can be considered constant for all fibres, so that the factor $1/r'^2$ can be removed from the integral and the sum. Under these preconditions, the integral can be expressed as a function $y_n(t)$, now depending only on time. Equation (20) becomes:

$$U(t) = \frac{1}{4\pi\sigma r'^2} \sum_{n=1}^{N} y_n(t) * a_n(t). \tag{21}$$

With synchronous stimulation of all muscle fibres, and the same shape of the curve of all $y_i(t)$, because of the refractory period of the cells, no interference phenomena can occur. In this case, the potential can be expressed by

$$U(t) = \frac{1}{4\pi\sigma \, r'^2} \, k \cdot (t) \cdot (y(t) * a_n(t)). \tag{22}$$

where $k(t)$ is the number of muscle fibres stimulated. If the potential function described in (7) is rectified and filtered, applying the same arguments as for the strength of the mechanical contraction, the spike train $a(t)$ can be replaced by the stimulation frequency $\omega(t)$ and we obtain, with constant c_3 for the intensity of the EMG, an expression having the form:

$$I(t) = \frac{c_3}{4\pi\sigma \, r'^2} \cdot k(t) \cdot \omega(t). \tag{23}$$

A comparison of (23) and (13) shows that for constant r', the force of contraction and the intensity of the EMG (IEMG) are proportional to each other:

$$f(t) = c_4 \cdot I(t). \tag{24}$$

In addition, using (11)

$$p(t) = p_0 + c_5 \cdot I(t). \tag{25}$$

applies. Accordingly, in the special case under consideration, both the intensity of contraction and the relative intrauterine pressure can be determined from a measurement of the myo-electric potential.

Of course, the question may now be asked as to what practical significance this result has. Although, in the derivation of (24) and (25), so many approximations were made that one might not expect the result to be quantitatively correct, experimental investigations show that at least qualitatively it does in fact conform to the physiological situation. The reason for including this derivation here, however, is to make it clear that, at least in principle, it *is* possible to find a fixed relationship between the EMG, and the intensity of contraction of the uterus. This would seem all the more important since, to date, reports in the literature have all denied this possibility. This is possibly the result of the fact that, formerly, it has always been the EMG itself, but not its intensity, that has been evaluated.

We must now investigate the nature of the relationship between $f(t)$ and $I(t)$ without the restricting conditions assumed in the derivation of (24) and (25), in particular in the case of an external recording of the myopotentials. In this connection, we must first examine the question as to whether the dependence of the intensity of the EMG on the force of contraction of the muscle changes when the assumption of synchronous stimulation of the muscle fibres—which is certainly not really the case—is dropped. For the mathematical determination of the relationship then applicable, the statistics of muscle excitation, and the geometry of the individual muscle fibres, must be known. On account of the complexity, perhaps even the impossibility, of this computation, no attempt was made to adopt this approach. Instead, experiments were performed to discover whether the linear relationship between $f(t)$ and $I(t)$ is preserved.

According to the literature (Person and Libkind, 1967; De Luca and Forrest, 1973), this is not the case for all muscles; sometimes, there is a square law relationship ($f(t) \propto I^2(t)$). With respect to the myometrium, however, we have been able to confirm the linear relationship already found by Milner-Brown and Stein (1975) for a number of other muscles. Accordingly, therefore, the assumption of synchronicity of fibre stimulation made in the derivation of the relations (24) and (25), does not result in a qualitative falsification of the result.

A further simplification, whose influence on the results has to be investigated, is the assumption of a spherical uterus and potential measurement in the centre of the sphere. Only under the above-mentioned conditions do the contributions of the individual dipole vectors in the overall potential, have identical weight. Both in the case of potential measurement outside of the centre of the sphere, and also in a change in the geometry of the uterus, a non-uniform weighting of the individual sources results. In accordance with Eqn (6), the influence of such sources that are closer to the measuring point becomes more marked, while those potentials originating in more distant muscle fibres become more attenuated. The result of this is that on moving the measuring point to a given part of the myometrium, mainly the activity of the muscle fibres in the immediate neighbourhood is recorded. Thus, by appropriately siting the electrodes, the local activity of individual regions of the uterus, or the spread of the contractions can be picked up. In this manner, motility disturbances, such as incoordination for example, can also be recognized.

For the global, non-invasive determination of the uterine activity, the measurement of the myopotentials at a single point is not adequate. For there is no point outside of the uterus that is equidistant from all the muscle fibres. Owing to the large spatial extension of the uterus, this condition is not even approximately fulfilled. Nevertheless, the uterine activity *can* be determined globally, if the potential is picked up simultaneously at a number of points. By appropriately siting the measuring electrodes and summating the individual potentials a measuring signal is obtained which can be used in Eqns (24) and (25) to give adequately accurate results. The degree of this approximation depends upon the number of measuring points and on their sites. Clinical investigations have shown that, in the majority of cases, the measurement of potentials at two points on the maternal abdomen is sufficient to ensure a result that is representative for the whole uterus.

3.5 Interference Potentials

The external recording of the uterine EMG is made difficult by super-imposed strong interference signals arising in the maternal ECG, the fetal ECG and the EMG of the abdominal wall musculature. The amplitudes of the interference signals are usually greater than those of the useful signal. Further possible interference components, such as electrode offset potentials, electromagnetic interference and noise potentials, are not considered here, since they can be avoided or suppressed by designing suitable measuring systems. In passing it might be mentioned that the bio-electrical interference signals are not included in the measurements indicated in the literature. One reason for this is, that in most measure-ments reported bipolar electrodes were used. With these the potential difference between two electrodes located close together is determined, so that the contributions to the signal of more distant sources, e.g. the heart, are strongly attenuated. The second reason is that the frequency response of the amplifiers employed only allowed signals below 2 Hz to be measured. In this low-frequency range, however, the interference signals mentioned have only a very small power density, so that their contribution to the measured signal is small. An analysis of the frequency spectrum of the uterine EMG, however, shows that it extends to about 250 Hz and has its greatest power density above 2 Hz. Accordingly, the recording of the EMG should be carried out in this frequency range. The low-frequency range below 2 Hz is not suitable for routine measurement since it is here that movement artefacts are strongly seen. The use of closely spaced bipolar measuring electrodes is reasonable only for the pick-up of local muscle activity.

For the analysis of the intensity of the EMG, the interfering components must be suppressed prior to rectification and low-pass filtering. The maternal ECG can be subtracted from the original signal using a pro-cedure described by Nagel and Schaldach (1980b). The R-peaks of the maternal ECG (MECG) are easily detectable by means of threshold detectors on account of their prominence in the abdominal signal. Through the exponential averaging of succeeding segments of the signal, all con-taining the maternal QRS complex in the same phase position, a reference signal corresponding to one interval of the MECG is obtained. The other signal components are suppressed in the reference since they are statistically independent of the MECG. Subtraction of the reference from the abdominal mixed signal after a special scaling operation, results in the complete elimination of the MECG. The fetal ECG can be eliminated in the same manner although, as practical experience shows, this is not necessary because of its small amplitudes; its influence on the labour (uterine activity) curve is negligible. A simplification of the procedure is achieved by limiting the potential measurement to the frequency range from *c.* 150 to 250 Hz. Since, here, the amplitude of the uterine EMG is considerably greater than that of the fetal and maternal ECG, signal separation is not necessary. When measurements were carried out in this restricted frequency range, no changes were observed in the labour curve.

The only interfering component that cannot be eliminated from the mixed signal, but can merely be reduced by appropriately positioning the electrodes, is the EMG of the abdominal wall musculature. This fact, however, is not necessarily a disadvantage. Its contribution to the labour curve is so characteristic that it can clearly be distinguished from the intensity of the uterine EMG. Furthermore, it can also show the behaviour of the mother under the stresses of labour, e.g., during expulsive contractions. Over and beyond this, it can be observed that the contractions of the abdominal wall muscles also lead to an increase in intrauterine pressure. Thus, the additional recording of the activity of the abdominal wall musculature is desirable rather than undesirable for the practical application of the measuring procedure described. In any case, the contractions of the abdominal wall muscles also strongly affect the external mechanical pressure recording.

3.6 Movement Artefacts

According to Eqn (6), the uterine myo-electric potential is dependent upon the measuring point. Thus it is to be expected that movements—either changes in the position of the patient, or changes in the geometrical state occurring during uterine contractions—will influence the potential of a measuring electrode affixed to the maternal abdomen. Depending upon the movement and the position of the electrode, or on the transmission path of the bio-electric signals, their amplitude decreases or increases, so that this effect also modulates the EMG, and can thus falsify the labour curve. The error can be eliminated by having the signal amplifier compensate for the fluctuations in amplitude. It is, of course, not desirable to adjust the amplitudes of the EMG to a constant level, since important information about its intensity would then be lost, for the amplitude changes caused by contraction would also be eliminated.

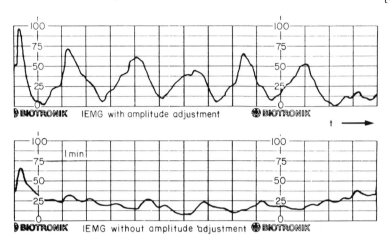

FIG. 4   Influence of amplitude adjustment of the EMG on the uterine contraction curve.

A surprising but simple solution is the use of the maternal ECG as a calibration signal for amplitude control. Experimental experience has shown that the amplitude fluctuations of the MECG correspond quite accurately to those of the EMG. From this it may be deduced that possible measuring errors can be avoided by employing automatic gain control that ensures the constant amplitude of the MECG within the original signal. In Fig. 4, the effects of adjusting the amplitude of the EMG to the contraction curve is represented by simultaneously recording the contraction curve with and without amplitude adjustment, using the same pick-up electrode.

4.0 COMPARISONS OF RECORDINGS OF THE UTERINE ELECTRICAL AND MECHANICAL ACTIVITY

Below, a number of examples of the recording of uterine activity are described, and these are intended to show the degree of conformity and also the differences between tocography and the recording of uterine activity via the EMG.

There is overall good agreement between the theory and actual measurement. Figure 6 shows a comparison of contraction curves measured mechanically and myographically. For the detection of the EMG, an electrode was placed at the isthmus and another at the fundus of the uterus. The potential reference point was obtained from a third electrode applied to the thigh. In contrast to the externally measured pressure (upper curve),

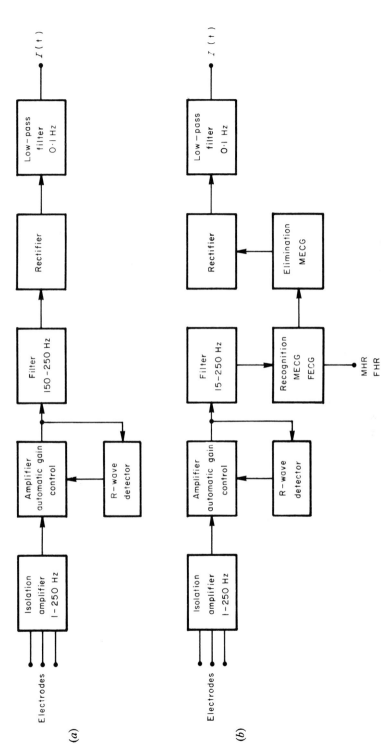

FIG. 5   Block diagrams of the electronic circuit for recording electrical uterine activity. (a) The simpler arrangement, in which only the high frequency components of the EMG are processed. (b) This circuit permits the recording of the labour curve from the complete EMG and also the maternal and fetal heart rates from the respective ECGs. Before producing the envelopes of the EMG, the maternal ECG is eliminated.

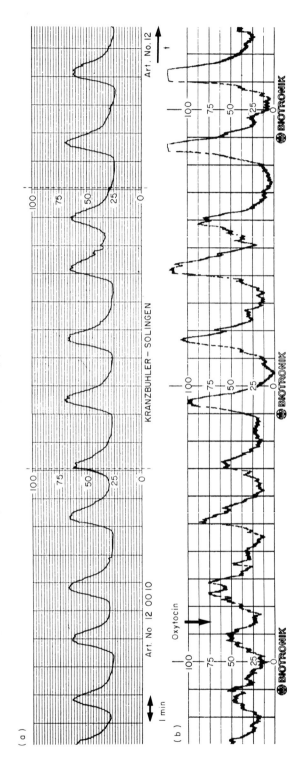

Fig. 6 Uterine contraction (labour) curve, recorded simultaneously with an external pressure pick-up (a) and from the electro-myogram (b).

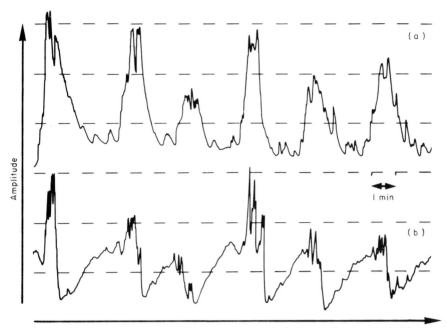

FIG. 7 Uterine contraction (labour) curve, recorded simultaneously with an external pressure pick-up (a) and from the electromyogram (b).

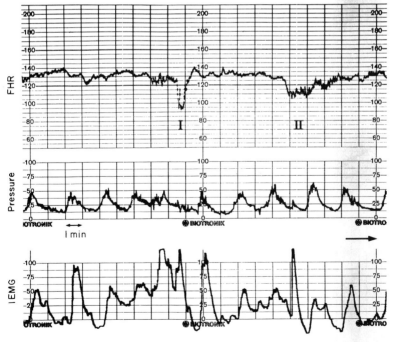

FIG. 8 Cardiotocogram with uterine contraction curve measured mechanically (external) and electromyographically.

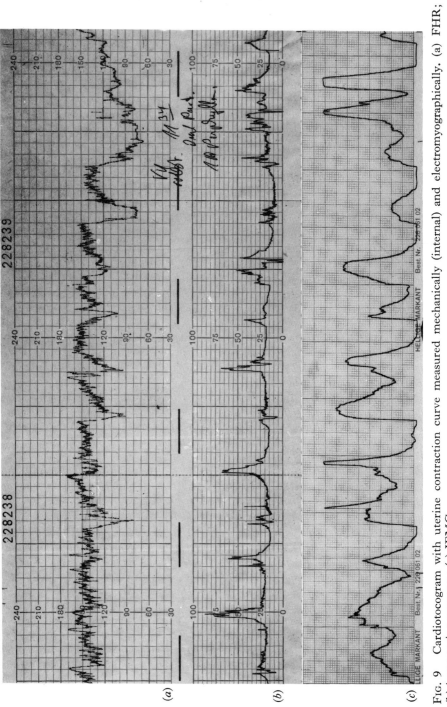

Fig. 9 Cardiotocogram with uterine contraction curve measured mechanically (internal) and electromyographically. (a) FHR; (b) intrauterine pressure; (c) IEMG.

the IEMG clearly reveals the increasing intensity of uterine contractions in response to infusion of oxytocin, the start of which is marked on the curve.

In Fig. 7, the myographic contraction curve shows that a new uterine contraction is stimulated immediately at the end of the previous contraction. The contraction is subjectively recognized only when a minimum intensity has been reached, and the activation of the abdominal musculature (expulsion contraction) is triggered. The externally recorded pressure curve does not contain this information about the course of labour.

The recordings shown in Fig. 8 reveal clear differences in the course of the uterine contraction curve. While the external pressure recording provides no explanation for the marked frequency drop at I and II of the FHR, the IEMG clearly reveals the powerful muscular contractions that may be considered as the cause.

A comparison between the intrauterine pressure and the IEMG shows that, in general, the two measuring procedures agree well. The invasive pressure recording technique, however, is less sensitive and considerably more susceptible to interference (Fig. 9).

5.0 CONCLUSIONS

All in all, uterine myographic recording has proved valuable in clinical trials. On account of the high information yield and the simplicity of handling, in particular for the simultaneous recording of fetal and maternal heart activity in addition to the uterine contraction curve, the procedure described is highly suitable for clinical routine work. The combined recording of the electrical activity of the myometrium and the intrauterine pressure will, it may be presumed, provide more information about the behaviour of the uterine musculature than either technique alone.

References

Bode, O. (1931). Das Elektrohysterogramm. *Arch. Gynaek.* **146**, 123.

Bozler, E. (1942). The action potentials accompanying conducted responses in visceral smooth muscle. *Am. J. Physiol.* **136**, 553.

Cibils, L. and Hendricks, C. H. (1969). Uterine contractility on the first day of the puerperium. *Am. J. Obstet. Gynecol.* **103**, 238.

Clamann, H. P. (1969). Statistical analysis of motor unit firing patterns in a human skeletal muscle. *Biophys. J.* **9**, 1233–1251.

Clason, S. (1934). Versuche mit obstetrischer Elektrographie. *Acta Obstet. Gynec. Scand.* **14**, 131.

Crodel, W. (1927). Wehenmessung durch die Bauchdecke. *Arch. Gynaek.* **132**, 23.

De Luca, C. J. and Forrest, W. J. (1973). Some properties of motor unit action potential trains recorded during constant force isometric contractions in man. *Kybernetik* **12**, 160–168.

Dill, L. V. and Maiden, R. M. (1946). The electrical potentials of the human uterus in labor. *Am. J. Obstet. Gynecol.* **52**, 735.

Faust, U. (1965). Über den gegenwärtigen Stand der Theorie des Ekgs. *Hippokrates* **15**, 573–585.

Frey, E. (1933). Der Hysterotonograph. Eine neue Apparatur mit automatischer Schreibvorrichtung für die klinische Wehenkontrolle. *Zbl Gynaek.* **57**, 545.

Halliday, E. C. and Heyns, O. C. (1950). Progress in the study of the modes of action of the human uterus during pregnancy and labor. *A. Afr. Med. J.* **24**, 571.

Jung, H. (1972). Zur Physiologie des Uterus–Muskels unter Berücksichtigung zellulärer und neurohumoraler Regelvorgänge bei der Ruhigstellung des schwangeren Uterus. "Perinatale Medizin" Bd. II, hrsg. von E. Salingund, F.-J. Schulte. Thieme, Stuttgart, 72.

Kehrer, F. A. (1867). "Beiträge zur vergleichenden und experimentellen Geburtskunde" Bd. II, Gießen, 132 and 170.

Larks, S. (1956). Electrical correlates of the human uterus in labor. *Fedn Proc.* **15**, 116.

Larks, S. (1960). "Electrohysterography", Charles C. Thomas, Springfield, Illinois.

Lowi, S. (1933). Über einen neuen wehenzeichnenden Apparat (Tokograph). *Zbl. Gynäk.* **57**, 554.

Mestwerdt, G. (1944). Über den Nachweis von Aktionsströmen am menschlichen Uterus. *Z. Geburtshilfe Perinatol.* **127**, 44.

Milner-Brown, H. S. and Stein, R. B. (1975). The relation between the surface electromyogram and muscular force. *J. Physiol.* **246**, 549–569.

Müller and Liechty (1954). Elektrohysterographie. *Geburth. Frauenheilkd.* **14**, 670.

Nagel, J. (1979). Analyse bioelektrischer Potential in der Perinatologie, Dissertation Universität Erlangen-Nürnberg.

Nagel, J. and Schaldach, M. (1980a). Recording of uterine activity from the abdominal lead EMG. *In* "Fetal and Neonatal Physiological Measurements" (Ed. P. Rolfe), p. 177, Pitman Medical Limited, Tunbridge Wells, UK.

Nagel, J. and Schaldach, M. (1980b). Processing the abdominal fetal ECG using a new method. *In* "Fetal and Neonatal Physiological Measurements" (Ed. P. Rolfe), p. 9, Pitman Medical Limited, Tunbridge Wells, UK.

Person, R. S. and Libkind, M. S. (1967). Modelling of interference bioelectrical activity. *Biofizika* **1**, 127–134.

Rech, W. (1934). Ein neues Verfahren zur selbsttätigen fortlaufenden Registrierung der Wehentätigkeit. *Arch. Gynäk.* **157**, 458.

Rübsamen, W. (1920). Die externe Hysterographie als klinisch-experimentelle Testmethode für die Bestimmung der Wertigkeit von Wehenmitteln (Chinin und Hydrastis-Kotarninpräparate). *Arch. Gynäk.* **122**, 461.

Sanderson, A. C., Kozak, W. and Calvert, T. (1973). Distribution coding in the visual pathway. *Biophys. J.* **13**, 218–244.

Schatz, F. (1872). Beiträge zur physiologischen Geburtskunde. *Arch. Gynäk.* **3**, 58, 174.

Steer, C. M. (1954). The electrical activity of the human uterus in normal and abnormal labor. Parts I and II. *Am. J. Obstet. Gynecol.* **68**, 867.

Steer, C. M. and Hertsch, G. J. (1950). Electrical activity of the human uterus in labor. The electrohysterograph. *Am. J. Obstet. Gynecol.* **59**, 25.

Sureau, C. (1955). Etude de l'activité électrique de l'utérus au cours de la gestation et du travail. Thèse, Paris.

Sureau, C. (1956). Etude de l'activité électrique de l'utérus au cours du travail. *Gynéc. Obstét.* **55**, 153.

Sureau, C. (1964). La contraction utérine. *Ann. Anestésiol. Franc.* **5**, No spécial, 7.

Sureau, C., Chavinié, J. and Cannon, M. (1965). L'électrophysiologie utérine. *Bull. Féd. Gynéc. d'Obstét.* **17**, 79.

Wolfs, G. M. J. A. and van Leeuwen, M. (1979). Electromyographic observations on the human uterus during labour. *Acta Obstet. Gynecol. Scand. Suppl.* 90.

4. MONITORING RESPIRATORY ACTIVITY IN INFANTS—A NON-INTRUSIVE DIAPHRAGM EMG TECHNIQUE

M. J. O'Brien, L. A. van Eykern and
H. F. R. Prechtl

*Department of Developmental Neurology,
University Hospital, Groningen, The Netherlands*

1.0 INTRODUCTION

Interest in developmental respiratory physiology and behaviour and in disorders of respiratory control in infancy has led to the development of minimally intrusive methods to monitor continuously aspects of respiratory activity over prolonged periods in sick and healthy infants (Rolfe. 1975). For instance, nasal and oral thermistor and pneumotachographic methods are used to measure air flow qualitatively or quantitatively (Prechtl *et al.*, 1968; Buhl, 1968; Palm *et al.*, 1971; Fleisch, 1925; Rigatto and Brady, 1972; Gordon and Thompson, 1975; Brouillette and Thach, 1980). Respiratory movements have been monitored using strain gauges (Milner, 1970), linear displacement transducers (Konno and Mead, 1967), magnetometers (Rolfe, 1971), movement-sensing mattresses (Lewin, 1969; Smith and Scopes, 1972), radar (Caro and Bloice, 1971), electrode belt (Kelly *et al.*, 1979) and impedance pneumography (Pallet and Scopes, 1965; see also Volume 1 of this series for an excellent review of this subject by Baker, 1979). The impedance technique is the one most widely used

NON-INVASIVE MEASUREMENTS: 2   *Copyright©1983 by Academic Press Inc. (London) Ltd.*
ISBN 0 12 593402 5   *All rights of reproduction in any form reserved*

for apnoea monitoring in neonatal intensive care. Under stable measurement conditions and with careful application, magnetometer and impedance methods can be calibrated to provide a representation of volume changes during spontaneous breathing. Calibration is tedious, however, and has to be repeated if the patient changes position (Ashutosh *et al.*, 1974), which probably explains why minimal clinical use is made of this possibility (Rolfe, 1975). These flow and movement sensing methods are ineffective as respiration monitors for ventilated patients as they not only sense respirations initiated by the patient but also those externally imposed by the respirator. The only way the clinician can judge the ventilatory efforts of the baby itself is by direct observation. The eye of a trained observer is indeed a superb measuring instrument but few clinicians have the time to watch babies for long enough to make an informed judgement of the quality of these efforts, and even the human eye may have trouble distinguishing the spontaneous from the imposed breath in a sick infant on a ventilator. This is especially so if the infant's spontaneous rate of breathing is close to the programmed rate of the ventilator. A measure of breathing activity which does not suffer this disadvantage would clearly be useful in a clinical setting.

Several groups have successfully applied surface electromyography in neonatal respiratory studies in recent years. Patterns of activity in diaphragm, intercostal, abdominal and other accessory respiratory muscles have been studied in newborns during different behavioural states and in different postures in six-hour duration polygraphic studies (Prechtl *et al.*, 1977, 1979; O'Brien *et al.*, 1979b). The potential value of respiratory muscle EMG as a form of respiratory monitoring in newborns (O'Brien and van Eykern, 1979) and in diagnosis of neonatal diaphragmatic fatigue (Muller *et al.*, 1979a, c) has been indicated. The accumulating evidence that the diaphragm myo-electric signal measured with surface electrodes contains physiologically meaningful information makes this measurement attractive for both research and clinical purposes.

2.0 THE PHYSIOLOGICAL BASIS OF ELECTROMYOGRAPHY

2.1 The Motor Unit

Lidell and Sherrington (1925) introduced the concept of the motor unit as the basic functional unit of the neuromuscular system. A good recent review of the properties of motor units is given by Buchthal and Schmalbruch (1980). A motor unit consists of an anterior horn cell in the spinal cord, its axon and a group of muscle fibres innervated by that axon. The number of muscle fibres per motor unit varies greatly depending on the animal species and the muscle examined. Edström and Kugelberg (1968) used the depletion of glycogen in muscle fibres produced by repetitive stimulation of a single motor nerve to study the size and arrangement of individual motor units. This approach has confirmed that fibres belonging to one motor unit share physiological and histochemical properties, and that the fibres of a motor unit are scattered over a large area. There is considerable overlap between the fibres of different motor units. Buchthal *et al.* (1959) using electrophysiological methods found that in humans the average diameter of the territory of motor units was 5–7 mm in upper limb muscles and 7–11 mm in lower limb muscles. A territory of this size contains space for the fibres from 15 to 30 different motor units. The average number of muscle fibres per motor unit varies between 110 and 1720 in different human muscles (Buchthal and Schmalbruch, 1980). Muscles that are finely controlled generally have low innervation ratios, e.g. few muscle fibres per motor unit. In this regard it is of interest that innervation ratios of 25–83 have been reported in the diaphragm in different species (Krnjević and Miledi, 1958; Campbell *et al.*, 1970).

When a motor neuron fires, a neural impulse is transmitted via axonal branches to the muscle fibres innervated by that cell. As a result the individual muscle fibres contract after a few milliseconds delay and each generates a minute action potential. Ekstedt (1964) recorded single muscle fibre action potentials in normal human muscles and found that they were characterized by a median voltage of 5·6 μV and a duration of 470 μs. The individual temporally dispersed muscle fibre potentials summate to form a composite motor unit action potential which in normal muscle ranges in amplitude from a few microvolts to 10 mV (De Luca, 1979). The size of the motor unit potential is directly related to the size of the motor unit. Motor unit size also determines recruitment order under most circumstances. This has become known as the size principle following the work of Henneman *et al.* (1965).

As shown by Adrian and Bronk (1929), the force generated by the individual motor unit is graded by means of changes in its frequency of discharge. In humans the rate of motor unit discharge rarely exceeds $30 \, s^{-1}$ (Buchthal and Schmalbruck, 1980) except within the first second of a maximal contraction when motor units may fire briefly at $150 \, s^{-1}$ (Marsden *et al.*, 1971). The force of whole muscle contraction is graded by three mechanisms (De Luca, 1979): (1) increase in the number of activated motor units, (2) increase in firing frequency of the active units, and (3) synchronization of the activity of the different motor units. At the beginning of a force varying contraction, for instance, the smallest motor units are recruited first at a firing rate of $5–6 \, pulses \, s^{-1}$. As force increases up to 30% of maximum voluntary contraction (MVC), progressively larger units are recruited. The firing rate increases also as a secondary factor. Increase in firing rate becomes the dominant factor above 30% MVC. When a muscle fatigues during both constant-force and force-varying isometric contractions, synchronization of the discharges of motor units occurs.

2.2 The Properties of the Surface Electromyogram

It is now generally agreed that the surface electromyogram represents the global level of excitation of the muscle (Maton, 1976). Moore (1967) studied the rules of combination and summation of motor unit action potentials using a synthetic waveform to represent the physiological potential. He assumed a random distribution of impulses and found that the root mean square (RMS) of the resultant wave was linearly related to the square root of the number of impulses summed, and that if two or more groups of impulses were added up randomly that the RMS was proportional to the square root of the sum of the individual mean squares (i.e. variances). Clamann (1969) studied motor unit firing statistics in the brachial biceps muscle in humans and found that Moore's assumption of randomness was justified. He concluded that the firing pattern could be modelled by a stochastic renewal point process, the underlying probability density being Gaussian. This has since been confirmed by other workers also (e.g. Freund *et al.*, 1973). Dick *et al.* (1974), like Moore, investigated how such random signals would combine. Using an analogue computer and a standard biphasic wave they confirmed Moore's finding of a linear growth in variance with increased numbers of active units. Variance was also linearly related to firing frequency. Biro and Partridge (1971), in a lucid paper, extended Moore's work to include physiological waves. They studied the summation of motor unit potentials recorded from the triceps sura muscle of anaesthetized cats. With an increase in the number of contributing identical units, the probable distribution of voltage became indistinguishable from a Gaussian distribution; in other words the global EMG is a noise-like signal. Like Moore, they showed that variance was proportional to the number of (identically distributed) active units, and that when units of non-identical variance were summed, variance of the combined signal was equal to the sum of the individual variances. Biro and Partridge reasoned that the same rule of linear summation of motor unit potentials should also apply to potentials from fibres within a unit, and that the variance contributed by an individual unit should therefore be proportional to the number of fibres in that unit. Since Håkansson (1957) had found that fibre voltage was proportional to the square of fibre circumference, Biro and Partridge concluded that the variance of the EMG can be interpreted as representing the total weighted fibre activity, i.e. it represents both the cross-sectional area of the active fibres and their firing rate.

In summary, the global EMG as measured with surface electrodes can, with certain restrictions, yield reliable descriptors of neuromuscular excitation.

2.3 The Relationships between Electrical Activity and Mechanical Action of Muscles

Since the work of Lippold (1952), demonstrating a linear relationship between the integrated surface electromyogram and the isometric tension in the human adult gastrocnemius muscle, many studies have confirmed that relationships exist between the surface EMG and biomechanical variables, such as work (Bouisset and Goubel, 1973) or force of isometric contraction (Moritani and De Vries, 1978). The exact relationship between muscle electrical activity and muscle force is dependent on the muscle state, on intrinsic muscle properties such as force/length and force/velocity ratios (which vary between different muscles) and on extrinsic loads which determine the mechanical conditions of the contraction. One of the most important factors to consider is the effect of fatigue. Fatigue causes a reduction in muscle force at the same time as an increase in the average half-wave rectified, integrated, or RMS surface EMG amplitude (Edwards and Lippold, 1956; De Vries, 1968; Lloyd, 1971; De Luca, 1979; Petrofsky, 1980). Since Piper (1912) it has been known that fatigue is associated with a relative increase in low frequency components in the surface EMG and this has been quantified either as the ratio of low (c. 40 Hz) to high (> 150 Hz) frequency power (Kaiser and Petersén, 1965; Stålberg, 1966; Lindström et al., 1970) or in terms of changes in the centre frequency of the EMG power spectrum (Petrofsky et al., 1975).

Fatigue has recently been detected in diaphragmatic EMG signals in adults (Gross et al., 1979) and in newborns (Muller et al., 1979a), and it has been suggested that monitoring of EMG frequency changes may be of benefit in directing the process of weaning infants from ventilators (Muller et al., 1979c). These changes in newborns were detected using surface electrodes. As discussed in the following section it has only recently been realized that surface recording of diaphragmatic EMG activity is possible.

3.0 ELECTROMYOGRAPHY OF THE DIAPHRAGM

3.1 Background

Taylor (1960) was perhaps the first to recognize that it is possible to record diaphragmatic EMG activity using surface electrodes. In the course of studying the activity of intercostal muscles in a group of 80 volunteer male servicemen, he recorded simultaneously from surface and bipolar needle electrodes in the seventh right intercostal space in the mid-clavicular line in order to test whether Campbell (1954) might have been mistaken in attributing inspiratory activity recorded with surface electrodes in the lower intercostal spaces to intercostal muscle instead of to the diaphragm. To quote Taylor,

> with the subject breathing quietly in the supine position, the needle in the intercostal (a single layer at this point) showed no activity, while the surface electrodes picked up distant inspiratory discharges. In deep breathing the needle revealed marked expiratory discharges only. The surface electrode, however, showed activity in both phases, that in inspiration being clearly from a much more distant source (diaphragm) than that in expiration (intercostals).

Green and Howell (1959) probably made the same error as Campbell. With surface electrodes placed in the lower (6th–9th) intercostal spaces in six healthy young male subjects, they recorded inspiratory activity which carried over into expiration—probably during quiet breathing—in the semi-recumbent posture. They clearly did not even entertain the possibility that they might be recording diaphragmatic and not intercostal activity although that now seems very likely to have been the case. The same could be said of Sempik (1977), who measured inspiratory activity from electrodes placed over the 7th or 8th intercostal space and attributed it, on Campbell's (1955) authority, to intercostal muscles. Koepke *et al.*, (1958) and Taylor's (1960) findings that during quiet breathing the lower intercostals normally show no inspiratory activity have not been disproved. Both showed that these muscles only become activated with progressively deeper breathing. In a study of phrenic nerve conduction, Newsom Davis (1967) found that electrical activity in the diaphragm, evoked in response to stimulating the phrenic nerve in the neck, could be readily recorded from surface electrodes over the 7th, 8th and 9th intercostal spaces on the anterolateral aspect of the chest wall. He made no mention, however, of findings during spontaneous breathing.

   The significance of these several observations as to the feasibility of recording diaphragmatic activity from skin surface electrodes seems to have been neglected until recently. Another approach became more popular in human adult studies, namely that of recording diaphragmatic activity via oesophageal bipolar electrodes (Draper et al., 1959; Agostini et al., 1960a, b; Petit et al., 1960; Lourenco and Mueller, 1967). This same approach was taken by Finkel (1975) in newborn infants. He recorded external intercostal EMG by means of surface electrodes, and diaphragm EMG by an oesophageal probe containing silver electrodes in newborn infants in different behavioural states. Interest in the recording of diaphragmatic activity from externally placed electrodes in infants is recent. Hagan et al. (1976) successfully recorded diaphragmatic EMG activity in newborn infants from surface electrodes in the 7th intercostal spaces in the midclavicular line and Prechtl et al. (1977) reported findings using a non-intrusive surface electromyographic technique developed for polygraphic study of newborn respiratory muscle activity. They adopted a new electrode placement for diaphragm recording, positioning them bilaterally at the costal margin in or just outside the midclavicular line. Curzi-Dascalova et al. (1979) placed electrodes over the right 7th and 8th intercostal spaces in the anterior and middle axillary lines in an infant with a *paralysed* diaphragm requiring assisted ventilation and recorded clear inspiratory intercostal activity. This finding emphasizes the need for caution in interpreting the origin of activity recorded in that area, a point which will be discussed later. This finding presumably corresponds to that of Koepke et al. (1958) and Taylor (1960) mentioned above regarding the recruitment of lower intercostals with deeper breathing.

3.2 Relationships between Diaphragm EMG Activity and Respiratory Control System Output

Pitts (1942) demonstrated that phrenic nerve activity was closely related to the tidal volume. Katz et al. (1962) found in humans that integrated diaphragmatic EMG activity (needle electrodes) was linearly related to tidal volume during rebreathing. These authors pointed out the need to establish control values before each experiment and emphasized that relationships with volume could only be expected under conditions of unobstructed breathing. Lourenco et al. (1966) showed in dogs that during unobstructed breathing, tidal volume was proportional to both phrenic nerve and diaphragm integrated activity whereas with breathing obstructed diaphragmatic activity was still highly significantly related to phrenic nerve activity. Eldridge (1975), dissatisfied with tidal volume and minute volume as indicators of the output of the respiratory control system, turned to airway occlusion pressure as a measure of respiratory muscle force output and investigated in cats the relationships between, on the one hand, several descriptors of phrenic nerve and external intercostal EMG activity, and, on the other hand, peak airway pressure generated during occlusion of the airway. Peak and average rates of inspiratory phrenic nerve activity and peak external intercostal muscle activity showed highly significant linear correlations with peak airway pressure (r values = 0·974, 0·973 and 0·915 respectively). Altose *et al.* (1975) showed in dogs that both mechanical loading and hypercapnia increased total inspiratory, diaphragmatic (and intercostal) electrical activity. However, since mechanical loading also increased inspiratory duration whereas hypercapnia reduced it, the average rate of diaphragmatic electrical activity was unaffected by loading but was increased by hypercapnia. Lopata *et al.* (1977) compared three methods of processing diaphragmatic EMG signals measured from oesophageal leads in humans. They found that the slope and peak values of moving average and moving variance EMG measures were highly correlated to changes in P_ACO_2 and ventilation during CO_2 rebreathing, and to mouth pressure measured during transient airway occlusion. As Derenne *et al.* (1978) point out in a review, respiratory muscles seem to function as flow generators, acting in the face of varying respiratory system loads to maintain constant flow. Under constant conditions, a proportional relationship exists between appropriate diaphragm EMG descriptors and force output of the inspiratory muscles. This electromechanical relationship is variable, however, and thus in order to understand fully the significance of a change in respiratory neural excitation, a knowledge of the muscle state, the mechanical conditions and the ventilatory demand are necessary.

4.0 TECHNICAL CONSIDERATIONS RELATING TO SURFACE EMG OF THE DIAPHRAGM IN INFANTS

4.1 The Measurement System for Recording in a Polygraphic Context

In infant studies, the surface EMG of the diaphragm is usually recorded in a polygraphic context, i.e. in conjunction with other signals. Such studies entail the non-intrusive measurement of multiple (electro-) physiological variables during the spontaneous cycling of behavioural states (Prechtl *et al.* 1968). Undistorted signals of high signal-to-noise ratio, with a minimum of movement artefacts or line frequency interference, are an essential prerequisite for this type of research. Recording in any of the following modes may be required:

(a) unipolar, where signals are measured as a voltage difference between individual electrodes in active regions and one or more electrodes in inactive or assumed inactive regions;

(b) average reference, where the signals are measured as a voltage difference between individual electrodes in active regions and the calculated mean voltage of all these electrodes;

(c) bipolar, where the signals are measured as a voltage difference between two electrodes both in active regions.

The system should offer the user operational flexibility and should be based on a signal measurement technique which deals with unwanted signal components in an acceptable way, i.e. without having to rely on high-pass, notch- and low-pass filters.

The most practical answer is to perform impedance transformation and preliminary signal amplification in a compact modular preamplifier to which electrode signals are transported by short flexible cables. The rest of the equipment can then be set up wherever it is most convenient.

4.2 Preamplifier Design Considerations and Principles

4.2.1 *Safety*

Complete isolation of the subject from earth potential is necessary to minimize the possibility of electrical hazards. Care is necessary to ensure that leakage current, for instance at the line frequency, from the subject via the equipment to earth is maintained below those levels which may cause ventricular fibrillation. This is achieved by using isolated power supplies and isolation amplifiers in all signal lines. Safeguard circuitry at the input stage prevents damage to the electronic circuits during procedures such as defibrillation and electrical cauterization.

4.2.2 *Modular Design*

A modular control-free input stage provides operational flexibility and ease of maintenance.

4.2.3 *Signal-to-Noise Ratio*

If electrode impedance is low the input voltage noise is the main factor determining the signal-to-noise ratio. If electrode impedance is high, as for instance when the effective electrode area is small, or becomes high during a recording due to deteriorating electrode contact, then the input noise current becomes a critical factor. Since the product of current and impedance is voltage, an increase in electrode impedance will significantly increase the input noise voltage. A low input noise current as well as low input noise voltage are thus both necessary to produce a high signal-to-noise ratio. Amplifying the signal in the input stage of the preamplifier will prevent noise contributions at subsequent stages from significantly affecting the signal-to-noise ratio. The amount of amplification possible is limited by the fact that the electrode offset potential is also being amplified. Too much amplification would result in a latch-up in the amplifier output stage.

4.2.4 *Artefact Prevention*

Line frequency interference and movement artefacts greatly diminish signal quality. Line frequency interference is very easy to pick up but less easy to reject. It is caused mainly by the capacitive coupling of the subject's body, as well as of the electrode cables, to the surrounding power lines and mains-powered equipment. If the electrical field is homogeneous and no current flows through the body of the subject to earth, all electrodes exhibit the same amount of line frequency interference which will be cancelled out if the common mode rejection ratio of the input stage is sufficiently high. Homogeneity of the electrical field at the measuring site can be promoted by ensuring that mains transformers and power lines are at a distance of at least 2 m from the measuring site. Leakage current from the subject to earth can be minimized by isolating the preamplifier from the mains supply by means of a highly isolated power supply and using isolation amplifiers in all output signal lines.

The amount of line frequency interference coupled into the electrode cable is reduced to insignificance if shielded electrode cables are used. The cable capacitance resulting from the use of shielded cables can be minimized by connecting a low impedance version of the electrode signal to the shield to create a signal guard. The final factor to consider in connection with the prevention of mains interference is the frequency range of the amplifier. Since non-linearities always exist between the input stages of different amplifier channels, higher harmonics of the mains frequency can cause deterioration of the common mode rejection. This cannot be corrected in subsequent amplifier stages and if allowed to occur may contribute unpredictable components to the output signal in the course of a long recording. Even though the highest frequencies encountered in surface EMGs will not exceed 2 kHz, it is none the less essential that linear performance from DC to at least 5 kHz be specified to prevent this happening. The major part of the common mode rejection should take place in the input stage of the preamplifier as a further precaution against non-linearities in subsequent stages in the signal path. The common mode rejection is also influenced by differences between channels in the ratio, electrode impedance/input impedance. As electrode impedance may vary considerably, it is important that the input impedance be so high that electrode impedance differences cannot significantly influence the common mode rejection.

4.2.5 *Movement Artefacts*

The input impedance of an amplifier forms a load for the electrode offset potential. If this impedance is too low, a current flows through the input circuit until an equilibrium is reached. Disturbances at the electrode–skin interface associated with movements change the electrode offset potential. The time it takes to reach a new equilibrium varies between different electrodes, even if the offset step changes are equal in amplitude. This will be seen as an artefact even in a bipolar recording. If the input impedance is high enough, almost no current will flow and after the initial step change in offset voltage there will be no further noticeable change in electrode offset. Thus the changes in the output signal will be primarily determined by the difference in the electrodes' offset voltage steps.

A second source of movement artefact originates in the so-called input bias current. This is a constant current flowing in the input stage of an amplifier and will cause a voltage drop across the electrode impedance. Changes in electrode impedance caused by movements alter this voltage and therefore cause changes in output signal during movements. To minimize movement artefacts the resistive part of the input impedance should be more than 1 GΩ ($10^9\Omega$) and the input bias current less than 50 pA (5×10^{-10} A) (Zipp and Ahrens, 1979).

4.2.6 *Signal Transport*

The output signals from the preamplifier must be insensitive to interference pick-up and ground loops during transport to the main amplifier. In order to achieve this while still retaining the choice of recording modes mentioned, the information from each electrode signal is best transported as a voltage difference between two signal lines. The line frequency interference picked up by the signal lines will then appear as an in-phase voltage at the main amplifier where, being a common mode signal, it will be cancelled in its differential input stage.

The only type of amplifier system in which all these features can be combined is one based on the average reference principle.

4.3 Amplifier Based on a Modification of the Average Reference Principle

This recording technique was introduced in electrocardiography by Wilson *et al.* (1934) and applied to electro-encephalography by Goldman

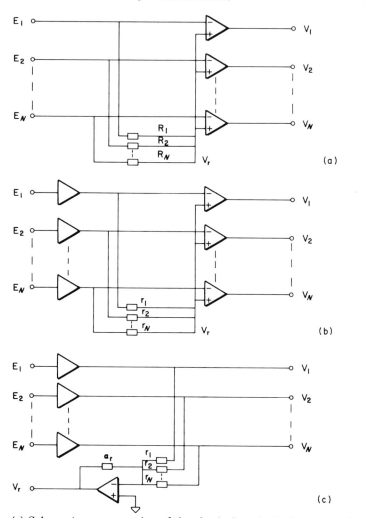

FIG. 1   (a) Schematic representation of the classical method of average reference recording. (b) An application of the average reference technique incorporating impedance transformation. (c) A further stage in the development of the technique. Feedback of the average reference, via the subject, to the input stage has been added.

(1950) and Offner (1950). The advantages of the classic method, shown diagrammatically in Fig. 1a are: (i) the common mode signals are rejected at the input stage (ii) the information is available as a voltage difference between output channels for unipolar and bipolar recording, and (iii) the electrode signal can be re-obtained for guarding. Disadvantages are related to loading of the electrode by the summing resistors and the additional noise generated in the (large) summing resistors. Several systems based on this recording principle are commercially available. None has all the features that have been discussed. One lacks common mode rejection at the input stage (Fig. 1b); in another modification (Fig. 1c) the reference signal is fed back to an electrode and thus to the subject. This can give rise to unstable performance and is not totally safe. The main shortcoming of these and of the many other amplifier systems sold for use in the measurement of bio-electric signals using surface electrodes is that they do not allow true electrode cable guarding. It does not seem to be fully appreciated that the electrode cable is the main source of line frequency interference pick up. Faraday cages, which are very expensive, do not solve the problem of line frequency interference, being only really effective against radio-frequencies.

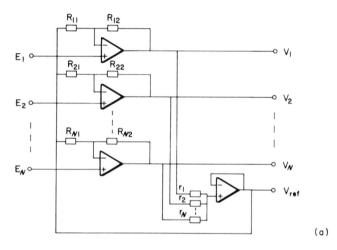

(a)

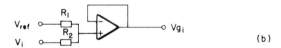

(b)

FIG. 2   (a) Circuit diagram of an average reference recording method which incorporates impedance transformation, amplification and feedback of the average reference directly to the input stage. (b) The electrode cable guard (V_{gi}) is obtained by weighted adding of the output signal (V_i) and the average reference signal (V_{ref}).

The reference amplifier shown in simplified diagrammatic form in Fig. 2a was developed by one of the authors (L. A. van E.) for use in polygraphic recordings. In addition to other advantages it incorporates signal guarding. After impedance transformation and amplification of the electrode signal the average reference is formed by means of "smaller" summing resistors. The average reference is fed back to the input stage after impedance transformation to prevent loading of the common point. Signals for the three recording modes can be obtained by simple subtraction of selected reference amplifier output lines. Guard signals are re-obtained from the output signals (V_i) and the average reference signal (V_{ref}) by means of weighted adding as shown in Fig. 2b. Tests showed that best performance was achieved by guarding the input signal with a low impedance version of itself containing both common and specific components. A system based on operational amplifiers (Type 52J, Analog Devices) has an input impedance of more than 1 GΩ, input bias current < 3 pA, an RMS noise voltage of 10 nV Hz$^{-\frac{1}{2}}$, an output impedance of 2 Ω, a frequency range from DC to at least 10 kHz, an amplification factor of 50 and common mode rejection at the line frequency of more than 100 dB. The preamplifier unit is modular. It is isolated from the mains supply by a highly isolated power supply and incorporates isolation amplifiers in all output lines.

This system has been used routinely in infant and animal polygraphic recordings since 1975 and has particularly benefited developmental EEG research (Vos *et al.*, 1977; Gramsbergen *et al.*, 1980), and neonatal respiratory muscle EMG research in humans (Prechtl *et al.*, 1977, 1979). The multichannel preamplifier systems in operation allow great flexibility in the types of measurement possible, incorporating not only average reference amplifier modules but also differential amplifier modules and a number of special purpose modules for measurements of temperature, sound, pressure, etc.

4.4 Electrode and Electrode Cables

4.4.1 *Electrodes*

Because of the very high input impedance of the system described, electrode impedance does not influence signal quality, so the size of the electrode can be tailored to the size of the subject or to suit the wishes of the experimenter. The main criteria governing choice of electrode are (a) the need to ensure a stable skin–electrode contact, both electrical and mechanical, and (b) patient acceptability and comfort. Indirect contact or "floating" electrodes, in which electrolyte is interposed between the skin and the electrode surface, provide a more stable electrical contact between skin and electrode than direct contact electrodes where the stability of the contact can be easily disrupted by movements of the skin or of the electrode. We have found re-usable sintered silver/silver chloride indirect contact electrodes (manufactured by Hellige GmbH) to be satisfactory for polygraphic work. These electrodes contain 2 mm diam. electrode pellets set in a shallow Perspex chamber containing two holes for injection of electrode paste. When used for long recordings in warm environments these holes have the disadvantage that they permit evaporation of the electrode paste. For use with babies we therefore seal the holes. Electrode paste is applied before fixing the electrode to the skin by means of double sided adhesive collars. We also deepen the electrode chamber by adding a 1 mm thick ring of Perspex to the base of the electrode chamber. Provided the electrode remains attached to the skin, stable respiratory muscle EMG signals can be recorded using this technique for several days. In a clinical intensive care unit setting we have experimented with disposable tin foil disc electrodes (made by Hoffman la Roche). These have an electrode diameter of 18 mm and are used normally for such measurements as ECG and transthoracic impedance. The contact with the skin is made through a thin layer of electrode paste. We have recorded quite satisfactory respiratory muscle EMG signals using these electrodes with our preamplifiers. Smaller versions would probably be just as effective and might be more suitable for tiny preterm babies.

4.4.2 *Electrode Cables*

As already discussed, it is highly desirable when measuring bio-electrical signals such as the EMG to use shielded and guarded electrode cables. Re-usable electrodes are not normally supplied with shielded cables. We have found the most practical solution to be to leave the last 10 cm of unshielded cable attached to the electrode and to connect this via 1 mm male and female gold-plated connectors to an 80 cm long flexible shielded cable which goes to the preamplifier module. This cable is composed of a highly flexible uninsulated shielded cable (Filotex C.I.P.) threaded through medical polyvinyl chloride (PVC) tubing, 0·8 mm i.d., 1·6 mm e.d.

4.5 Signal Detection, Processing and Interpretation

4.5.1. *Signal Detection*

We have investigated diaphragmatic EMG activity using surface electrodes during the course of 94 consecutive 6-h polygraphic recordings. This represents a heterogeneous group of infants, polygraphically investigated for a variety of reasons. Most were full-term newborns or preterm infants recorded at term conceptional age, but a small number of older infants

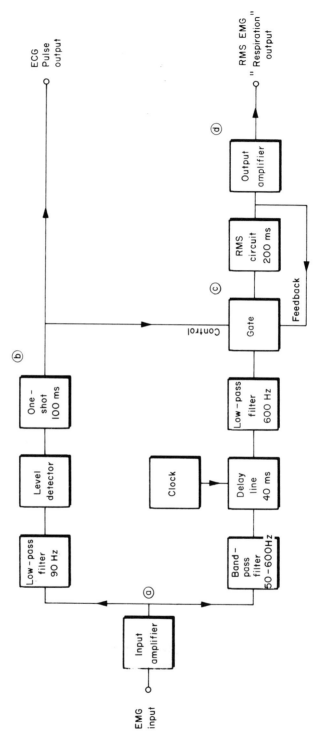

EMG input

Input amplifier

(a)

Low-pass filter 90 Hz

Level detector

One-shot 100 ms

(b)

ECG Pulse output

Band-pass filter 50-600Hz

Clock

Delay line 40 ms

Low-pass filter 600 Hz

Gate

Control

(c)

RMS circuit 200 ms

Feedback

Output amplifier

(d)

RMS EMG "Respiration" output

(A)

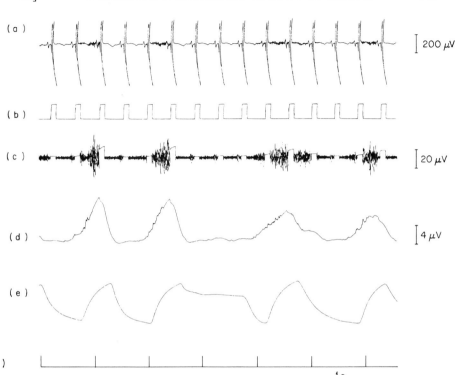

FIG. 3   Functional diagram of the EMG processing technique (A) with examples of signals recorded from strategic points in the signal path (B): (a) the composite raw signal at the output of the input amplifier containing low amplitude diaphragm EMG and high amplitude ECG components; (b) normalized pulse derived after detection of the QRS complex; (c) the amplified diaphragm EMG signal after gating out the QRS complex and filling in the ongoing RMS value via a feedback loop into the gate; (d) the RMS output signal showing good interpolation during gating and almost no residual heart beat components. The signal has a respiration wave form; (e) a reference respiration signal recorded via a nasal thermistor.

was also studied. The EMG of the diaphragm (measured from more than one electrode derivation simultaneously in 10 infants) was recorded in conjunction with EMGs of other respiratory muscles: in all infants EMG activity from at least one upper intercostal derivation was measured; in 27 infants EMGs from abdominal muscles were also recorded. A reference respiratory signal was recorded in all cases from a thermistor placed at the entrance to one nostril, together with a standard ECG signal. Depending on the other purposes of the recording, signals such as EEG, EOG and other non-respiratory EMGs were also measured.

The position of the recording electrodes for the diaphragm EMG was chosen somewhat intuitively. In order to avoid contamination from possible intercostal activity, electrodes were placed below the costal margin on the right and left sides of the body just outside the nipple lines. It seemed unlikely that during quiet breathing there would be much risk of contamination from abdominal muscles (Campbell, 1958) and the site seemed acceptably close to the insertion of the costal fibres of the diaphragm along the costal margin. This choice of electrode site proved fortunate, as bipolar measurements consistently yielded a clear respiratory EMG signal, peaking in inspiration (see Fig. 3). In five infants a comparison was made between the above derivation and one based on electrodes placed in the 7th intercostal space in the anterior axillary line on either side of the body. In three infants electrodes were attached in the 7th and 9th intercostal space in the anterior axillary line on the right side of the chest to compare unilateral and bilateral recording. In two infants a variety of other unilateral bipolar combinations were evaluated. The respiratory EMG signal recorded from the bilaterally placed subcostal or costal electrodes was always larger than that recorded simultaneously at other sites. It is not completely clear why this is so. Possibly at this site the relationship between electrode and electrically active muscle is least influenced by factors such as change in posture, movements of skin, alterations in lung volume, etc. but this is speculative. The chance that the EMG signal will be diminished by in-phase common components is minimal with widely separated electrodes. In the differential amplifier the measurement, in effect, represents the vectorial sum of two random signals. Unilateral electrodes are likely to measure a greater common EMG signal and thus a smaller difference signal especially if the distance between the electrodes is small (Vigreux *et al.*, 1979). Electrodes in intercostal spaces such as the 7th are probably at a somewhat greater distance from the diaphragm than the costal electrodes. Workers in Toronto have also successfully recorded diaphragm EMG from several sites, unilateral and bilateral, using surface electrodes, over a wide range of subject ages (Tusiewicz *et al.*, 1977; Muller *et al.*, 1979a; Fleming *et al.*, 1979). Macklem *et al.* (1978) placed surface electrodes in the 6th and 7th intercostal space in the anterior axillary line and measured diaphragm EMG in three adults.

Electrodes placed near the costal margin anterolaterally right and left allow pick-up of a signal derived from both hemidiaphragms. The separate activity of each hemidiaphragm can be recorded either in the bipolar mode by appropriately placed unilateral electrodes or in the unipolar mode using the common average reference technique. Figure 4 shows diaphragmatic activity recorded in this way from four electrode sites, two at the costal margin anteriorly and two on the posterior chest.

4.5.2 *Signal Processing*

(i) *ECG artefact elimination.* As shown in Fig. 3 the measured signal, a, contains a low amplitude respiratory EMG component and a high amplitude ECG component. The ECG component prevents a satisfactory quantitation of the EMG signal either in the time or frequency domains. There is no ideal way to separate the ECG and the EMG components. Bandpass filtering is useless since the ECG and EMG frequency spectra overlap (Schweitzer *et al.*, 1979).

One approach to this problem is to subtact an average ECG complex from the composite signal. Bickford *et al.* (1971a, b), who were concerned with ECG contamination of EEG signals in cases of suspected brain death, described a computer technique whereby an ensemble averaging procedure, using the "R" wave to trigger the computer, generated an average ECG waveform which was then subtracted from the ECG contaminated signal. This technique was adapted by Barlow and Dubinsky (1980) for real-time use with up to 16 channels of clinical EEG. Nagel and Schaldach (1979) reported success using a similar technique to eliminate maternal ECG complexes from abdominal lead fetal ECG signals (see also Chapter 1 in this volume). These authors achieved a 60% increase in the detection of fetal ECG complexes compared to conventional techniques. The main drawback with the averaging approach is that short-term modulation in the amplitude of the QRS complex related to modulations in depth of breathing, and abrupt reductions in ECG amplitude during body movements reduce the efficacy of the procedure. Such changes are quite marked in the ECG contaminating diaphragm EMG signals in infants and would certainly result in a residual artefact—after subtraction—that would occasionally exceed the low amplitude EMG activity measured during quiet breathing.

Lourenco and Mueller (1967) attempted to minimize the effect of the ECG by clipping the signal above a chosen level and subtracting the integrated normalized ECG from the total integrated signal. This method is only adequate when the EMG does not exceed the ECG in voltage, e.g. at rest (Evanich *et al.*, 1976).

Bruce *et al.* (1977) described a non-linear digital technique for processing respiratory EMG signals, whereby the signal was first bandpass filtered (50–500 Hz) and then sampled at 1000 Hz and full wave rectified. It was then passed through a digital voltage window. This set the signal to zero whenever it exceeded a preset threshold voltage, a procedure which the authors called "blanking". Portions of the ECG signal not exceeding the voltage threshold are not eliminated by this technique, whereas it suffers the same disadvantage as the clipping method of being ineffective if the EMG voltage exceeds that of the ECG. Mucke and Buchholz (1977)

Respiration   Insp

RR   60   0

HR   150

Cardiotach   100 150   80

Intercostal — Bipolar   Right anterior

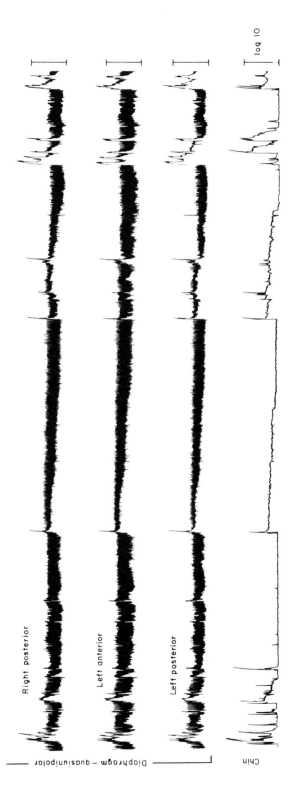

FIG. 4 Quasi-unipolar diaphragm EMG recording. The specific electrode activity at symmetrical sites on the lower lateral chest, front and back, was measured in a 5-day-old infant using the common average reference technique. The average value of the four diaphragm electrode signals was used as reference. The processed EMG signals are displayed logarithmically on a compressed time scale to facilitate visual perception of tonic activity changes. The record shows an epoch of behavioural state 1 bounded by incomplete epochs of state 2 (see Table I for state definitions). Abbreviations: Respiration = nasal thermistor respiration signal; RR and HR = respiration and heart rate signals from a moving window event counter; bipolar intercostal EMG signal from electrodes placed parasternally right and left in an upper intercostal space; chin = moving average chin EMG.

developed a similar approach for non-respiratory EMG analysis based on analogue detection of the ECG by means of a comparator, the output of which is used to control on-line digitization of the EMG.

The technique shown schematically and illustrated in Fig. 3 allows more complete elimination of the ECG artefact. The input signal is delayed by a variable delay line in one signal path while the QRS complex is detected in a second circuit by a level detector. The output of the level detector goes to a one-shot or pulse stretcher which generates a standard pulse, the width of which is set to match the duration of the QRS complex. The output of the one-shot controls an electronic switch which gates the delayed signal for the duration of the standard pulse. Additional gates can be included if desired to eliminate the P and T waves, but this becomes self-defeating at high heart rates as too much signal is eliminated. The ECG signal used for detection of the QRS complex can either be separately recorded or can be that present in the EMG signal. In the latter case, the large ECG component in the signal recorded from widely separated electrodes is a positive advantage, since it minimizes the risk of missed detection due to the QRS component becoming submerged in the EMG under conditions of increased activity.

(ii) *EMG processing*. Following gating of the ECG component the raw EMG is suitable for visual inspection and further processing. The gating technique carries the disadvantage that not just the QRS complex is eliminated but also any superimposed EMG activity. For visual inspection of the signal the gates are not really a problem, since the patterns of respiratory EMG activity can still easily be appreciated and the observer has the ability to ignore the presence of the gates, a process made easier by time compressing the displayed signal. Visual inspection provides a check on signal quality and allows a rough qualitative description of the behaviour of the muscle over time. Biro and Partridge (1971) defined perception of signal standard deviation as a property of visual inspection of the raw EMG. Visual scanning of the signal represents a form of continuous moving time window analysis. These two aspects of visual analysis, standard deviation tracking and moving window processing, also characterize the EMG processing technique we have developed (see Fig. 3). The RMS value of the signal is calculated in a moving time window of about 200 ms by a so-called box-car averager. To compensate for the loss of EMG signal during the "R" wave gating the output of the RMS unit is fed back to the electronic switch and during the time that the signal is interrupted to delete the QRS complex, the ongoing RMS value is filled into the gate, as substitution for the RMS of the deleted EMG activity. Further low pass filtering of the RMS signal to smooth out the gated portion of the signal is made unnecessary by this step. It can of course be done for cosmetic reasons and a second order low pass filter with $f_0 = 6$ Hz is satisfactory for this purpose.

Muller *et al.* (1979c) adopted a somewhat similar approach to that just described, but with two key differences. Firstly, they relied on low pass filtering alone to smooth out the gated portion of the EMG. The low-pass filter used had a time constant of 0·1s. This degree of filtering has the disadvantage that as respiratory rate increases, there is increasing loss of signal resolution. This effect will already be significant at a rate of 60 min^{-1}. Secondly, these authors used the moving average (or mean absolute) value and not the RMS value as signal descriptor. In order to derive this value, the signal must first be full wave rectified. One of the consequences of this step is to alter drastically the amplitude distribution of the signal from a more or less symmetric form, with a mean value of zero, to an asymmetric form (whose mean value can then be calculated). It can be shown that if the voltage distribution of the raw signal is Gaussian, the RMS value will be larger than the mean absolute value by a factor of 1·25. Deviation from a Gaussian distribution alters this ratio, which is known as the form factor. We have compared the RMS and the mean absolute value amplitudes of infant diaphragm EMGs and have found that their peak inspiratory amplitudes generally differ by at least a factor of 1·5, which rises under certain circumstances to 2·5. This means that the form of the respiratory EMG signal deviates significantly from the Gaussian and the RMS provides much better signal resolution than the mean absolute value. The arguments cited earlier suggest that a measure such as signal variance (in the case of EMG, the square of the RMS) is closely related to global muscle fibre excitation. The RMS has the additional advantage of being the standard, statistically valid, measure of electrical power.

The gating technique causes slight errors in the estimation of the timing of onsets and peaks of inspiratory activity if these events coincide with a QRS complex. Under this circumstance, the RMS signal will always give the impression of an onset or a peak occurring at the end of the gate, irrespective of the precise instant within the gate time that the event actually occurs. If the gate is 100 ms, then the maximum possible timing error will be 100 ms. An error in the estimation of the slopes of the inspiratory activity is a further consequence of the gating technique under this specific circumstance. Just as with all forms of signal integration, the rectangular moving window technique introduces phase shifts which are dependent on the window length and the signal form and repetition rate. Most breaths are associated with a more or less diamond shaped pattern of increase to a peak followed by a decrease in EMG activity. It follows that the peak in the RMS signal will appear about $\frac{1}{2}$ a window length after the actual peak, i.e. about 100 ms delay if the time window is 200 ms.

4.5.3. Signal Interpretation and Validity

(i) *Phasic respiratory activity*. There are good reasons to believe that the measured RMS signal genuinely represents diaphragmatic activity during normal breathing and probably suffers little or no contamination from neighbouring accessory muscles of respiration when breathing is augmented. The inspiratory diaphragmatic EMG activity has an RMS value about 2·5 times that of intercostal activity recorded from electrodes placed parasternally over upper external intercostal muscles. In humans, inspiratory activity in lower external intercostals seems to be negligible during quiet breathing (Taylor, 1960; Koepke *et al.*, 1955, 1958), but even if lower intercostal activity were mobilized during augmented breathing, there is no reason to expect that its peak RMS amplitude would be larger than that measured in higher spaces. Since random signals such as the EMG of two independent muscles summate vectorially, the RMS value of a signal containing contributions from two sources is equal to the square root of the sum of their individual variances (Biro and Partridge, 1971). It follows that even if lower intercostal activity bore a $1:2$ relationship to diaphragmatic activity, it would still contribute only 11% to the RMS value of the combined signal.

The fibres of the external oblique abdominal muscle overlap the lower ribs and are thus within the pick-up range of electrodes at the costal margin. The electrodes are also near the lateral border of the rectus abdominus muscle. We have compared the EMG of the external oblique, measured from electrodes placed 2–3 cm apart in the right lower quadrant of the abdomen, and of the rectus muscle measured from electrodes placed either to the right and left of the umbilicus or over the belly of the muscle just above and below the umbilicus, with activity in the diaphragm signal in a variety of behavioural conditions. During quiet regular breathing low amplitude expiratory activity may infrequently be seen in the abdominal muscle signals, especially the rectus, superimposed on an elevated tonic level. This does not contaminate the diaphragm signal. At the start of a feeding (see Fig. 5a), significant expiratory abdominal activity is seen in some infants, but the diaphragm signal measured simultaneously remains strictly inspiratory. During behaviours such as crying, coughing, sneezing, defaecating and vomiting, an inspiration is followed by expulsive or paroxysmal expiratory abdominal muscle activity.

It is known that the diaphragm contracts synergistically with the abdominal muscle to increase intra-abdominal pressure. This has been shown electromyographically in man (Agostini *et al.*, 1960a) and in dogs (Jiménez-Vargas *et al.*, 1967; Monges *et al.*, 1978). Diaphragmatic EMG activity has also been demonstrated during forced expiration (Agostini and

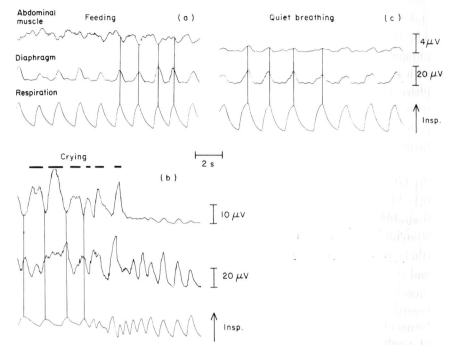

FIG. 5 Abdominal muscle (rectus abdominis) and diaphragm EMG activity in a newborn infant in three situations: (a) bottle feeding: abdominal activity is expiratory on an elevated tonic background while diaphragm activity is inspiratory; (b) crying: each cry, indicated by a line, is associated with a peak in abdominal activity preceded by an (inspiratory) peak in diaphragm activity; (c) quiet breathing: weak inspiratory activity in the abdominal muscle signal may represent distant diaphragmatic activity.

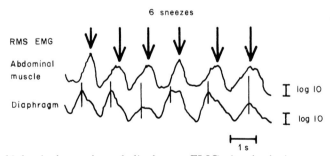

FIG. 6 Abdominal muscle and diaphragm EMG signals during sneezing in a newborn. The RMS EMG signals are displayed logarithmically. The sneezes are indicated by arrows. The phase lag between the peaks in the two signals, emphasized by the vertical lines drawn through the diaphragm signal peaks, is evidence that the intense expulsive abdominal muscle activity does not contaminate the "diaphragm" signal to the extent that diaphragmatic activity is obscured.

Torri, 1962; Delhez *et al.*, 1964) and during coughing and sneezing (Agostini *et al.*, 1960b). In our observations on infants during crying, sneezing (see Figs 5b and 6) and coughing, the background activity level of abdominal, diaphragm, and intercostal muscle signals tends to be elevated. The inspiratory phase of these acts is still clearly detectable in the diaphragm signal as a typical rapidly peaking increment in activity coincident with inspiratory air flow. At the beginning of the expulsive or expiratory phase, the diaphragmatic EMG activity plateaus or falls sharply, coincident with a rapid rise in abdominal muscle EMG activity. The fact that the reciprocal nature of these acts is often reflected in the EMG signals is further evidence of the validity of the diaphragm signal.

(ii) *Gross motor activity.* The diaphragm probably acts as a "fixator" (Macklem *et al.*, 1978) during gross movements to prevent sudden undesirable changes in pressure across the diaphragm. When infants make stretching movements, breathing is usually suspended for brief periods (there is breath-holding against a closed or partially closed upper airway) and the diaphragm probably co-contracts with abdominal and chest wall muscles. Disorganized breathing activity is evident in the polygraphic record, see Figs 7b and 8. The diaphragm EMG shows respiratory wave forms of varying amplitude superimposed upon an undulating background of moderate to high amplitude. The high cross-correlation that exists during continuous breathing between diaphragm and thermistor respiratory signals (Prechtl *et al.*, 1977, 1979) is absent and some diaphragmatic contractions are relatively or completely ineffective, being associated with little or no air flow. Electrodes placed over the cricothyroid muscle in the ventral neck area often show increased EMG activity, reflecting at least external laryngeal muscle contraction, in association with this type of breathing as illustrated in Fig. 8.

Based on our own work and that of others cited, we think that the signal recorded by the "diaphragm" electrodes during general movements still represents activation of the diaphragm during such movements, even though the measured signal may contain some contribution from other muscles.

(iii) *Tonic activity.* One of the interesting findings that emerged from our studies was that tonic activity, defined as the presence of a constant low level of background myo-electrical activity, was recorded in diaphragm signals during certain behavioural states (Prechtl *et al.*, 1977, 1979). Finkel (1974) had made similar observations in infants using oesophageal electrodes. This activity appears to originate in the diaphragm. It certainly does not represent contamination from neighbouring abdominal muscles as

(a)   State I

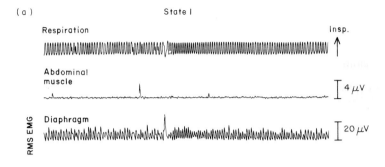

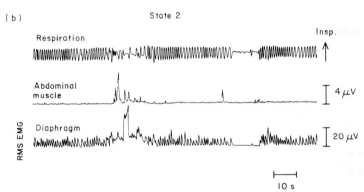

FIG. 7 Abdominal muscle and diaphragm EMG activity (a) during regular breathing in state 1, and (b) during irregular and interrupted breathing in state 2. As shown in (b) breathing may be interrupted either in association with increased muscle activity during a movement or as a result of a pause in muscle activity (i.e. central apnoea).

we found in simultaneous recordings that tonic activity in these muscles was much less consistently present than in the diaphragm and was almost confined to appearance in the supine position (O'Brien *et al.*, 1979b). The other possible source could be adjacent intercostal muscles. The level of diaphragmatic tonic activity is higher than that simultaneously measured from electrodes placed over upper intercostal muscles. There is no reason we are aware of to expect higher levels of tonic activity in lower compared with upper intercostal muscles—indeed the opposite has been found in cats (Duron, 1973). Harding *et al.* (1979) found that posturally dependent tonic activity recorded from electrodes implanted in the peripheral diaphragm in the lamb, could be abolished or markedly reduced

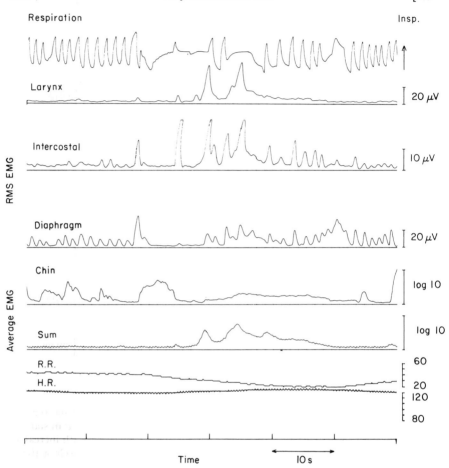

FIG. 8   Piece of polygraphic recording showing disordered breathing during a general movement. The latter is indicated by increased activity in the "sum EMG" signal—this signal represents the summed activity of 6 neck and limb muscles. Several inspiratory efforts indicated by the diaphragm signal are ineffective, probably because of upper airway closure (note the increased laryngeal activity).

by blockade of the lower four intercostal nerves but not by phrenectomy. This lamb experiment, interesting though it is, does not invalidate our observations made in human infants. Independent confirmation of the presence of tonic activity in the diaphragm in non-REM sleep in humans was obtained by Muller *et al.* (1979b) who studied five adults and five premature infants. In one of the adults tonic activity was still recordable in the diaphragm signal following blockade of the lower six intercostal muscles by mepivacaine. These authors considered that the most striking evidence for tonic diaphragmatic activity is the cephalad displacement of the paralysed diaphragm in pediatric patients. Fink *et al.* (1960), using needle electrodes, also recorded continuous low-grade "background" activity in the diaphragm at the onset of apnoea in adults undergoing anaesthesia and reported that this activity invariably disappeared if the $P CO_2$ fell more than 2 mmHg below the apnoeic value.

To summarize, electrodes placed near the costal margin anterolaterally right and left allow reliable bipolar or unipolar measurement of diaphragmatic EMG activity in infants and, perhaps less consistently, in adults. The signal probably validly represents changes in both tonic and phasic diaphragmatic activity under most circumstances. The diaphragm has important non-respiratory functions and participates, along with abdominal, thoracic and upper airway muscles, in many complex acts. Electrodes that record its activity unequivocally during quiet breathing will not fail to pick up its augmented activity, but the recorded signal will reflect the ratio of this activity to that of other contracting muscles within the detection volume of the electrodes (Vigreux *et al.*, 1979).

5.0 EXPERIENCE USING THE METHOD OF SURFACE DIAPHRAGMATIC EMG MEASUREMENT IN INFANTS

5.1 Polygraphic Studies of Breathing in Newborn Infants

The polygraphic studies already mentioned not only allowed testing and development of the method of recording and processing diaphragm (and other respiratory muscle) EMG signals, but also permitted observations of respiratory motor activity in newborn infants under naturalistic conditions. The categorization of newborn behaviour into five behavioural states (Prechtl and Beintema, 1964; Prechtl, 1974) provided the conceptual and physiological framework within which the 6-h polygraphic recordings

TABLE I
Vectors of behavioural states

	Eyes open	Respiration regular	Continuous gross movements	Vocalization
State 1	−1	+1	−1	−1
State 2	−1	−1	−1	−1
State 3	+1	+1	−1	−1
State 4	+1	−1	+1	−1
State 5	0	−1	+1	+1

+ = true   − = false   0 = true or false.

were analysed. These states are recognized on the basis of the maintenance with time of relatively stable constellations of behaviours, which are either visually or electrophysiologically recorded. As shown in Table I, four aspects of behaviour are sufficient to define five mutually exclusive states.

5.1.1. *The influence of body position and state on tonic and rhythmic respiratory EMG activity of various muscles*

We have been interested in analysing EMG activity of diaphragm, intercostal, abdominal, neck and laryngeal muscles to define their respective roles and interrelationships during spontaneous behaviour in different body positions, during specific behaviours such as feeding, and during respiratory pauses.

(i) *Tonic activity.* Schloon *et al.* (1976) extensively studied averaged EMG activity of chin, neck and limb muscles in newborns and showed that in state 2 tonic activity was never maintained for more than 30 s. During states 1 and 3 and during feeding, sustained tonic activity was generally present in the chin. If present in the other muscles its duration was short and its expression varied greatly. The probability of its occurrence apparently diminished according to a cephalocaudal sequence. Prechtl *et al.* (1977, 1979), studying respiratory muscles, found that in state 2 tonic activity in the diaphragm was, on occasion, briefly present following general body movements, whereas in state 1 tonic activity was present throughout about 80% of the state time and was more consistent in expression than in the chin. High levels of tonic diaphragm activity were also present in states 3, 4 and 5.

O'Brien *et al.* (1979b) compared the duration of state 1 tonic activity in supine and prone positions in a study involving both hypotonic and normotonic newborns, in whom the EMGs of diaphragm, upper intercostal and abdominal muscles were measured. In diaphragm and intercostal muscles the percentage state 1 duration was lower in hypotonic than in normotonic infants, but was not different in prone than in supine in either group. Abdominal muscle tonic activity, on the other hand, was strikingly influenced by the body posture. It was very rare in prone but was not infrequently present in supine normal infants. This suggests that postural reflexes play a role in triggering and maintaining tonic activity in abdominal muscles in state 1, whereas diaphragm and intercostal tonic activity may be primarily an epiphenomenon of that state, generated spontaneously and maintained by descending supraspinal facilitatory mechanisms. Further research is needed to determine whether postural loads can modify this activity once it is present.

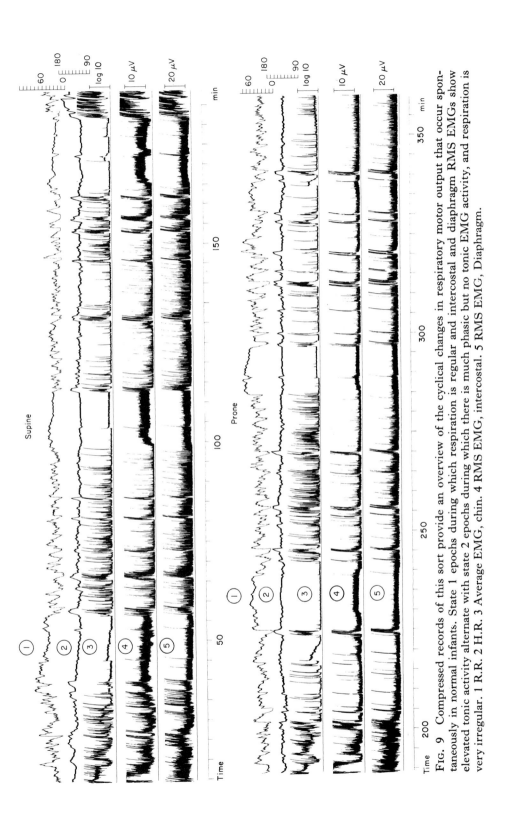

FIG. 9 Compressed records of this sort provide an overview of the cyclical changes in respiratory motor output that occur spontaneously in normal infants. State 1 epochs during which respiration is regular and intercostal and diaphragm RMS EMGs show elevated tonic activity alternate with state 2 epochs during which there is much phasic but no tonic EMG activity, and respiration is very irregular. 1 R.R. 2 H.R. 3 Average EMG, chin. 4 RMS EMG, intercostal. 5 RMS EMG, Diaphragm.

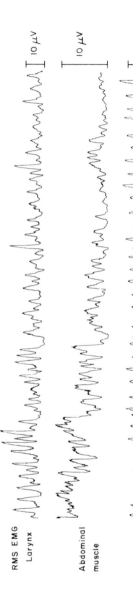

Feeding

RMS EMG
Larynx

Abdominal
muscle

Diaphragm

Respiration

10 µV

10 µV

10 µV

Insp

10 s

Fɪɢ. 10 Patterns of EMG activity during bottle feeding. The laryngeal EMG signal reflects sucking and swallowing activity. Abdominal muscle activity is complex: phasic activity, probably representing rhythmic expiratory and sucking activity, is superimposed on an initially high tonic activity level—none of this activity contaminates the diaphragm signal. Occasional breaths are partially obstructed, indicating that during feeding, control of ventilation is exercised at the level of the upper airway.

During active feeding tonic activity is usually present in diaphragm and intercostal EMGs and may also be prominent but for shorter periods in abdominal muscles (see Figs 5a and 10). In the latter case it seems to reflect the active sucking posture (Casaer et al., 1973), one component of which is extension of the legs at the knee and slight hip flexion, a posture requiring contraction of abdominal muscles.

(ii) *Respiratory activity*. The compressed record illustrated in Fig. 9 shows cyclical reciprocal changes in diaphragm and intercostal RMS EMG activity during sleep in a normal infant who was placed after the first feeding in a supine position and after the second prone. This record is reasonably typical, though we have found that the range of interindividual differences may be quite large. The amplitude of diaphragmatic inspiratory EMG activity is about 20% higher in state 2 than in state 1, whereas upper intercostal muscle activity is variable and sometimes absent in state 2. Hagan et al. (1976) related the measured reduction in intercostal activity in REM sleep (contained in state 2) to paradoxical inspiratory chest movement. Cross-correlation of the two RMS signals shows that the coupling between two muscles is different in the two sleep states. Maximum correlation coefficients are higher and more stable in state 1 than in state 2. Intercostal activation clearly precedes diaphragmatic activation in state 1, whereas in state 2 this phase lead is either slight or non-existent. Diaphragm respiratory activity is about 1·4 times higher in supine than in prone whereas there is no individually consistent difference in the level of intercostal activity (unpublished observations). The physiological significance, if any, of this observation requires further study but it is interesting in view of Martin et al. (1979) finding that transcutaneous PO_2 levels in a group of preterm infants were higher in prone than in supine, and Hutchison et al. (1979) study on ventilation and lung mechanics in preterm infants in which the prone position resulted in an improvement in ventilation compared to supine. The prone posture may cause this improvement through abdominal loading as discussed by Fleming et al. (1979).

Abdominal muscle EMGs occasionally show an expiratory activity superimposed on tonic activity during quiet breathing in the supine position. This expiratory modulation is not always evident when the muscle is tonically active. In the absence of tonic activity a very low amplitude inspiratory modulation whose significance is not clear may or may not be observed (see Figs 5 and 7).

(iii) *Respiratory activity during feeding*. As shown in Fig. 10, simultaneous measurement of air flow, and laryngeal and respiratory muscle EMGs during feeding reveals that rhythmical inspiratory activity continues in the diaphragm (and intercostals) even during active feeding. In some infants inspiratory air flow is regularly cut off to permit swallowing.

Feeding therefore causes a form of physiological upper airway obstruction. As shown in Fig. 10, at the beginning of a feeding session abdominal EMGs often show tonic activity with superimposed expiratory (? accessory sucking/swallowing) activity. This fades as sucking becomes less intense.

5.2 Monitoring Diaphragmatic EMG Activity in Newborns under Intensive Care

Since the polygraphic experiments show that it is possible to measure diaphragmatic EMG activity for at least 6 h without deterioration in signal quality, and since the processing technique allows recognition of the inspiratory activity associated with each breath, the question arises as to whether this measurement could be used for monitoring respiratory activity in a clinical intensive care setting. Bock and Liesegang (1971) reported encouraging preliminary experience in this regard using needle electrodes to monitor intercostal muscle activity in adults. O'Brien *et al.* (1979a) drew attention to the potential clinical usefulness of diaphragmatic EMG monitoring using surface electrodes in infants.

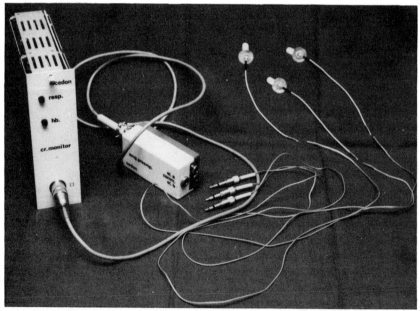

FIG. 11   EMG processor module, preamplifier and flexible shielded electrode cables disconnected from terminal non-shielded electrode bearing cable. The three electrodes, one of which serves as common electrode, are fixed to adhesive collars ready for application to the skin.

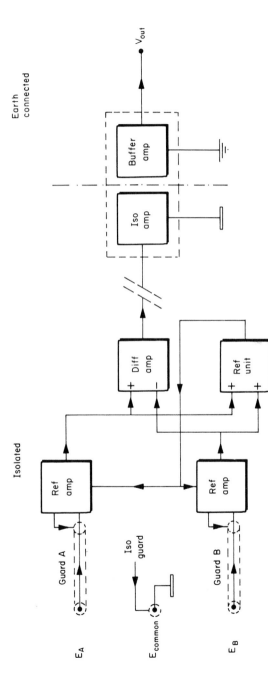

Fig. 12 Block diagram of the preamplifier based on the average reference principle developed for monitoring of diaphragm EMG activity in an intensive care setting.

A prototype EMG signal detection and processing unit, based on the technique shown in Fig. 3, which could form the basis for a clinical respiration and heart rate monitor has been developed (see Fig. 11). The unit consists of a preamplifier and an EMG processing module. The preamplifier, based on the common average reference principle, contains two reference amplifiers and a differential amplifier (see Fig. 12). Each reference amplifier uses a high quality operational amplifier for impedance transformation and signal amplification, and a guard amplifier. Via the reference unit the mean value of the output signal is fed to the input amplifiers to reduce common mode signal components at the input stage. The guard signals are formed by adding a portion of the output signal of the input amplifier and the reference signal. The guard signals represent a low impedance version of the input (electrode) signals. The diaphragm EMG is obtained by subtracting the output signal of the two reference amplifier channels in a differential amplifier.

A simpler preamplifier based on the commonly known differential amplifier concept could also be used with the exception that a real signal guard could then not be made.

Protection circuits are included to protect the infant against high guard currents in the case of a faulty electrode cable and against too high electrode currents in case of damage to the input amplifier.

The preamplifier is connected to the processing module via an isolation amplifier which also provides an isolated power-supply for the preamplifier. The shield of the common electrode cable as well as the shield of the connection cable to the preamplifier is connected to the guard signal of the isolation amplifier to reduce line frequency interference to a minimum. The diaphragm EMG signal is coupled to the isolation amplifier via a smooth bandpass filter to reduce the noise bandwidth for low frequencies and to prevent aliasing in the isolation amplifier at higher frequencies. The output of the isolation amplifier is AC coupled to the input of the processor by means of a first order high pass filter to prevent distortion of the QRS complex. The time constant of this filter is fairly high to remove low frequency signal components and amplifier drift from the EMG signal prior to processing.

Electrodes and electrode cables are the same as those used in polygraphic studies, though disposable tinfoil electrodes of the type widely used in impedance monitoring are also being evaluated.

Detection and gating of the ECG complex and estimation of the RMS value take place in the processing module according to the method already outlined and as shown in Fig. 3. The module's available outputs are a standardized ECG pulse, the ungated and gated EMG signals and the RMS (respiration) signal.

Diaphragm EMG measurements have been made using the prototype unit in 11 infants in a newborn intensive care unit. They ranged in birthweight between 750 and 3300 g and in gestational age between 26 and 40 weeks. Seven of the infants were receiving intermittent positive pressure ventilation at the time the diaphragm EMG measurements were made—one was weaned off the ventilator during the recording session. All infants were being routinely monitored using an impedance based system (Hewlett-Packard). The outputs of the prototype unit, the impedance respiration signal and in some infants a transcutaneous PO_2 signal, were recorded polygraphically in ten of the 11 infants. No attempt was made to influence the clinical management in this essentially exploratory investigation. The number of hours of EMG recording per infant varied between 3 and 64, with a total recording time of about 250 hours.

Despite the intensive care environment, in which electrophysiological measurement is notoriously difficult, reliable diaphragm EMG signals were recorded in these infants with little difficulty. It proved possible to record impedance and EMG signals simultaneously if the same common electrode was used for both measurements. However, this did cause an unpredictable increase in line frequency interference in the impedance system's ECG signal. A small amount of line frequency interference was picked up by the EMG system if the respirator was very close to the incubator. This caused no deterioration in the quality of the respiratory signal, merely a DC baseline shift. This interference is probably picked up by the short length of unguarded cable attached to the electrodes and could probably be eliminated by use of an electrode cable guarded along its whole length. EMG signal quality remained stable as long as electrode fixation to the skin was secure. It was found that the small adhesive electrode collars, which are normally satisfactory for polygraphic recording, were not always robust enough, and tended to become detached or displaced in infants requiring much handling. This problem is fundamentally the same as it is for impedance measurement with the exception that we found the EMG signal to be less affected by an electrode becoming partially detached than was the impedance signal.

Figure 13 shows impedance and RMS EMG respiration signals and a heart rate signal—derived from the EMG processor—in a 36-week gestation 2665-g birthweight infant with incomplete nasal obstruction. This illustration shows the typical appearance in impedance and EMG signals of normal breaths, a sigh, and a central respiratory pause or apnoea. Figure 14 shows a longer piece of recording taken from a 30-week gestation, 1190-g birthweight infant who was in good condition. Three respiratory pauses, central in type, are very clearly defined in the RMS EMG signal, and slightly less well so in the impedance signal. This

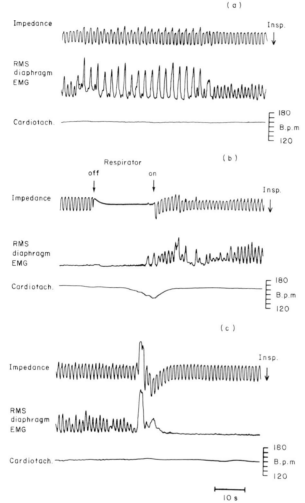

FIG. 15   The pattern of endogenous breathing activity of an artificially ventilated infant revealed by the diaphragm RMS EMG technique. For description of figure, see text.

illustration is quite typical of the quality of the signals recorded in all infants.

Figure 15 demonstrates the unique advantage of the diaphragmatic EMG measurement. This infant was on a ventilator and the impedance signal reflects the chest movement induced by the ventilator. The diaphragm signal, however, shows only the infant's own inspiratory efforts, and in Fig. 15a the infant makes a series of large inspiratory efforts at exactly half the frequency of the ventilator and then switches to a rate equal to that of the machine. In Fig. 15b complete absence of endogenous respiratory effort is evident for half the record. The assisted ventilation was discontinued for about 30 s and during this period both signals are silent. Towards the end of this time, the infant begins to breathe spontaneously. After the first spontaneous breath, which is barely visible in the impedance signal, the ventilator was switched on again and the impedance signal again reflects the machine-induced chest movement while the infant continues to breathe irregularly and more or less independently of the respirator. In Fig. 15c the infant stops breathing spontaneously again after a brief general movement. This ability to record the endogenous diaphragmatic contractions was confirmed in each of the seven infants on ventilators.

6.0 FUTURE PROSPECTS

There are several potential benefits to be expected from EMG-based clinical monitoring of respiration. A monitoring system combining surface measurement of diaphragm EMG activity and transcutaneous measurement of blood gases would provide the intensive care physician with non-intrusively obtained information on (a) the central patterning of respiratory neural output, (b) the state and effectiveness of the respiratory pump, and (c) the chemical drive and the adequacy of oxygenation. Rational and timely provision, alteration and withdrawal of ventilatory assistance would be promoted by such monitoring.

Patient–ventilator competition may gravely compromise the effectiveness of ventilatory assistance. Mitchell *et al.* (1979) have described an algorithm based on measurement of respiratory pressure and gas flow to detect and quantitate this. The EMG technique might lend itself to a simpler solution to this problem based on on-line measurement of the phase relationships between the endogenous respiratory activity and the imposed ventilation.

Sudden switches in respiratory rate occur in spontaneously breathing healthy infants, as shown in Fig. 7a, and also in ventilated infants as evident in Fig. 15a. Our impression is that in this area much of interest and clinical importance awaits discovery.

Before leaving the topic of assisted ventilation, the diaphragm EMG signal may prove to have yet another useful function, namely as the trigger signal for patient-synchronized intermittent mandatory ventilation (Aoki *et al.*, 1978). Since muscle electrical activity precedes air flow, the delay introduced by the EMG processing is short enough to allow the ventilator, triggered on the leading edge of the processed signal, to augment the subjects own inspiratory flow.

Further research is necessary to define the optimal way to extract information on muscle state from the EMG. It is possible that a measure such as the form factor, which as has been mentioned is the ratio of the RMS value to the mean absolute value, may prove to be a very sensitive and useful measure in this regard.

The EMG technique has other applications also. Combined with air flow monitoring, it constitutes a most reliable way to distinguish between different types of respiratory rhythm disturbance such as between centrally and peripherally located causes of apnoea (see Figs 7b, 8, 13, 14 and 15).

It may prove useful also in evaluating infants with upper airway dysfunction, since we have seen in three infants that changes in the amplitude and form of the inspiratory EMG activity appeared to reflect changes in the resistance to breathing.

Surface EMG measurements will probably be used increasingly in studies in all age groups of respiratory control in health and disease as signal processing techniques improve and as the physiological significance of the processed signals becomes better appreciated. Already some workers positively endorse the use of quantitative EMG measurements combined with occlusion pressure and ventilation measurements as a means of assessing the neuromuscular–mechanical components of the respiratory control system (Evanich et al., 1976; Lopata et al., 1978). Others, such as Sharp et al. (1976), are slightly more reserved, placing a careful and sensible emphasis on the limitations of EMG measurements.

Using the EMG technique described it will be possible to study the development of the respiratory and non-respiratory function of important muscles even in the smallest preterm infants. The results of such study will contribute to further improving the respiratory care of infants.

Acknowledgement

This research is supported by grant no. 13-51-91 from the Foundation for Medical Research Fungo, which is subsidized by the Netherlands Organization for the Advancement of Pure Research (ZWO).

Note Added in Proof

Since this chapter was written, it has become apparent that when EMG is used simply to detect diaphragmatic respiratory activity, RMS processing of the signal is unnecessarily complex. The technically simpler though statistically less correct technique of moving-window averaging following full-wave rectification is adequate for this purpose and greatly simplifies the design of a monitoring system intended for clinical use.

References

Adrian, A. D. and Bronk, D. W. (1929). The discharge of impulses in motor nerve fibers. II. The frequency of discharge in reflex and voluntary contractions. *J. Physiol.* **67**, 119–151.

Agostoni, E. and Torri, G. (1962). Diaphragm contraction as a limiting factor to maximum expiration. *J. Appl. Physiol.* **17**, 427–428.

Agostoni, E., Sant'Ambrogio, G. and Del Portillo Carrasco, H. (1960a). Electromyography of the diaphragm in man and transdiaphragmatic pressure. *J. Appl. Physiol.* **15**, 1093–1097.

Agostoni, E., Sant'Ambrogio, G. and Del Portillo Carrasco, H. (1960b). Elletromiografia del diaframma e pressione transdiaframatica durante la tosse, lo sternuto ed il riso. *Atti Acad. Naz. Rend.* **28**, 493–496.

Altose, M. D., Stanley, N. N., Cherniack, N. S. and Fishman, A. P. (1975). Effects of mechanical loading and hypercapnia on inspiratory muscle EMG. *J. Appl. Physiol.* **38**, 467–473

Aoki, N., Shimizu, H., Kushiyama, S., Katsuya, H. and Isa, K. (1978). A new device for synchronized intermittent mandatory ventilation. *Anesthesiol.* **48**, 69–71.

Ashutosh, K., Gilbert, R., Auchincloss, J. H., Erlebacher, J. and Peppi, D. (1974). Impedance pneumograph and magnetometer methods for monitoring tidal volume. *J. Appl. Physiol.* **37**, 964–966.

Baker, L. E. (1979). Electrical impedance pneumography. *In* "Non-invasive Physiological Measurements" (Ed. P. Rolfe), Vol. 1, pp. 65–94, Academic Press, London and New York.

Barlow, J. S. and Dubinsky, J. (1980). EKG-artifact minimization in referential EEG recordings by computer subtraction. *Electroenceph. Clin. Neurophysiol.* **48**, 470–472.

Bickford, R. G., Billinger, T. W., Sims, J. K., Fleming, N. I., Hoffman, R. S. and Aung, M. H. (1971a). A computer and enunciator system for estimation of brain death. *Proc. San Diego Biomed. Symp.* **10**, 117–123.

Bickford, R. G., Sims, J. K., Billinger, T. W. and Aung, M. H. (1971b). Problems in EEG estimation of brain death and use of computer techniques for their solution. *Trauma* **12**, 61–95.

Biro, G. and Partridge, L. D. (1971). Analysis of multiunit spike records. *J. Appl. Physiol.* **30**, 521–526.

Bock, W. J. and Liesegang, J. (1971). Eine neue Methode zur Atmungsüberwachung. *Biomed. Technik.* **16**, 48–50.

Bouisset, S. and Goubel, F. (1973). Integrated electromyographical activity and muscle work. *J. Appl. Physiol.* **35**, 695–702.

Brouillette, R. T. and Thach, B. T. (1980). A self-retaining nasal flowmeter for preterm infants. *J. Appl. Physiol.* **48**, 569–571.

Bruce, E. N., Goldman, M. D. and Mead, J. (1977). A digital computer technique for analyzing respiratory muscle EMGs. *J. Appl. Physiol.* **43**, 551–556.

Buchthal, F. and Schmalbruch, H. (1980). Motor unit of mammalian muscle. *Physiol. Rev.* **60**, 90–142.

Buchthal, F., Erminio, F. and Rosenfalck, P. (1959). Motor unit territory in different human muscles. *Acta Physiol. Scand.* **45**, 72–87.

Buhl, D. (1968). New approach to understanding the operation of thermistors in gas chromatography. *Analyt. Chem.* **40**, 715–726.

Campbell, E. J. M. (1954). The muscular control of breathing in man. Ph.D. thesis, University of London.

Campbell, E. J. M. (1955). An electromyographic examination of the role of the intercostal muscles in breathing in man. *J. Physiol.* **129**, 12–26.

Campbell, E. J. M. (1958). "The Respiratory Muscles and the Mechanics of Breathing", Lloyd Luke (Medical Books), London.

Campbell, E. J. M., Agostoni, E. and Davis, J. N. (Eds) (1970). "The Respiratory Muscles. Mechanics and Neural Control", Lloyd Luke (Medical Books), London.

Caro, C. G. and Bloice, J. A. (1971). Contactless apnoea detector based on radar. *Lancet* **ii**, 959–961.

Casaer, P., O'Brien, M. J. and Prechtl, H. F. R. (1973). Postural behaviour in human newborns. *Aggressologie* **14**, No. B, 49–56.

Clamann, H. P. (1969). Statistical analysis of motor unit firing patterns in a human skeletal muscle. *Biophys. J.* **9**, 1233–1251.

Curzi-Dascalova, L., Radvanyi, M. F., Moriette, G., Morel-Kahn, F. and Korn, G. (1979). Respiratory variability according to sleep states during mechanical ventilation: A polygraphic study in a baby with bilateral diaphragmatic paralysis. *Neuropädiat.* **10**, 361–369.

Delhez, L., Troquet, J., Damoiseau, J., Pirnay, F., Deroanne, R. and Petit, J. M. (1964). Influence des modalités d'exécution des manoeuvres d'éxpiration forcée et d'hyperpression thoraco-abdominale sur l'activité électrique du diaphragme. *Arch. Int. Physiol. Biochim.* **72**, 76–94.

De Luca, C. J. (1979). Physiology and mathematics of myoelectric signals. *IEEE Trans. Biomed. Engin.* **26**, 313–325.

Derenne, J.-Ph., Macklem, P. T. and Roussos, Ch. (1978). The respiratory muscles: mechanics, control and pathophysiology. Part I. *Am. Rev. Resp. Dis.* **118**, 119–134.

De Vries, H. A. (1968). Method for evaluation of muscle fatigue and endurance from electromyographic fatigue curves. *Am. J. Phys. Med.* **47**, 125–135.

Dick, D. E., Meyer, J. R. and Weil, J. V. (1974). A new approach to quantitation of whole nerve bundle activity. *J. Appl. Physiol.* **36**, 393–397.

Draper, M. H., Ladefoget, P. and Whitteridge, D. (1959). Respiratory muscle in speech. *J. Speech. Res.* **2**, 16–27.

Duron, B. (1973). Postural and ventilatory functions of intercostal muscles. *Acta Neurobiol. Exp.* **33**, 355–380.

Edström, L. and Kugelberg, E. (1968). Histochemical composition, distribution of fibres and fatiguability of single motor units. *J. Neurol. Neurosurg. Psychiat.* **31**, 424–433.

Edwards, R. G. and Lippold, O. C. J. (1956). The relation between force and the integrated electrical activity in fatigued muscle. *J. Physiol.* **132**, 677–681.

Ekstedt, J. (1964). Human single muscle fiber action potentials. *Acta Physiol. Scand.* **61**, Suppl. 226, 1–96.

Eldridge, F. L. (1975). Relationship between respiratory nerve and muscle activity and muscle force output. *J. Appl. Physiol.* **39**, 567–574.

Evanich, M. J., Lopata, M. and Lourenco, R. V. (1976). Analytical methods for the study of electrical activity in respiratory nerves and muscles. *Chest* **70**, 158S–162S.

Fink, B. R., Hanks, E. C., Holaday, D. A. and Ngai, S. H. (1960). Monitoring of ventilation by integrated diaphragmatic electromyogram. *Am. Med. Assoc.* **172**, 1367–1371.

Finkel, M. L. (1975). Activity of respiratory muscles in newborn infants during wakening and sleep. *J. Evol. Biochem. Physiol.* **1**, 92–95.

Fleisch, A. (1925). Der Pneumotachograph: ein Apparat zur Geschwin eils Registrierung der Atemluft. *Arch. Ges. Physiol.* **209**, 713–722.

Fleming, P. J., Muller, N. L., Bryan, M. H. and Bryan, A. C. (1980). The effects of abdominal loading on rib cage distortion in premature infants. *Pediatrics* **64**, 425–428.

Freund, H. J., Dietz, V., Wita, C. W. and Kapp, H. (1973). Discharge characteristics of single motor units in normal subjects and patients with supraspinal motor disturbances. *In* "New Developments in Electromyography and Clinical Neurophysiology" (Ed. J. E. Desmedt), Vol. 3, pp. 242–250, Karger, Basel.

Goldman, D. (1950). The clinical use of the "average" reference electrode in monopolar recording. *Electroenceph. Clin. Neurophysiol.* **2**, 209–212.

Gordon, D. H. and Thompson, W. L. (1975). A new technique for monitoring spontaneous respiration. *Med. Instrum.* **9**, 21–22.

Gramsbergen, A., Vos, J. E. and Mooibroek, J. (1979). The ontogeny of the EEG in the rat: development of coherence functions. *In* "Ontogenesis of the Brain", Vol. 3, Proceedings of the International Symposium Neuro-ontogeneticum Tertium, Praga, 1979. Charles University, Prague.

Green, J. H. and Howell, J. B. L. (1959). The correlation of intercostal muscle activity with respiration air flow in conscious human subjects. *J. Physiol.* **149**, 471–476.

Gross, D., Grassino, A., Ross, W. R. D. and Macklem, P. T. (1979). Electromyogram pattern of diaphragmatic fatigue. *J. Appl. Physiol.* **46**, 1–7.

Hagan, R., Bryan, A. C., Bryan, M. H. and Gulston, G. (1976). The effect of sleep state on intercostal muscle activity and rib-cage motion. *Physiologist* **19**, 214.

Hagan, R., Bryan, A. C., Bryan, M. H. and Gulston, G. (1977). Neonatal chest wall afferents and regulation of respiration. *J. Appl. Physiol.* **42**, 362–367.

Håkansson, C. H. (1957). Action potentials recorded intra- and extracellularly from the isolated frog muscle in Ringer's solution and in air. *Acta Physiol. Scand.* **39**, 291–312.

Harding, R., Henderson-Smart, D. J., Johnson, P. J. and McClelland, M. E. (1979). Posturally related tonic activity recorded from the peripheral diaphragm in awake and sleeping lambs. *J. Physiol.* **292**, 57P.

Henneman, E., Somjen, G. and Carpenter, D. O. (1965). Functional significance of cell size in spinal motoneurons. *J. Neurophysiol.* **28**, 560–580.

Hutchison, A. A., Ross, K. R. and Russell, G. (1979). The effect of posture on ventilation and lung mechanics in preterm and light-for-date infants. *Pediatrics* **64**, 429–432.

Jiménez-Vargas, H., Asirón, M., Voltas, J. and Onaindia, J. (1967). Electromiografia de músculos respiratorios en la tos, en los reflejos de la glotis y en el vómito. *Rev. Esp. Fisiol.* **23**, 65–74.

Kaiser, E. and Petersen, I. (1965). Muscle action potentials studied by frequency analysis and duration measurement. *Acta Neurol. Scand.* **41**, 19–42.

Katz, R. L., Fink, B. R. and Ngai, S. H. (1962). Relationship between electrical activity of the diaphragm and ventilation. *Proc. Soc. Exp. Biol. Med.* **110**, 792–794.

Kelly, D. H., O'Connell, K. and Shannon, D. C. (1979). Electrode belt: a new method for long-term monitoring. *Pediatrics* **63**, 670–673.

Koepke, G. H., Murphy, A. J., Rae, J. W. and Dickinson, D. G. (1955). An electromyographic study of some of the muscles used in respiration. *Arch. Phys. Med. Rehab.* **36**, 217–222.

Koepke, G. H., Smith, E. M., Murphy, A. J. and Dickinson, D. G. (1958). Sequence of action of the diaphragm and intercostal muscles during respiration: I. Inspiration. *Arch. Phys. Med. Rehab.* **39**, 426–430.

Konno, K. and Mead, J. (1967). Measurement of the separate volume changes of rib cage and abdomen during breathing. *J. Appl. Physiol.* **22**, 407–422.

Krnjevic, K. and Miledi, R. (1958). Motor units in the rat diaphragm. *J. Physiol.* **140**, 427–439.

Lewin, J. E. (1969). An apnoea alarm mattress. *Lancet* **ii**, 667.

Lidell, E. G. T. and Sherrington, C. S. (1925). Recruitment and some other factors of reflex inhibition. *Proc. R. Soc. Lond. Ser. B* **97**, 488–518.

Lindström, L., Magnusson, R. and Petersén, I. (1970). Muscular fatigue and action potential conduction velocity changes studied with frequency analysis of EMG signals. *Electroenceph. Clin. Neurophysiol.* **10**, 341–356.

Lippold, O. C. (1952). The relation between integrated action potentials in a human muscle and its isometric tension. *J. Physiol.* **117**, 492–499.

Lloyd, A. J. (1971). Surface electromyography during sustained isometric contractions. *J. Appl. Physiol.* **30**, 713–719.

Lopata, M., Evanich, M. J. and Lourenco, R. V. (1977). Quantification of diaphragmatic EMG response to CO_2 rebreathing in humans. *J. Appl. Physiol.* **43**, 262–270.

Lopata, M., Evanich, M. J., Onal, E., Zubillaga, G. and Lourenco, R. V. (1978). Airway occlusion pressure and respiratory nerve and muscle activity in studies of respiratory control. *Chest* **73**S, 285S–286S.

Lourenco, R. V. and Mueller, E. P. (1967). Quantification of electrical activity in the human diaphragm. *J. Appl. Physiol.* **22**, 598–600.

Lourenco, R. V., Cherniack, N. S., Malm, J. R. and Fishman, A. P. (1966). Nervous output from the respiratory center during obstructed breathing. *J. Appl. Physiol.* **21**, 527–533.

Macklem, P. T., Gross, D., Grassino, A. and Roussos, C. (1978). Partioning of inspiratory pressure swings between diaphragm and intercostal-accessory muscles. *J. Appl. Physiol.* **44**, 200–208.

Marsden, C. D., Meadows, J. C. and Merton, P. A. (1971). Isolated single motor units in human muscle and their rate of discharge during maximal voluntary effort. *J. Physiol.* **217**, 12P–13P.

Martin, R. J., Herrell, N., Rubin, D. and Fanaroff, A. (1979). Effect of supine and prone positions on arterial oxygen tension in the preterm infant. *Pediatrics* **63**, 528–531.

Maton, B. (1976). Motor unit differentiation and integrated surface EMG in voluntary isometric contraction. *Eur. J. Appl. Physiol. Occup. Physiol.* **35**, 149–157.

Milner, A. D. (1970). The respiratory jacket. *Lancet* **ii**, 80.

Mitchell, R. R., Kunz, J. C., Lamy, M. and Fallat. R. J. (1979). On-line frequency measurement of patient-ventilator fighting. *Comput. Biomed. Res.* **12**, 433–443.

Monges, H., Salducci, J. and Naudy, B. (1978). Dissociation between the electrical activity of the diaphragmatic dome and crura muscular fibers during esophageal distension, vomiting and eructation. An electromyographic study in the dog. *J. Physiol.* **74**, 541–554.

Moore, A. D. (1967). Synthesized EMG waves and their implications. *Am. J. Phys. Med.* **46**, 1302–1316.

Moritani, T. and De Vries, H. A. (1978). Reexamination of the relationship between the surface integrated electromyogram (EMG) and force of isometric contraction. *Am. J. Phys. Med.* **57**, 263–277.

Mucke, R. and Buchholz, Ch. (1977). Die Eliminierung EKG-überlagerter EMG-Abschnitte bei der on-line-Auswertung der bioelektrischen Muskelspannung. Z. Gesam. Hyg. Grenzgeb. **5**, 290–292.

Muller, N., Gulston, G., Cadem, D., Whitton, J., Froese, A. B., Bryan, M. H. and Bryan, A. C. (1979a). Diaphragmatic muscle fatigue in the newborn. J. Appl. Physiol. **46**, 688–695.

Muller, N., Volgyesi, G., Becker, L., Bryan, M. H. and Bryan, A. C. (1979b). Diaphragmatic muscle tone. J. Appl. Physiol. **47**, 279–284.

Muller, N., Volgyesi, G., Bryan, M. H. and Bryan, A. C. (1979c). The consequence of diaphragmatic muscle fatigue in the newborn infant. J. Pediatr. **95**, 793–797.

Nagel, J. and Schaldach, M. (1979). Registration of uterine activity from the abdominal lead EMG. In "Book of Abstracts, Foetal and Neonatal Physiological measurements Conference", The Biological Engineering Society, Oxford, 3.6.

Newsom Davis, J. (1967). Phrenic nerve conduction in man. J. Neurol. Neurosurg. Psychiat. **30**, 420–426.

O'Brien, M. J. and van Eykern, L. A. (1979). Monitoring the newborns breathing by surface electromyography-research and clinical aspects. In "Book of Abstracts. Foetal and Neonatal Physiological Measurements Conference", The Biological Engineering Society, Oxford, 6.6.

O'Brien, M. J., van Eykern, L. A. and Prechtl, H. F. R. (1980). Diaphragmatic intercostal and abdominal muscle tonic EMG activity in normal and hypotonic newborns. In "Ontogenesis of the Brain", Vol. 3, Proceedings of the international symposium neuro-ontogeneticum tertium, Praga, 1979. Charles University.

Offner, F. F. (1950). The EEG as potential mapping: The value of the average monopolar reference. Electroenceph. Clin. Neurophysiol. **2**, 213–214.

Pallett, J. E. and Scopes, J. W. (1965). Recording respirations in newborn babies by measuring impedance of the chest. Med. Electron. Biol. Engin. **3**, 161–168.

Palm, D. G., Brodner, H. and Heller, K. (1971). Probleme der apparativen Atemüberwachung bei Früh-und Neugeborenen (Erfahrungen mit einem neuen Atemmonitor). Arch. Kinderheilk. **63**, 70–79.

Petit, J. M., Milic-Emili, G. and Delhez, L. (1960). Role of diaphragm in breathing in conscious normal man: An electromyographic study. J. Appl. Physiol. **15**, 1101–1106.

Petrofsky, J. S. (1980). Computer analysis of the surface EMG during isometric exercise. Comput. Biol. Med. **10**, 83–95.

Petrofsky, J. S., Dahms, A. R. and Lind, A. R. (1975). Power spectrum EMG during static exercise. Physiologist **18**, 91–95.

Piper, H. (1912). "Electrophysiologie menschlicher Muskeln", Springer, Berlin.

Pitts, R. F. (1942). Excitation and inhibition of phrenic motor neurones. J. Neurophysiol. **5**, 75–88.

Prechtl, H. F. R. (1974). The behavioural states of the newborn infant (a review). Brain Res. **76**, 185–212.

Prechtl, H. F. R. and Beintema, D. J. (1964). "The Neurological Examination of the Full-term Newborn Infant", Clinics in Developmental Medicine No. 12, Heinemann, London.

Prechtl, H. F. R., Akiyama, Y., Zinkin, P. Kerr Grant, D. (1968). Polygraphic studies of the full term newborn infants. I. Technical aspects and qualitative analysis. In "Studies in Infancy", (Eds M. C. O. Bax and R. C. MacKeith) Clinics in Developmental Medicine, No. 27, pp. 1–25, Heinemann, London.

Prechtl, H. F. R., van Eykern, L. A. and O'Brien, M. J. (1977). Respiratory muscle EMG in newborns: A non-intrusive method. *Early Hum. Develop.* **1**, 265–283.

Prechtl, H. F. R., O'Brien, M. J. and van Eykern, L. A. (1979). Neonatal breathing in different states of sleep and wakefulness. *In* "Central Nervous Control Mechanisms in Breathing" (Eds C. von Euler and H. Lagercrantz), pp. 443–455, Pergamon Press, New York.

Rigatto, H. and Brady, J. P. (1972). A new nosepiece for measuring ventilation in preterm infants. *J. Appl. Physiol.* **32**, 423–424.

Rolfe, P. (1971). A magnetometer respiration monitor for use with premature babies. *Biomed. Engin.* **6**, 402–404.

Rolfe, P. (1975). Monitoring in newborn intensive care. *Biomed. Engin.* **10**, 399–404.

Schloon, H., O'Brien, M. J., Scholten, C. A. and Prechtl, H. F. R. (1976). Muscle activity and postural behaviour in newborn infants. *Neuropädiat.* **7**, 384–415.

Schweitzer, T. W., Fitzgerald, J. W., Bowden, J. A. and Lynne-Davies, P. (1979). Spectral analysis of human inspiratory diaphragmatic electromyograms. *J. Appl. Physiol.* **46**, 152–165.

Sharp, J. T., Druz, W., Danon, J. and Kim, M. J. (1976). Respiratory muscle function and the use of respiratory muscle electromyography in the evaluation of respiratory regulation. *Chest* **70**, 105S–154S.

Sempik, A. K. (1977). An electromyographic study of intercostal muscle activity in man during breath-holding. *J. Physiol.* **276**, 10–11P.

Smith, J. E. and Scopes, J. W. (1972). A new apnoea alarm for babies. *Lancet* **ii**, 7775, 545–546.

Stålberg, E. (1966). Propagation velocity in human muscle fibers *in situ*. *Acta Physiol. Scand.* **70**, Suppl. 287.

Taylor, A. (1960). The contribution of the intercostal muscles to the effort of respiration in man. *J. Physiol.* **151**, 390–402.

Tusiewicz, K., Moldofsky, H., Bryan, A. C. and Bryan, M. H. (1977). Mechanics of the rib cage and diaphragm during sleep. *J. Appl. Physiol.* **43**, 600–602.

Vigreux, B., Cnockaert, J. C. and Pertuzon, E. (1979). Factors influencing quantified surface EMGs. *Eur. J. Appl. Physiol.* **41**, 119–129.

Vos, J. E., Lammertsma, A. A. and van Eykern, L. A. (1977). Ordinary and partial coherences of bipolar and quasi-unipolar derivations of infant electro-encephalograms. Random signal analysis. *IEEE Conf. Publ.* No. 159, 154–160.

Wilson, N. F., Johnston, F. E., Macleod, A. G. and Barker, P. S. (1934). Electrocardiograms that represent the potential variations of a single electrode. *Am. Heart J.* **9**, 447–458.

Zipp, P. and Ahrens, H. (1979). A model of bioelectric motion artefact and reduction of artefact by amplifier input stage design. *J. Biomed. Engin.* **1**, 273–276.

5. NON-INVASIVE ASSESSMENT OF GASTRIC ACTIVITY

R. H. Smallwood and B. H. Brown

Department of Medical Physics and Clinical Engineering,
Royal Hallamshire Hospital, Sheffield, England

1.0 INTRODUCTION

The first non-invasive recording of the electrical activity of the human stomach was made by Alvarez in 1922—this was the first *in vivo* recording of electrical activity from any part of the human intestine. Alvarez succeeded in making a single recording from "a little old woman whose abdominal wall was so thin that her gastric peristalsis was easily visible". We now know that it is much easier to record the electrical activity from the stomach than from any other area of the gastro-intestinal (GI) tract. This is because the gastric electrical activity is very well co-ordinated—simultaneous recordings from any part of the electrically active area of the stomach wall will show that the frequency of the activity is the same from all parts of the stomach—about 3 cycles min^{-1} (0·05 Hz) in humans—with a constant phase shift between the records from different sites (Fig. 1). This is far from being the case throughout the remainder of the GI tract—12 cycles min^{-1} activity is present in the duodenum, with time-dependent phase shifts between different sites (Christensen *et al.*, 1964, 1966a, b; Duthie *et al.*, 1971); the activity in the remainder of the small intestine decreases progressively in frequency to 8 cycles min^{-1} at the terminal ileum (op. cit.; Waterfall *et al.*, 1973); the right colon has activity

NON-INVASIVE MEASUREMENTS: 2   *Copyright©1983 by Academic Press Inc. (London) Ltd.*
ISBN 0 12 593402 5

Fig. 1   Human gastric electrical activity at 3 cycles min^{-1} recorded from a spike electrode in the mucosal surface of the gastric musculature.

at 10 and 3 cycles min^{-1}, the transverse colon at 7 and 3 cycles min^{-1}, and the rectum at 6 and 3 cycles min $^{-1}$ (Taylor *et al.*, 1974a, b, 1975). Furthermore, the colonic activity is not present all the time and the anal canal has activity at 16 and 1 cycles min^{-1} (Wankling *et al.*, 1968; Kerremans, 1968).

Most of the attempts to record gastro-intestinal activity non-invasively in humans have concentrated on gastric electrical activity. (In this context, non-invasive means that probes do not puncture the skin and are not inserted into body cavities, though it will be necessary to discuss many invasive measurements in this review.) The nomenclature concerning electrical signals is confusing. The electrical signals, recorded directly from the muscular wall of the stomach or small intestine, show two characteristic types of activity: *slow waves* (also called pacesetter potential, control activity, or basic electrical rhythm) and *fast activity* (also called spike potentials, action potentials, response activity and spike burst). The *slow waves* are present all the time (both *in vivo* and *in vitro*), and are characteristic of smooth muscle—no activity is present in unstimulated striated muscle. The frequencies quoted above are the slow wave frequencies. Muscular contractions are accompanied by *fast activity*, which consists of bursts of oscillations 50 to 200 ms long lasting for up to 3 s. The fast activity corresponds to the muscle action potentials in striated muscle. It is often difficult to distinguish higher frequency components of the slow wave from genuine fast activity, and there is no universally accepted definition of fast activity.

The signals can be recorded, *in vivo*, from both the serosal and mucosal surface of the muscular wall of the gut, and from electrodes placed on the skin surface, and a wide variety of names have been applied to the different recordings. The most widespread and suitable terms for recordings from surface electrodes are *electrogastrogram* (EGG) for recordings of gastric electrical activity and *electro-enterogram* as the generalized term for surface recordings from the GI tract. Electromyography is traditionally applied to the recording of muscle action potentials from striated muscle, and is best avoided.

2.0 HISTORICAL REVIEW

The recording and analysis of gastro-intestinal activity from surface electrodes is but a small part of the total body of work on the smooth muscle of the GI tract. Readers who are interested in the wider aspects of the subject are referred to the comprehensive review by Daniel and Chapman (1963), the reviews by Kohatsu (1970) and Duthie (1974), the monograph by Brown (1976) and the bibliography in Smout (1980), the volumes on smooth muscle edited by Bulbring et al. (1970) and by Bulbring and Shuba (1976), and the proceedings of the biennial symposia on gastro-intestinal motility (Daniel, 1974; Vantrappen, 1975; Duthie, 1978; Christensen, 1980).

This review concentrates on measurements using surface electrodes. Two main lines of work can be distinguished—the mainstream, which concentrates on the clinical and/or physiological uses of surface electrode recording; and the considerable use which has been made of surface electrode recording of gastric electrical activity in psycho-physiological research (i.e. the effect of psychological stimuli on physiological parameters). It is perhaps hardly surprising that the psycho-physiological research is almost unmentioned in the mainstream literature, but it is astonishing that the advances in recording, analysis and interpretation of surface electrode recording reported in the mainstream literature have been ignored by the psycho-physiological researchers until very recently, with the result that one is left with grave doubts about much of this research.

As mentioned above, the first recording of human gastric activity was made by Alvarez (1922), who was unable to repeat his feat. It is indicative of the apparent lack of connection between the slow waves and gastric pathology that the next in vivo recording on humans was not made until 1936, though it had been possible to record and extract diagnostic information from the similar amplitude ECG since the late nineteenth century. Goodman (1943) measured the potential difference between a pair of calomel electrodes, one within the stomach and one on the arm. He thus measured the sum of the trans-mucosal potential difference and the slow wave. Several papers in the intervening period refer to work on acute or isolated preparations on dogs and rabbits (Richter, 1924; Puestow, 1932a, b; Berkson et al., 1932; Berkson 1933; etc.). In a further short report, Goodman et al. (1951) used silver–silver chloride electrodes, and recorded frequencies of 3–12 cycles min^{-1} from 63 normal stomachs—the 12 cycles min^{-1} was presumably duodenal activity which can be recorded from the stomach when close to the pylorus. He also mentions the response of the trans-mucosal potential to milk—a test used by many early workers.

Fukushima et al. (1951) also recorded the sum of trans-mucosal potential and slow wave using platinum electrodes. The usual 3 cycles min^{-1}

activity was found in 25 records from ten normal subjects. In 50 cases of gastric cancer, the rhythm was irregular, and similar recordings were obtained from three cases of gastric ulcer. In three cases of gastric spasm, the rate was unaltered but the amplitude increased during spasm—the description is consistent with the change being a movement artefact. In 12 cases of duodenal ulcer, the period was irregular. This paper is typical of many papers published during the following 15 years, in which a reasonable number of patients have been studied, but the analysis is completely subjective, and no attempt is made to establish statistically significant differences between groups.

In 1952 Martin and Morton described measurements of the combined trans-mucosal potential and slow wave (called by them DC and AC—a terminology used frequently by subsequent workers) and described a number of cases. Ingram and Richards (1953) used an electrometer amplifier and saturated KCl junction electrodes to measure trans-mucosal and slow wave potentials, and also made surface measurements on 45 subjects. They were not able to record simultaneously from both mucosa and skin, and do not appear to have recorded slow wave activity at the surface. They state that the 3 cycles min^{-1} activity of the stomach is abolished during sleep—a common early observation that is difficult to explain. Morton and Martin (1953) give a further anecdotal account that is mainly concerned with the trans-mucosal potential difference.

In a much more detailed paper, Morton (1954) analysed his records by a "scatter diagram of potentials", a "total activity gradient" and a "regularity score" which separated carcinomas, but does not describe the calculation of these indices. Goodman et al. (1955) analysed records (from luminal Ag–AgCl electrodes) from 200 subjects, both visually and by scoring 26 characteristics of the records—again, no details of the method are given. They were clearly convinced that their technique was diagnostic, even though their method classified 60 control subjects as 26 normal, 26 benign disease and eight carcinomas, and 46 gastric carcinomas as 26 cancer, 19 benign disease and one normal. We can take our leave of early measurements of the slow waves in conjunction with measurement of the trans-mucosal potential difference by mentioning two further papers. Colcher et al. (1959) recorded regular activity from normals, and irregular activity from "dumpers". Goodman et al. (1959) claimed to have recorded 18–20 cycles min^{-1} from the fundus and 18–22 cycles min^{-1} from the small intestine. The separation of potential difference recording from slow wave recording commences with the careful work by Katzka et al. (1955), who used calomel electrodes to demonstrate that the measured trans-mucosal potential is related to gastric physiologic activity rather than stomach contents.

A related non-invasive technique which turns up now and again is first described by Wenger *et al.* (1955), who used a magnetometer to detect a $2\frac{1}{2}$-g magnet swallowed by the patient. Two-hour recordings from the stomach correlated well with simultaneous pressure recordings from a balloon. The method was further described in Wenger *et al.* (1957). The alnico magnet was 3/16″ diam. by $\frac{1}{2}$″ long and coated with polystyrene. The recording system had an effective sensitivity of 0.2 mGs cm^{-1} and the peak-to-peak signal was 1 mGs. Further experiments, in dogs, show a correlation between pressure, strain gauge and magnetometer recordings of motility (Engel and McFall, 1959). Further recordings of gastric motility were made in 180 normal subjects by Wenger *et al.* (1961). Alternatives to the swallowed magnet are pressure or pH sensitive radio pills. These have not been used in conjunction with surface electrical recordings and will not be considered further.

The use of surface electrode recording of gastric electrical activity in psycho-physiological research commences with the work of Davis *et al.* (1957). A DC amplifier was used to record from EEG electrodes placed on the skin. The DC potential of the electrodes was backed-off by inserting a DC voltage source in series with the electrodes, and signals of 100–500 mV [*sic*—this appears to be a misprint for μV] were recorded, with frequencies up to 30–40 cycles min^{-1}. A variation of Wenger's technique, using a $\frac{1}{4}$″ ball bearing which was swallowed and a mine detector was also used. Anecdotal examples of the changes in the DC level in response to psychological stimuli are given. The methodology and terminology for much of the psycho-physiological research is established in a further paper by Davis *et al.* (1959) in which monopolar recordings were made from 1″ square chlorided silver discs. The records were divided into 30 s [*sic*] segments, and the "amplitude", "displacement" and "frequency" were measured. The effects of feeding, visual stimuli and inflation of a gastric balloon were recorded. The authors appear to believe that the 3 cycles min^{-1} electrical activity is a measure of gastric motility, a confusion which persists throughout the psycho-physiological literature.

The first modern surface electrode recording of gastric electrical activity and its relationship to disease is reported in Sobakin *et al.* (1962). They used monopolar surface recording in a large number (800–900) of normals and patients. Results from patients with gastric ulcer, pyloric stenosis, other gastric emptying disorders and gastric cancer, are quoted. It is particularly unfortunate, considering the large numbers, that there is no statistical analysis to support the optimistic conclusion that the technique is "completely suitable for objective pathophysiological studies of the displacements of the motor system of the stomach during digestion". These authors also appear to believe that the electrical activity is a result of motility, as they claim the method gives "registration of peristaltic action during digestion". There is an extensive Russian language literature, very little of which is available in translation, e.g. Krasilnikov and Fishzon-Russ (1963) made surface recordings from 59 healthy subjects.

Several papers on psycho-physiological research appear over the next few years. Similar techniques are used in all of them to test the response to different stimuli, and the same doubts also apply to all of them, e.g. measurements of frequency from a 30 s period of a noisy 3 cycles min^{-1} waveform; measuring DC offset changes from Ag–AgCl surface electrodes which do not have a potential which is independent of surface effects, etc. Davis and Berry (1963) placed electrodes over the upper left and upper right abdominal quadrants, and claimed to be recording gastric and duodenal activity. Subjects performing a noise avoidance task were claimed to show increased gastric activity. Stern (1964) examined the effects of visual and auditory stimulation, and quantified "motility" in terms of the amplitude of 1 min stretches of signal. Fedor and Russell (1965) and Stern (1966) examine the effects of response contingent stimulation, and Russell and Stern (1967), in a review article, define the measured parameters in more detail, but still claim that surface electrodes record gastric motility.

The use of modern data analysis techniques is first reported in a series of papers by Nelsen and co-workers. This is also, perhaps, the first work in this field in which one can have confidence. In 1965, Nelsen and Wallace described recordings from serosal electrodes, with on-line calculation of the instantaneous frequency and the mean and variance of the frequency. Further work using implanted electrodes in dogs (Nelson et al., 1967) demonstrated the transitory change in gastric frequency following feeding. In 1967, Nelsen reported the use of phase-lock techniques to track the gastric slow waves recorded from surface electrodes. Simultaneous recordings from implanted electrodes were performed on three patients, and surface electrode recordings only from a further 12, with only one failure to record activity from the surface electrodes. Kohatsu (1968) and Nelsen and Kohatsu (1968) report further work with cutaneous electrodes. Kohatsu recorded from 12 female and 18 male patients using Ag–AgCl surface electrodes. Nine of the patients also had implanted electrodes, and a further 20 normal subjects had surface electrode recordings. The records were analysed by eye and by autocorrelation. There was a 1:1 relationship between the internal and cutaneous records. Ninety per cent of the cutaneous records were useable, and a significant drop in frequency was seen following a meal. The further work reported in Nelsen and Kohatsu (1968) confirms in more detail these encouraging results.

At about the same time, Christensen and his co-workers reported many careful measurements of the electrical activity of the small intestine using an intraluminal wick electrode (Christensen *et al.*, 1964, 1966a, b). They demonstrated that there is a frequency gradient in the small intestine in normal subjects, and showed statistically significant changes in frequency with fever, hypo- and hyperthyroidism, gastric hypersecretion, diarrhoea, intestinal malabsorption, gastric ulcer and chronic pulmonary insufficiency, and suggested that the metabolic rate is a major factor governing the rate of the duodenal "pacemaker".

Oi *et al.* (1967) used mucosal electrodes to show a variation in gastric frequency with acidity, and a paper by Condrea *et al.* (1967) claims to measure motility using surface electrodes.

There is a very considerable literature in French on surface electrode recording in a large number of patients with many different diseases. The majority of the reports are anecdotal, with no statistical analysis, though there are one or two outstanding exceptions. It will be convenient to consider all of these papers together, though this does take us out of the chronological sequence. Many of the French references are to very short and uninformative abstracts, and most of these are omitted. Martin *et al.* (1967) made 6–24 h recordings from a rectangular array of stainless steel discs placed near the left iliac fossa. The amplifier time constants were equivalent to high pass filtering at 6·4 or 12·6 cycles min^{-1} (i.e. 2 or 4 times the gastric slow wave frequency). The recordings from 56 subjects showed a variety of frequencies, including electrical silence during sleep. Martin *et al.* (1970) recorded from 12 stainless steel electrodes. They give diagrams of the amplitude (in three bands, normal, high and low) versus time in two patients, and suggest that the method is useful for diagnosing psychosomatic disease. In 1971, Martin and Thillier (1971a) reported recordings from a rosette of surface electrodes and from limb leads, and claimed that the electrical axis of the stomach, small intestine and colon could be calculated in an analogous manner to the Einthoven triangle for ECGs. In a further report (Martin and Thillier, 1971b) this thesis was elaborated, with 300 recordings for 24 h. It was claimed that, by suitable selection of the electrical axis, recordings could be made selectively from the stomach, small intestine and colon. Martin *et al.* (1971) recorded simultaneously from internal and external electrodes in the rat, but added little to our knowledge of the propagation from the muscle to the skin. Martin *et al.* (1971) gave a case history in epithelioma, and Combe *et al.* (1972a) give a number of case histories in infants. The electrical axis reports were developed further by Martin and Thillier (1972) to give a vector-gastrogram. There are three more abstracts from the same group (Martin *et al.*, 1972; Thouvenot *et al.*, 1972; Combe *et al.*, 1972b). Of rather more interest is Thouvenot *et al.* (1973), in which the specification of the electrodes and recording system is discussed and evidence for the identity of the electrical activity recorded from the muscle and from the skin is presented. (This paper and the following paper are both in English.) It was claimed by Thillier and Bertrand (1975) that the stomach acts as an electrical dipole and electrical activity is therefore only recorded from the surface when the stomach contracts. The lack of surface activity during fasting (a finding contrary to that of other workers) was said to be due to the surface activity corresponding to the fast activity which triggers the contractions. Finally, an impressive paper by Tonkovic et al. (1975) describes spectral analysis of surface electrode recordings, and gives the percentage incidence of activity in different frequency bands.

Returning now to the chronological order, McIntyre et al. (1969) made recordings from the serosal surface of the stomach wall at operation in 205 patients: 112 normal, 59 duodenal ulcer, 14 gastric ulcer, 12 gastric carcinoma and eight other. They were unable to find any significant difference in the frequency, amplitude or conduction velocity between the four main groups.

One further easily accessible example of the use of the electrogastrogram in Russia is available as a British Library translation of Furman and Kazakov (1970), in which they report on 108 patients. Amplitude and frequency distributions are given for their different groups, and show considerable overlap.

In two reports (in German), Schulz and his colleagues describe their recording method and give mean frequency and the standard deviation for a number of disease states (Schulz et al., 1973, 1975).

The totally isolated line of psycho-physiological research is continued by Lilie (1974) in a volume called "Bioelectric Recording Techniques". This is essentially a repeat of the description of Russell and Stern, with the addition of frequency analysis, and there is no reference to any of the reliable modern papers on analysis of the electrogastrogram. Stevens and Worrall (1974) compare the mechanical [sic] activity from the stomach in cats with the electrical activity recorded with surface electrodes, and analyse their records by eye and by Fourier and correlation techniques. They claim that the surface electrical activity reflects the electrical and mechanical activity of the stomach, and find segments of record in which surface electrical activity is present in the absence of mechanical activity.

In 1975, Brown et al. reported surface electrode recordings from 16 normal subjects both before and after a meal. Several simultaneous recordings were also made from mucosal and surface electrodes. Fourier analysis and correlation techniques were used to establish that the surface electrical activity was indeed a reflection of the gastric electrical activity. A completely objective method of analysis was described by Smallwood et al. (1975), which worked well with gastric electrical activity, but was less successful with colonic activity. A further series of recordings in 77 patients and normal subjects undergoing a gastric emptying test has been described by Smallwood (1976). Recordings were made both before and after the patients underwent either highly selective vagotomy or truncal vagotomy and pyloroplasty. Significant differences in frequency were found between pre-operative and post-operative records, but there were insufficient numbers in the groups to demonstrate a difference between operations. Smallwood also attempted to repeat the work of Martin and Thillier (1971), in which recordings from different pairs of limb leads were said to record selectively from different parts of the GI tract. It was not surprising, in view of the volume conductor properties of the body, to find that the gastric electrical activity could be recorded from all the electrode combinations other than right arm to left arm. The same reference also includes a preliminary report on the use of an integrated circuit phase-lock system for tracking cutaneous gastric activity.

In a further paper in English (Sobakin and Privalov, 1976), with a Russian bibliography, the Russian group claim that recordings made from different positions on the anterior abdominal wall reflect the electrical activity of different parts of the stomach. Autocorrelations of the electrical activity recorded from different parts of the stomach show different frequencies, but the authors do not state whether the records were made simultaneously.

The psycho-physiological line is continued with three papers by Walker and co-workers, in which the classical analysis of Russell and Stern is used. In Walker and Sandman (1976) the "phasic" changes in the EGG are said to reflect responses to environmental stimuli, whereas the "tonic" changes (i.e. changes in DC level) are said to indicate chronic gastro-intestinal disturbance. Spectral analysis is used in Walker and Sandman (1977), but the spectra do not show any peaks, even at the gastric frequency. A trial of voluntary control of gastric electrical activity is reported in Walker et al. (1978) using 36- and 90-s trial periods. Control of the EGG was achieved in the longer trials.

Nelsen's use of phase-lock techniques to track the cutaneous gastric activity is continued in a two part paper by Smallwood (1978a, b) in which an integrated circuit realization of the phase-lock loop and the exponentially mapped past (EMP) statistics calculation is described. The first paper describes in detail the design and performance of the system, and the second paper describes the operation of the system with the signals previously analysed by Brown et al. (1975). A method of attaching levels of significance to the EMP variance of the frequency is proposed, and short-term variations in the gastric frequency are demonstrated. Technical details of the system are described in further papers (Smallwood 1978c, d, e, f).

The first accounts of psycho-physiological research that are well able to stand comparison with mainstream research are those of Holzl (1979) and Holzl et al. (1979). The second paper contains a critical review of previous psycho-physiological research. A combination of surface electrode recording and "magneto-gastrography" (Wenger's magnet measurement of motility) was used, and both types of record were frequency analysed in 10-min stretches (the length of the stress period). A significant increase in signal power was found during the stress periods.

The Sheffield group have published several papers on the methodology of the analysis of electrical activity (Stoddard et al., 1979; Smallwood et al., 1980b; Kwok et al., 1980). These papers are principally concerned with the analysis of serosal and mucosal recordings of colonic slow waves using autoregressive modelling, fast Fourier transforms and visual analysis, but the methods and problems are directly relevant to the analysis of cutaneous gastric signals.

The dissertation by Smout (1980) contains two relevant papers (Smout et al., 1980a, b), and considers recordings directly from the gastric muscle and from cutaneous electrodes in both dog and man. The literature is reviewed, and the author discusses at length the origin of the cutaneous signals. This is further discussed in the next section.

3.0 ORIGIN OF SURFACE ELECTRICAL ACTIVITY

It is quite clear from the historical review that there has been some uncertainty about the origin of the electrical signals that can be recorded from the skin overlying the stomach. There are two main possibilities: the signals are the result of the mechanical activity of the stomach; or the signals originate from the electrical activity of the stomach. It is sometimes not clear which explanation is implied, as the term "motility" is used rather loosely. *Motility* refers to the mechanical activity of the stomach, which causes the contents to be mixed and moved down the stomach. Human gastric motility has been measured *in vivo* by several methods: direct X-ray visualization of a radiopaque "meal"; pressure measurement using open-ended catheters, which are usually perfused at a low rate; pressure measurement using air-filled balloons; pressure measurement using a free or tethered radio-pill; movement of a small magnet; movement of pressure-sensitive radio-pill; and, quite recently, the use of a gamma camera following a radioactive meal to follow stomach wall movements. In chronic animal experiments, the contraction of the muscle has been measured using strain gauges attached directly to the muscle. *Transport* refers to the movement of the contents along the length of the gastrointestinal tract, and has also been measured by several methods: X-ray visualization of radiopaque markers; counting non-absorbable markers in the stool; labelling a meal with a radio-isotope and following the progress with a suitable X-ray counter or imaging system; breath hydrogen tests. It has not been claimed that surface electrical activity is an indicator of transport.

The human gastric electrical activity can be recorded *in vivo* by several types of electrode: intraluminal electrodes of various types, the most satisfactory of which is the calomel wick electrode; mucosal electrodes, which are usually fine spikes attached to a nasogastric tube and held in place by suction; serosal electrodes, which are fine wires inserted in the serosal surface of the muscle at operation—the connecting wires are brought out through a surgical drain and the electrodes are removed with the drain a few days after operation.

Although it is possible to infer that the surface electrical signals originate as electrical signals from the gastric muscle, it is clearly more satisfactory to record the surface and internal signals simultaneously in order to establish their identity.

Alvarez recorded simultaneous electrical and mechanical signals from many different animal preparations, and obviously believed that the electrical signals he recorded from the surface (Alvarez, 1922) originated as the electrical activity of the gastric muscle. Confusion then reigns for about the next 40 years, particularly as some authors possibly use "motility" in a wider sense to mean both electrical and mechanical activity. Several authors (e.g. Ingram and Richards, 1953) comment that the 3 cycles min^{-1} activity disappears during sleep, which suggests that the electrical signals that were being recorded were in fact movement artefacts, i.e. they were indeed measuring motility! Nelsen (1967) performed simultaneous recordings from implanted and cutaneous electrodes in three patients, and showed that the frequencies were identical. Kohatsu (1968) recorded from implanted and surface electrodes, and showed that the two electrical signals had the same frequency. This is confirmed by Nelsen and Kohatsu (1968). Thouvenot *et al.* (1973) state that the cutaneous electrical activity is identical to simultaneously recorded electrical activity from the muscle. Stevens and Worrall (1974) correlated the mechanical activity of the cat stomach with surface electrical recordings, and noted that the electrical activity persists in the absence of mechanical activity.

Brown *et al.* (1975) recorded from both cutaneous and mucosal electrodes, and considered in detail the origin of the cutaneous activity. They examined four possible sources of 3 cycles min^{-1} activity: electrode artefacts; a 3 cycles min^{-1} oscillator other than the stomach; mechanical artefacts due to gastric motility; and the electrical signals from the gastric muscle.

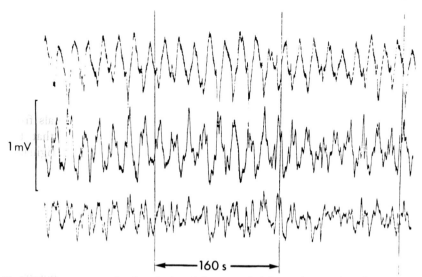

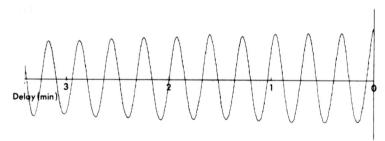

Fig. 2 Human gastric electrical activity recorded from three pairs of cutaneous electrodes arranged as shown in Fig. 7. There is a very clear rhythm at about 3 cycles min^{-1} in the top two traces. The rhythmic nature of the lower trace is partially obscured by noise. (From Brown *et al.*, 1975.)

Electrode artefacts should be independent of electrode siting, and recordings were therefore made from electrodes on the arm, which gave very low signal levels with no consistent frequency components. Activity at such a low frequency is only likely to originate from smooth muscle. The only known source of 3 cycles min^{-1} activity (other than the stomach) is the colon, and the activity is only present for about 20% of the total recording time from the colon (Taylor *et al.*, 1974b). The third possibility, that the mechanical activity of the stomach gives rise to an artefact that is recorded as an electrical signal at the surface, is not consistent with the regularity of the surface electrical signal. There is a one-to-one relationship between

Fig. 3 The autocorrelation function plotted against time delay for the upper trace in Fig. 2. (From Brown *et al.*, 1975.)

the slow waves and the gastric contractions when the latter are present, but contractions are not present all the time (Monges et al., 1969). The regularity of the signal can be established by calculating the autocorrelation function for the signal—the rate of decay of the autocorrelation function is inversely proportional to the regularity. Figure 3 shows the autocorrelation calculated for the upper trace in Fig. 2, which demonstrates that the surface electrical signal is very consistent—it can easily be shown that the missing contractions in the mechanical signal give rise to a rapidly decaying autocorrelation function. Furthermore, the amount of mechanical activity would be expected to increase after a meal, but the percentage of time for which a cutaneous electrical signal was obtained did not increase after feeding. Several simultaneous recordings were made from mucosal electrodes and surface electrodes (Fig. 4). Frequency analysis of 20-min stretches of data (giving a frequency resolution of 0.05 cycles min^{-1}) showed that the frequency of internal and external signals were identical, and cross-correlation of the mucosal and cutaneous signals showed excellent correlation between the signals (Fig. 5). The mucosal electrode probe also contained a fine open-ended tube through which pressure recordings could be made. Cutaneous electrical activity at 3 cycles min^{-1} was recorded even in the absence of pressure waves within the stomach (Fig. 6). It was concluded that the cutaneous electrical activity originated from the gastric electrical activity, and was independent of the mechanical activity of the stomach.

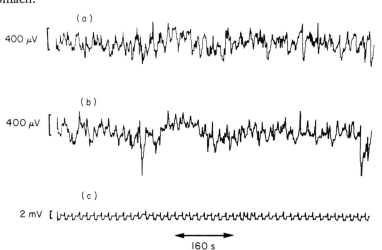

FIG. 4   Gastric electrical recordings from cutaneous electrode pairs placed as for the upper and lower pairs in Fig. 7 (upper two traces) and from a mucosal suction electrode (lower trace). No pressure waves were present during this recording. (From Brown et al., 1975).

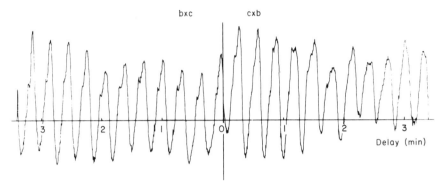

FIG. 5   The cross-correlation function plotted against time delay for internal signal b and external signal c in Fig. 4. (From Brown et al., 1975.)

Smout et al. (1980a, b) made simultaneous recordings from serosal and cutaneous electrodes and from extraluminal force transducers in chronic dog experiments. The records were analysed in 1024 s (17·07 min) blocks. Out of a total of 530 blocks, 95·8 % had a cutaneous signal which was sufficiently noise-free to be analysable. In the absence of mechanical activity, a small sinusoidal signal at 5 cycles min^{-1} (the dog gastric frequency) was obtained for 91 % of the time. Spectral analysis of the cutaneous and serosal signals showed the frequencies to be identical. In the presence of tachygastria (increased gastric frequency), the tachygastric frequency was recorded from both serosal and cutaneous electrodes. The amplitude of the cutaneous signal increased when mechanical activity was present. The impedance between the electrodes was measured continuously, and was unchanged by the mechanical activity, which produced no visible movement of the electrodes. Vigorous movement of the electrodes did not produce any artefact on the cutaneous signals. The authors conclude that the cutaneous signal is generated by the electrical activity of the stomach, and that it consists of slow wave activity (during mechanical quiescence) and a summation of slow wave and fast activity (during mechanical activity) but is not influenced by the mechanical activity itself. A model was proposed which takes into account the contribution of both slow and fast activity in the generation of the cutaneous signal.

4.0 APPLICATION OF GASTRIC ELECTRICAL ACTIVITY RECORDING

There are three inter-related areas in which the non-invasive recording of gastric electrical activity may be useful: exploration of the physiology of

the slow waves is diagnostic. Unfortunately, all the evidence is anecdotal—there is usually no attempt to analyse the recordings in an objective repeatable manner, and no statistical analysis has been performed on the results. The work by Goodman et al. (1955), quoted above, is typical. They analysed records from 200 subjects, and performed both a visual analysis and also a more objective analysis in which 26 characterisitics of the records were scored. No details of these characteristics are given, no figures are produced for the relationship between the scores and the diagnosis, and no statistical results are quoted. Their classification of the subjects (given above) does not inspire confidence in the method. A very small number of papers describe recordings from different patient groups with statistical analysis of the results. The conclusions of McIntyre et al. (1969) are typical: they were unable to find any significant difference in frequency, amplitude or conduction velocity in their recordings at operation. The only significant changes in slow wave frequency with disease are those reported for the small intestine by Christensen et al. (1964, 1966a, b). Unfortunately, there is evidence to suggest that cutaneous recording of small intestinal electrical activity would have a very low success rate.

The apparent lack of association between slow wave frequency and disease may be due to the fact that gastric contractions and slow wave frequency are not related. Although the gastric contractions have a one-to-one relationship with the slow waves, the contractions do not occur for every cycle of the slow waves. A change in motility may therefore be due to the average number of contractions altering, whilst the slow wave frequency is unchanged. It would be interesting (and perhaps useful!) to relate some form of motility index to disease states. The recent work of Smout et al. (1980a, b) suggests that the amplitude of the cutaneous slow waves (in the dog) is related to the presence of fast activity, which in turn causes contractions of the gastric musculature. Confirmation of this relationship in humans would provide a non-invasive means of relating motility to disease. The ability to monitor simultaneously electrical activity and motility might also be of value in further exploring the physiology of the gastro-intestinal tract. The one well-documented change in slow wave frequency in the stomach is the dip following feeding. The significance of this reproducible change in frequency is unknown. Could stimulated changes in frequency be the basis of diagnostic tests (as suggested 15 years ago by Nelsen)? Recent work in which circadian rhythm display techniques have been applied to slow wave frequency measurement (see below) shows that the average gastric frequency is stable, but that there are considerable short-term frequency changes. Some of these appear to be related to the migrating complexes of Szurszewski (1969), which have been extensively studied in fasting subjects.

The morphology and function of these complexes in the normal GI tract is unknown, but could possibly be explored using these techniques. Other short-term frequency changes appear to be periodic and unrelated to the complexes—what is their function (if any)? It is perhaps worth restating at this point that the number of possible explanations of a change in the cutaneous recordings is large, and any hypothesis as to their origin must be proven by simultaneous recording of cutaneous electrical activity, mucosal (or serosal) electrical activity, and motility.

The psychological aspects of psycho-physiological research are outside the competence of the authors of this article, and sufficient has already been said about the physiological aspects of most of the published research related to electro-gastrography. Holzl et al. (1979) give the rationale behind the experiments. There is evidence that stress can lead to ulcer formation—this is well-established in animal experiments, but the extrapolation to man has been challenged, though of course there is ample anecdotal evidence that stress and ulcers are related. Data from human subjects is therefore required, and will hopefully lead to behaviourally orientated therapies for stress-induced illness. It is clearly not possible to expose human subjects to controlled stress and look for the development of ulcers, but it is possible to expose subjects to relatively short periods of stress and look for changes in gastric function which may favour ulcer formation. The use of non-invasive tests is clearly to be preferred, particularly as the invasive tests impose a considerable psychological stress on the subject.

5.0 RECORDING TECHNIQUES

5.1 Electrodes

Several different materials have been used as cutaneous electrodes for recording gastric electrical activity. The commonest are stainless steel, platinum black and chlorided silver. The properties of electrodes made from these materials are discussed by Geddes (1972). The ideal electrode would have a low impedance at the very low frequencies of interest, a stable potential difference between electrode pairs, and low noise. Although several electrode materials would fulfill these requirements, a discussion of the advantages of the different types is academic, owing to the ready availability of excellent chlorided silver electrodes intended for use with patient monitoring systems. The impedance of the Ag–AgCl electrode is almost independent of frequency at the low frequencies of interest, and it is essentially non-polarizable. A well-chlorided pair of electrodes will have a noise level of less than 10 μV peak-to-peak. It has been shown that movement artefacts at electrodes are caused by disturbing the charge double-layer at the electrode–electrolyte interface, so that movement artefacts can be reduced by a cup-shaped electrode which moves the interface away from the skin. It has also been shown that the noise level from a pair of electrodes separated by an electrolyte decreases with time after the electrolyte is applied, so that, for the best results, the electrolyte should be applied to the electrode perhaps an hour before the recording commences. Commerically available Ag–AgCl electrodes take advantage of most of these findings—they have a reasonable area (about a square centimetre) to reduce the impedance; they are disposable so that the chloride layer does not wear off (which would increase the noise level by orders of magnitude); they are recessed, and fairly free from movement artefact; and they are pre-jelled, so that the noise level should be minimal immediately on application. They should be applied to skin which has been cleaned and slightly abraded by a pad soaked in alcohol—the stratum corneum has a high impedance, but is only a few cells thick, and may therefore be easily removed with light abrasion.

5.2 Electrode Position

Gastric electrical activity can be recorded from a pair of electrodes placed anywhere on the body, with the stomach lying more or less in between them, as shown by Smallwood's (1976) recordings from limb leads. In practice, it is useful to define monopolar and bipolar recording systems, and to attempt to standardize electrode positions which give the best signals and success rate. Monopolar recordings measure the potential between an indifferent electrode, placed a long way from the source of the potential, and an active electrode close to the source. Bipolar recordings measure the potential between a pair of electrodes placed close to the source. Monopolar recordings measure the potential change at a point, but also unavoidably pick up any other signal source between the electrodes, and therefore tend to give a poor signal-to-noise ratio. Bipolar recordings have a higher signal-to-noise ratio, but cannot distinguish a change in potential at one electrode from a change of opposite sign at the other electrode. In the event of the potential change being the same at both electrodes, no signal would be recorded. If the iso-electric lines for a particular interfering source can be plotted, then the bipolar electrodes can be orientated to minimize the interference.

Two possible arrays of electrodes are the linear array used by Brown *et al.* (1975; Fig. 7), and the rosette array used by Smout (1980; Fig. 8) and several others. Brown and co-workers only recorded between horizontal pairs of electrodes. It is, of course, possible to record between any of

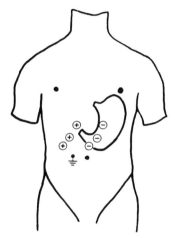

FIG. 7   The position of surface electrodes on the trunk used by Brown and Smallwood. Three pairs were placed over the gastroduodenal area. The cathode of the upper pair was halfway between navel and left nipple, and the anode 8 cm to the right. The electrode separation for the other channels was also 8 cm.

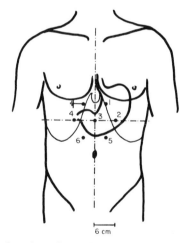

FIG. 8   The position of surface electrodes used by Smout. Electrodes 2, 3 and 4 were placed on a transverse line mid-way between the lower end of the sternum and navel. Electrode 3 is on the median plane, and all other electrodes are 6 cm from electrode 3. (From Smout, 1980.)

the 15 combinations of the six electrodes, and this has been done by Smout for the rosette configuration. The data recording problem can be reduced by recording from each electrode with respect to, say, electrode 1 (or with respect to a separate indifferent electrode) using amplifiers matched for gain and phase response, and then subtracting appropriate pairs of records to give the desired bipolar recording. Smout (1980) concluded that the lower two electrodes gave little extra information and could be discarded. The three best bipolar pairs were different before and after a meal. Both Brown *et al.* (1975) and Smout (1980) showed a 3 cycles min^{-1} signal to be present for *c.* 80–90% of the recording time, so there would appear to be little to choose between the arrays. Smallwood (1976) also recorded from post-operative patients who had a large dressing along the midline. The electrodes were placed at the same vertical position, but separated horizontally so as to fall at the edge of the dressing. The success rate post-feeding was again about 80%. The rosette array in similar circumstances would have to lose the central electrode.

5.3 Amplifier Characteristics

The impedance of a typical disposable Ag–AgCl electrode will be < 10 kΩ at low frequencies on well-prepared skin, with a maximum impedance on poorly prepared skin of *c.* 100 kΩ. An amplifier with an input impedance of at least 10 MΩ is therefore desirable. Frequency analysis of cutaneous gastric signals recorded with a wide band amplifier shows that there are no frequency components of greater amplitude than the system noise level at frequencies above about 20 cycles min^{-1} (0·3 Hz). Components near this frequency can usually be shown to be movement artefacts due to respiration, but occasionally harmonics of the 3 cycles min^{-1} activity are present, as in Fig. 6. The upper frequency limit of the amplifier can therefore be placed at 1 Hz with no loss of information. Frequencies as low as 1 cycle min^{-1} (the "minute rhythm") have been observed in electrical and motility records (Golenhofen, 1970). However, the noise level in cutaneous recordings is frequently found to increase at frequencies < 1–2 cycles min^{-1} (i.e. the power level in the signal increases with falling frequency, but there are no significant peaks in the frequency spectrum), so that the low frequency cut-off is usually set at 1 cycle min^{-1} (0·016 Hz). The phase distortion from this high pass filter is minimized by limiting the roll-off to 6 dB per octave.

The signal level may only be a few hundred microvolts, so the peak-to-peak noise levels of the amplifier (referred to the input) should be ⩽ 10 μV. The common mode rejection ratio of the amplifier is not very critical, as any 50 Hz interference will be attenuated by the 1 Hz low pass filter.

The circuit diagram of a suitable isolated amplifier is shown in Fig. 9.

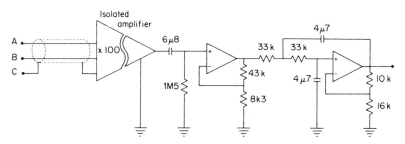

Fig. 9   Circuit diagram of an isolated amplifier with a bandwidth of 1–60 cycles min^{-1}, and an overall gain of 1000. The active electrodes are connected to A and B, and the isolated screen is connected, at C, to an electrode placed at a convenient position on the body. Point C must not be earthed.

5.4 Tape and Chart Recording

It is essential that a visible record of the signal is available when the recording is being made, in order that problems due to faulty electrodes, movement artefacts etc. can be corrected immediately. The signal bandwidth is very low, so that any type of chart recorder is suitable. The most commonly used types are multichannel galvanometric recorders. The paper speed should be sufficiently fast that it is possible to distinguish individual cycles of the slow waves. This is still a slow speed (of the order of 5 cm min^{-1}), and some types of recorder may not produce a satisfactory image. Ultraviolet recorders, for instance, may fog the paper at such low speeds. It may be necessary to purchase an accessory gear-box to achieve a sufficiently low speed.

All the signals should be recorded on magnetic tape, without any form of preprocessing, for future analysis. The bandwidth of any FM recorder at the lowest recording speed is more than adequate—the IRIG Intermediate Band specifies a response of DC to 300 Hz at 15/16 in. s^{-1}. Some recorders (e.g. Hewlett-Packard 3964 and 3968) have a minimum speed of 15/32 in. s^{-1}, which gives 24 h of recording on a 7 in. reel of triple-play tape. The tape-recorder is also used for data compression—the Racal Store 4 recorder has a maximum speed of 60 in. s^{-1}, so that a 4-h recording at 15/16 in.s^{-1} can be digitized in less than 4 min with replay at 60 in. s^{-1}. All good quality instrumentation recorders lock the tape speed to a crystal-controlled clock, so that frequencies measured on a signal replayed from tape will be as accurate as frequencies measured directly.

5.5 Practical Problems

In principle, the problems encountered when recording cutaneous gastric signals are no different from those found when making any other electrophysiological measurement. The skin should be cleaned and gently abraded with an alcohol-soaked pad before applying the electrodes. The electrode lead should be taped down, with sufficient slack lead between electrode and tape that any movement of the lead will not be transmitted to the electrode. The recordings should be done in an electrically quiet environment—in practice, this means in a room with the mains wiring in earthed metal conduit. As far as possible, the leads to the individual electrodes should be run close together and close to the skin to minimize wiring loops which will pick up radiated magnetic interference from transformers—if the room is too close to a substation transformer or its feeder cables, the magnetic interference may be so great as to make recordings impossible. The only solution is to move further away.

The subject should be calm, and be encouraged to relax and lie still to minimize artefacts due to both movement of the electrodes, and movement of the stomach with respect to the electrodes.

The cutaneous gastric signal is fairly sinusoidal, and any sharp transition in the waveform is likely to be due to a movement artefact or to a faulty electrode or connector. The system noise level should be less than the width of the line drawn by the chart recorder pen, so any thickening of the line will probably be due to excessive 50 Hz interference.

6.0 ANALYSIS TECHNIQUES

In this context, "analysis" is taken to mean "the provision of an objective description of the signal". There are three main variables that can be described: the frequency components that are present; the amplitudes of the components; and their time duration. Unfortunately, the signals are contaminated with noise of unknown spectral distribution, which makes an apparently simple task rather difficult. Several methods of analysis are outlined by Linkens (1978), and the problems of applying these methods to the analysis of colonic electric activity (which is even more intractable than cutaneous gastric activity) are described by Smallwood et al. (1980b).

6.1 Visual Analysis and Filtering

A visual record is essential—the frequent artefacts on cutaneous recordings are readily identified on the visual record, but may not be immediately obvious on, for instance, a frequency transform. An intelligent appraisal of the visual record is essential to the interpretation of any automated analysis. Manual determination of the frequency and amplitude of the signal is unreliable, due to the poor signal-to-noise ratio. It has also been found (Smallwood *et al.*, 1980b) that visual analysis is dominated by the largest amplitude component—harmonics or near harmonics of the fundamental frequency tend not to be noted—and higher frequency components are frequently masked by artefacts due to respiration.

Automatic recognition of artefact has not yet been applied to the analysis of gut signals—it should, for instance, be fairly easy to recognize gross movement artefacts by setting limits on the rate of change of the signal. Sophisticated artefact recognition techniques have been applied in the analysis of respiratory waveforms, which are similar to cutaneous gastric signals (Wilson *et al.*, 1982).

Bandpass filtering has been applied, with success, in a wide variety of situations where the signal is constant for very many cycles, e.g. in EEG and speech analysis. Filtering is less successful with cutaneous gastric signals. Gross movement artefact approximates to an impulse input to the filter, which gives as its output the impulse response, i.e. the filter "rings" at its resonant frequency. The movement artefact thus appears, to the untutored eye, as a transient burst of activity.

6.2 Fourier and Related Transforms

Fourier transformation is the classical method of frequency analysis (Oppenheim and Schafer, 1975; Rabiner and Gold, 1975; Randall, 1977) and, in the now readily available form of the fast Fourier transform (FFT), is frequently applied as a "black box" in the analysis of gastro-intestinal activity, with an often uncritical acceptance of the "objective" analysis. In practice, the output of the Fourier transformation is a plot of frequency versus amplitude—the frequency spectrum—which still requires interpretation.

The classical condition for performing a frequency analysis is that the data must be stationary, i.e. the spectral composition of the signal must not change during the analysis period. This condition is always violated by biological signals, without any disastrous consequences being apparent. In practice, there is a trade-off between transforming many cycles of the signal, and therefore getting good frequency resolution, and transforming a short length of signal, and so seeing changes in frequency.

The first step is to decide on the rate at which the analogue signal will be digitized. The maximum frequency present in the frequency spectrum will be $f_s/2$, where f_s is the rate at which the signal is digitized. Any signal of greater frequency than $f_s/2$ (the Nyquist frequency) will be reflected about $f_s/2$ and appear as a lower frequency signal. This process is known as aliasing, and is avoided by ensuring that the signal contains no information at frequencies above $f_s/2$. The signal is filtered by an analogue filter before digitization. The cut-off of a real filter is not infinitely steep, so it is usual to set the $-$ 3 dB point of the anti-aliasing filter at $f_s/3$. A suitable sampling rate for cutaneous gastric signals is 1 Hz.

The frequency resolution (i.e. the spacing of the points on the frequency spectrum) is given by $1/T$, where T is the time length of the signal being transformed. If the sampling rate was 1 Hz, and a 256 point transform was taken, the frequency resolution would be $(1/256)$ Hz $= (1/4\cdot26)$ cycles min^{-1} $= 0\cdot23$ cycles min^{-1}. The resolution of a particular frequency is also given by $(100/N)\%$, where N is the number of cycles of the waveform in the time T. For instance, for a 3 cycles min^{-1} signal and the same conditions as above, there would be $3 \times 4\cdot26 = 12\cdot8$ cycles in time T, and resolution would be $(100/12\cdot8)\%$ of 3 cycles min^{-1}, i.e. $7\cdot8\%$ at 3 cycles min^{-1}. This is equivalent to $(3 \times 7\cdot8/100)$ cycles min^{-1} $= 0\cdot23$ cycles min^{-1} as before. In practice, a Fourier transform of about 10 cycles of signal is the shortest length that is worthwhile. It is interesting that the human eye seems to need about the same length in order to recognize the signal in the presence of noise (Smallwood et al., 1980b).

The data to be transformed have been obtained by multiplying the complete signal by a "window" function which is zero outside the region of interest, and unity within the period to be transformed (i.e. unity for 256 points in the above example). This is equivalent, in frequency space, to convolving the Fourier transform of an infinite length of signal with the Fourier transform of the window, which is a $\sin x/x$ function. The ideal line spectrum therefore has "side lobes" which can conceal small components close to the main peak (Harris, 1978). It is possible to overcome this problem by using different window shapes, but only at the expense of frequency resolution. As the cutaneous gastric signals are noisy, and the main aim is to measure the frequency accurately, the straightforward rectangular window is usually used.

It is possible to place confidence limits on FFT spectra, but these are of limited use with cutaneous gastric data. The data are normally assumed to approximate to Gaussian noise, for which the amplitude of the peaks in the frequency spectrum are distributed as a Chi-square variable with two degrees of freedom. The confidence limits are tabulated (Blackman and Tukey, 1959) and give 95% confidence limits of $3\cdot00\,H$ and $0\cdot05\,H$ for a peak height of H, which is less than helpful. There are two methods for increasing the number of degrees of freedom, and thereby reducing the 95% confidence interval. Adjacent spectral peaks can be averaged or several consecutive spectra can be averaged. For the same frequency resolution, the same length of data is required for both methods, and they are both unsuitable for data in which the frequency may be changing. It is also clear from an examination of the spectra that they are not Gaussian—

they consist of a small number of sinusoidal components and a background noise spectrum. The shape of this noise spectrum has been determined for cutaneous gastric signals (Smallwood *et al.*, 1975). The amplitude of the noise was found to be greater at low frequencies, with an approximately lognormal amplitude distribution. Significant peaks in the spectrum could be automatically selected by comparing the signal spectrum with the mean and standard deviation of the noise spectrum.

The use of FFT analysis to determine the nature of periodic changes in amplitude of the signal (in this case, from the duodenum) has been described in Smallwood *et al.* (1980a).

The Walsh transform has also been used for the analysis of gastro-intestinal electrical signals (Linkens, 1978). The Walsh algorithm has the advantage that it requires no analogue-to-digital conversion of the signal, which is simply squared to give a digital signal, and that no multiplications are required, thus reducing the computation very greatly. It has the disadvantages that it generates spurious high frequency components and gives no amplitude information. The cheap and fast FFT processors that are available for use with microprocessors have made the Walsh transform obsolete in this application.

6.3 Phase-lock Techniques and EMP Statistics

Phase-lock techniques have been extensively used in, for instance, communications, frequency synthesis and speed control (Gardner, 1979). This was the first modern data analysis system to be applied to cutaneous gastric activity (Nelsen, 1967; Nelsen *et al.*, 1968), and the first technique to which statistical analysis was applied (Nelsen and Wallace, 1965). Nelsen had to use a pair of analogue computers, but it is now possible to implement a phase-lock system using a small number of integrated circuits (Smallwood, 1978a, b), and, indeed, to use a single integrated circuit phase-lock loop (Smallwood, 1978e).

The basic system consists of a phase detector (which has an output voltage proportional to the difference in phase between the input signal and the voltage controlled oscillator); a filter (which removes noise and determines the dynamic performance of the loop); and a voltage controlled oscillator (VCO). The error signal from the phase detector forces the frequency of the VCO to change until there is 90° phase shift between the input signal and the VCO output. There is then zero output voltage from the phase detector, and, clearly, the frequency of the VCO is identical to the input signal frequency. In practice, a finite voltage is required to drive the VCO, resulting in a phase error at the input to the phase detector, which can be made arbitrarily small by increasing the loop gain of the system.

When the system is locked (i.e. there is a one-to-one relationship between cycles of the input signal and VCO cycles), the VCO control voltage is proportional to the frequency of the input signal, and will track changes in frequency. Some means must be provided to indicate that the loop is locked. This is trivial with a noise-free signal, but is a statistical problem of some difficulty with a poor signal-to-noise ratio. The usual method is to use a quadrature phase detector to provide synchronous detection of the input signal. When the loop is not locked, the average value of the quadrature phase detector output will be zero. A lock/unlock decision can thus be made using the amplitude of the phase detector output.

The dynamic performance of the loop, and the lock detector signal, will depend on the level of the input signal. As the quadrature phase detector output gives a measure of the signal level, this can be used to control an automatic gain control at the loop input to maintain the signal level constant.

The mean and variance of the frequency can be calculated using the exponentially mapped past (EMP) technique of Otterman (1960), which gives a statistical analysis which is weighted in favour of the recent behaviour of the system. The EMP average frequency is given by:

$$\overline{f_\tau(0)} = \frac{1}{\tau}\int_{-\infty}^{0} f(t)e^{t/\tau}\,dt, \tag{1}$$

where τ is the weighting time constant, and the EMP variance on the frequency is given by:

$$\sigma_\tau^2(0) = \frac{1}{\tau}\int_{-\infty}^{0} [f(t) - \overline{f_\tau(t)}]^2\, e^{t/\tau}\,dt, \tag{2}$$

where $f(t)$ is the instantaneous frequency and $\overline{f_\tau(t)}$ is the average frequency. As $[f(t) - \overline{f_\tau(t)}]$ clearly has zero mean, and the variance is the mean square value of a zero mean process, the EMP standard deviation can be found by taking the RMS value of the difference between the mean and instantaneous frequencies. The technical details of the system are described in Smallwood (1978c, f). Although EMP statistics have been used in a number of applications, the problem of estimating statistical significance does not appear to have been tackled. Smallwood (1978b) therefore proposed a method by which a number of degrees of freedom could be assigned depending on the signal frequency and the averaging time constant. Values of EMP mean and variance which are separated by more than 3τ were taken as independent (as the contribution of the first measurement to the second would be $<5\%$ after three time constants), and Student's t test was used to calculate the significance.

Tests on tape-recorded cutaneous gastric signals showed that the system could track changes in frequency, and measurements were also made of the static and dynamic performance of the loop. Nelsen suggested, in a personal communication, that the replacement of the square wave VCO and switching phase detector by a sine wave VCO and multiplier would effect a 6 dB improvement in performance. The design of a suitable quadrature sine wave VCO is described in Smallwood (1978d). Smallwood has also suggested (1978b) that the loop filter bandwidth could be reduced to improve the performance. These improvements were incorporated and showed the expected improvement in performance when bench tested (as in Smallwood, 1978a). It appeared that a further limitation on the performance at high noise levels was clipping of the input signal when the loop was locked. The AGC circuit sets the noise-free signal to a constant level, and the noise therefore causes overloading with high noise levels. The dynamic range of the circuit as tested could probably be improved by 20 dB, which should improve the performance still further.

The great advantage of the phase-lock system is that it can track signals and provide statistical information in real time. It can only track a single frequency component, but this is not a serious disadvantage with the almost sinusoidal cutaneous gastric signals.

6.4 Intensity Modulated Raster Scanning

A further analogue technique (developed by Brown) which is extremely sensitive to small changes in frequency, is the intensity modulated raster scan, which is similar to methods which have been used to demonstrate changes in circadian rhythm. The spot on an oscilloscope screen is driven slowly in the X direction by the time base, and a sawtooth waveform is applied in the Y direction. The time base is adjusted so that the sawtooth ramps almost merge. The signal of interest is used to modulate the intensity of the beam. If the repetition rate of the signal and the sawtooth frequency are identical, the beam will be increased in intensity at the same position on each sawtooth, and a horizontal line will result. If the signal frequency is increased slightly, the line will move steadily downwards, and vice versa. In practice, several sawtooths stacked one above the other are used to achieve continuity when the trace reaches the top or bottom of the sawtooth (Fig. 10). The system is easy to set up, and produces remarkably clear traces showing the cycle-to-cycle variations in the frequency, even in the presence of a considerable amount of noise.

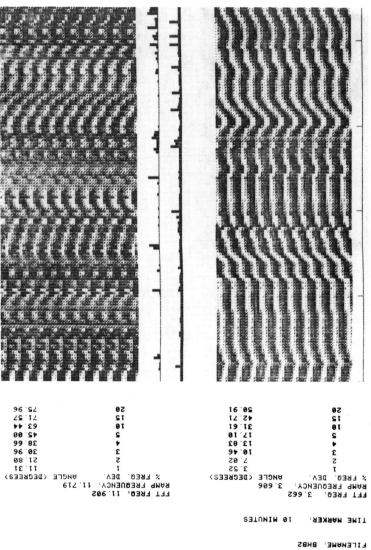

% FREQ. DEV.	ANGLE (DEGREES)
20	50.91
15	42.71
10	31.61
5	17.10
4	13.83
3	10.46
2	7.02
1	3.52

RAMP FREQUENCY, 3.606
FFT FREQ. 3.662

TIME MARKER, 10 MINUTES

FILENAME, BH82

% FREQ. DEV.	ANGLE (DEGREES)
20	75.96
15	71.57
10	63.44
5	45.00
4	38.66
3	30.96
2	21.80
1	11.31

RAMP FREQUENCY, 11.719
FFT FREQ, 11.902

Fig. 10 Intensity modulated raster display showing duodenal (above) and anal (below) frequency changes in slow wave electric activity. An increase in frequency corresponds to a decrease in gradient of the trace, with the calibration as shown on the left. A dip in gastric frequency of approximately 15% occurs 30 min from the start of the record. The two centre traces show the occurrence of duodenal and gastric spikes. (Diagram by courtesy of Dr A. J. Wilson, Department of Medical Physics and Clinical Engineering, University of Sheffield.)

6.5 Autoregressive Modelling

The fast Fourier transform will give a frequency spectrum, but requires many cycles of data for reasonable resolution, and estimation of the significance of the peaks is difficult. The phase-lock system can only track a single frequency component, and also requires several cycles to give a lock indication. The intensity modulated raster scan can also only track a single frequency, but can follow rapid changes in frequency. Autoregressive modelling (Datardina and Linkens, 1977; Linkens, 1978; Smallwood et al., 1980b) in effect does a least-squares fit of a large set of sine waves to the data, and is thus able to give an accurate estimate of frequency over a few cycles of data, and an indication of the significance of the result.

The basis of the method is that any signal can be represented as the output of a special filter (the autoregressive model) whose input is white noise (i.e. a signal whose spectrum has equal amplitudes at all frequencies). The flat frequency spectrum of the white noise is multiplied by the frequency response of the filter to give the spectrum of the signal, i.e. the coefficients of the filter frequency response must have the same relative amplitudes as the coefficients of the signal frequency spectrum. The modelled signal (the output of the filter) and the actual signal are tested for identity by subtracting one from the other, and then testing the residuals (the difference signal) to ensure that they are white noise. This is done by calculating the autocorrelation function of the residuals.

Two parameters must be selected: the number of data points (i.e. the number of cycles of signal) and the model order (i.e. the number of coefficients in the autoregressive filter). Both the number of points and the model order need to be increased for signals with more harmonics or more noise. The choice can be made either empirically or by applying optimization techniques (Kwok et al., 1980). Smallwood et al. (1980b) used a model order of 20 with colonic data digitized at 1 Hz (i.e. 20 samples per cycle at 3 cycles min^{-1}). The data contained frequencies of up to 12 cycles min^{-1}, and good results were obtained using 64 data points. It is often thought that the resolution of the FFT of a short length of data can be improved by "padding" the signal with zeros, e.g. a 64 point signal could be increased to 256 points by adding 192 zero amplitude points. Unfortunately, we cannot get something for nothing! The line spacing of the spectrum will be reduced, but will only serve to delineate the $\sin x/x$ shape of the transformed window, which has not been altered in length. The usable resolution is still determined by the number of cycles of the signal. An alternative explanation is that the amount of information in the input has not been increased, so that more information cannot be obtained from the output.

The form of the autoregressive filter can be expressed as a polynomial in z (the delay shift operator), which is factorized to give a set of roots from which the spectral components are calculated. The complex roots have a magnitude of approximately unity and a phase angle of $2\pi f$, where f is the spectral frequency. The actual magnitude of the spectral component is related to its significance. Trials on a typical set of gut data showed that significant rhythms have magnitudes lying between 0·95 and 1·05. The nature of the solution is such that a list of frequencies with their significance can be produced. With suitable artefact detection techniques, this would give a fully automated analysis system that had as output a list of significant frequencies—this is not possible using FFTs. The autoregressive modelling algorithm requires more computing power than the FFT algorithm, but this is no longer a serious limitation.

7.0 STATE OF THE ART

It is now well established that, in the narrow context of the electro-gastrogram, surface electrical signals arise from the slow wave electrical activity of the stomach. This electrical activity does not correspond to gastric motility as had been assumed to be the case by many earlier workers in this field. However, there is ample evidence that when motility is present the timing of this is controlled by the slow wave electrical activity.

After a meal the fast activity on the duodenal slow waves is correlated either with the slow waves in the antrum (Allen et al., 1964; Carlson et al., 1966)\ or with the fast activity in the antrum (Stoddard et al., 1973). This co-ordination is the best example of the function of electrical activity. Elsewhere it is as difficult to relate electrical activity to movement as it is to relate motility to transport of intestinal contents.

Ingestion of food interrupts the myo-electric complexes, which appear in the fasting state, and within a few minutes bursts of co-ordinated fast activity on gastric and duodenal slow waves are stimulated. This fast activity lasts for as long as three hours and the return to the fasting state may take eight hours or more. The frequency of the slow waves temporarily decreases after a meal and then returns to normal (Duthie et al., 1971; Nelsen et al., 1968; Smallwood et al., 1975).

Recent work has attempted to relate either extractable parameters (such as frequency and amplitude) in the electro-gastrogram to motility or to develop alternative methods of recording motility.

7.1 Recording of Migrating Myo-electric Complexes

Szurszewski in 1969 first described the presence of myo-electric complexes in the fasting subject; these complexes consist of periods of motility where

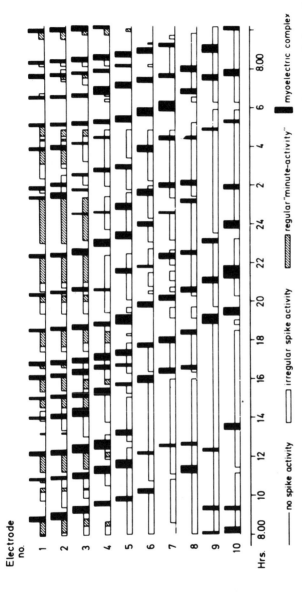

FIG. 11 Schematic representation of fasting myo-electrical activity in the entire human small intestine recorded continuously for 26 h. Bipolar electrodes were placed at intervals of 10 cm down a 5·6 mm diam. PVC tube. (From Fleckenstein, P., Krogh, F. and Øigaard, A. in Duthie (1978), pp. 19–28.)

every cycle of the slow wave is associated with fast activity and mechanical activity. It has been shown (see Duthie, 1978) that these complexes recur at intervals of about one hour and can be recorded at all levels along the stomach and small intestine. Conduction can be shown down the small intestine (see Fig. 11) and hence the term *migrating* myo-electric complex (MMC) has been adopted.

In view of the well-described time course of an MMC and its associated motility it is attractive to attempt to identify complexes from surface electrical activity. Two methods have recently shown transient frequency disturbances associated with duodenal myo-electric complexes. The first uses the intensity modulated raster display described in Section 6.4, the second uses the autoregressive modelling technique. Both methods enable small transient frequency changes, which would not be visible in Fourier transforms of data blocks, to be seen.

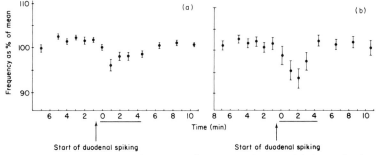

FIG. 12   (a) Mean duodenal slow wave frequency changes (± 1 standard error) for 19 human duodenal myo-electric complexes. (b) Mean gastric slow wave frequency changes for 20 human duodenal myo-electric complexes. In both cases the frequency changes have been related to the start of duodenal spiking.

If, in the intensity modulated raster display, the vertical ramp and fly-back display is running at the same frequency as a signal which is used to intensify and modulate the display then the points of maximum intensity will trace a straight line. If the modulating signal does not have a constant frequency then the line traced will have a variable gradient such that a positive gradient corresponds to a lower frequency than the ramp and a negative gradient a higher frequency (see Fig. 10). This type of raster display has been used to show transient frequency dips in gastric and duodenal electrical slow waves associated with duodenal MMCs (Fig. 12). These frequency changes have been shown in simultaneous serosal and surface electrical recordings and demonstrate the possibility of identifying MMCs from surface recordings (see Fig. 13).

The autoregressive modelling technique described above has also shown transient frequency changes in surface electrical recordings during the

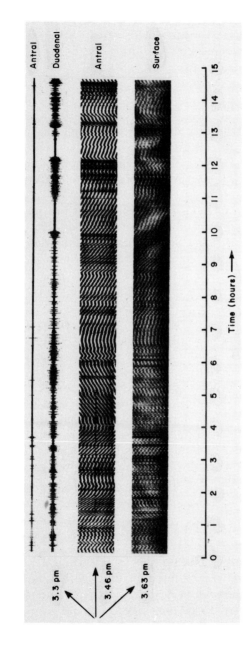

Fig. 13   Intensity modulated raster display showing that transient changes in antral slow wave frequency can be observed in the simultaneously recorded surface electrode activity. The changes between 5 and 11 min and the myo-electric complex at 13 min show clear parallel changes. This is not the case during the first 3 min where there appear to be different frequency changes. The top two traces show antral and duodenal spike activity.

presence of MMCs. However, data block lengths of 64 s do not enable shorter frequency transients to be seen.

Both the above methods have demonstrated gastric frequency dips of the order of 10% lasting for several minutes during the final phase of the myo-electric complex.

7.2 Measurement of Motility

The intensity modulated raster scan display described above enables the relatively large frequency transients associated with MMCs to be observed using surface recordings. It also shows the presence of continuous frequency perturbations. These perturbations may be associated with motility but this has not been established.

Currently two alternative methods of recording intestinal motility are being investigated. The first, described by Akkermans *et al.* (1980) uses the same technique of a radioactively labelled meal as has previously been used to make measurements of gastric emptying time (see Sheiner, 1975). A gamma camera is used to follow movements of the stomach wall by following the count rate variations in two or more "areas of interest" defined on the gamma camera image. This technique has been shown to give a useful recording of stomach wall movement and, in some cases, a measurement of the velocity at which mechanical contractions are conducted. However, the method is only suitable for recording a period of about 30 min and is an invasive procedure.

The second method currently being attempted is the use of a real-time ultrasound scanner to follow intestinal wall movements. This method has been shown to be feasible but no comprehensive study has yet been published. The technique holds considerable promise as a non-invasive technique but currently has the disadvantages of high cost, difficulty in separating organ depth changes from intestinal wall movements and the requirement for a full time skilled operator.

8.0 FUTURE DEVELOPMENTS

There have been many suggestions for the routine clinical use of the electro-enterogram, but with the exception of the reported usage in the USSR no significant penetration into medical practice has been reported elsewhere. Amongst the many suggestions have been the possible application of electrical stimulation via surface electrodes to overcome post-operative inhibition of intestinal electrical activity, which can be recorded via surface electrodes. Gastric emptying studies have shown that duodenal ulceration is associated with changes in the rate and pattern of emptying of solid meals. Identifiable patterns in the electro-gastrogram following a meal might have diagnostic application. There is some evidence of correlations of electrical activity and pathology in the large intestine. In the colon diverticular disease has been shown to change the frequency content of the slow wave electrical activity and there is some evidence that this might be recorded from surface electrodes.

A major obstacle to progress remains the inability to relate non-invasive recordings to intestinal motility. The best hope may be the use of direct and yet non-invasive methods of obtaining motility and in this context real-time ultrasound imaging is probably the most promising technique. The electro-gastrogram has certainly been shown to allow recording of gastric slow wave activity and there is a reasonable hope that further methods of analysis will allow inferential information on motility to be obtained. The following section makes brief mention of these techniques.

8.1 Methods of Analysis

Fast Fourier transforms, autoregressive modelling, phase-lock loop systems and raster displays have already been mentioned and examples given of their application to the electro-gastrogram. There are certainly transient frequency changes associated with feeding, myo-electric complexes and during mechanically quiescent periods. These changes can be seen from surface recordings and there is the need for a considerable amount of work to quantify and identify their origins. Many of these changes occur over only a few cycles of the slow wave activity and methods to be applied must be able to identify changes as small as 5% in frequency over periods of 1–2 min.

The surface electro-gastrogram is thought to be the result of signals which reinforce each other because the whole of the stomach has the same slow wave frequency. It may well be that transient frequency changes disturb the phase-locking of all areas of the stomach, in which case amplitude changes would be expected in association with frequency transients. There has been very little work in this area which might allow motility to be inferred from the frequency and amplitude changes of the electro-gastrogram. Simultaneous recordings from several electrode pairs might enable evidence of activity migration to be obtained.

Whilst this area of physiological investigation remains one with little clinical application it also remains as an area of poor physiological understanding and hence one of great interest. The hormonal, neuronal and myogenic control of intestinal smooth muscle almost undoubtedly has its origins at the cellular level and in particular the role of intercellular coupling of electrical activity is poorly understood. Whilst much of the physiological investigation will require work at the micro-electrode level there is still considerable scope for non-invasive *in vivo* investigation to elucidate the methods of control of intestinal motility.

References

Akkermans, L. M. A., Jacobs, F., Hong-Yoe, O., Roelofs, J.-M. M. and Wittebol, P. (1980). *In* "Gastrointestinal Motility" (Ed. J. Christensen), pp. 195–202, Raven, New York.
Allen, G. L., Poole, E. W. and Code, C. F. (1964). *Am. J. Physiol.* **207**, 906–910.
Alvarez, W. C. (1922). *J. Am. Med. Assoc.* **78**, 1116–1118.
Berkson, J. (1933). *Am. J. Physiol.* **104**, 62–66.
Berkson, J., Baldes, E. J. and Alvarez, W. C. (1932). *Am. J. Physiol.* **102**, 683–692.
Blackman, R. B. and Tukey, J. W. (1959). "The Measurement of Power Spectra", Dover, New York.
Brown, B. H. (1976). *In* "IEE Medical Electronics Monographs" (Eds D. W. Hill and B. W. Watson), Vol. 4, pp. 1–26, Peter Peregrinus, London.
Brown, B. H., Smallwood, R. H., Duthie, H. L. and Stoddard, C. J. (1975). *Med. Biol. Engng* **13**, 97–103.
Bulbring, E. and Shuba, M. F. (Eds) (1976). "Physiology of Smooth Muscle", Raven, New York.
Bulbring, E., Brading, A., Jones, A. and Tomita, T. (Eds) (1970). "Smooth Muscle", Edward Arnold, London.
Carlson, H. C., Code. C. F. and Nelson, R. A. (1966). *Am. J. Dig. Dis.* **11**, 155–172.
Christensen, J. (Ed.) (1980). "Gastrointestinal Motility", Raven, New York.
Christensen, J., Schedl, H. P. and Clifton, J. A. (1964). *J. Clin. Invest.* **43**, 1659–1667.
Christensen, J., Schedl, H. P. and Clifton, J. A. (1966a). *Gastroenterology* **50**, 309–315.
Christensen, J., Clifton, J. A. and Schedl, H. P. (1966b). *Gastroenterology* **51**, 200–206.
Colcher, H., Goodman, E. N. and Katz, G. M. (1959). *Am. J. Gastroent.* **31**, 408–418.
Combe, P., Thillier, J.-L., Fauchier, C. L., Martin, A. and Regy, M. (1972a). *Paediatrie* **27**, 483–495.
Combe, J., Thillier, J.-L., Martin, A., Fauchier, C. and Regy, J. M. (1972b). *Lyon Medical* **228**, 78.
Condrea, H., Manoach, M., Gitter, S. and Feller, N. (1967). *Arch. Surg.* **94**, 112–116.
Daniel, E. E. (Ed.) (1974). *Proceedings of the Fourth International Symposium on Gastrointestinal Motility*, Mitchell, Vancouver.
Daniel, E. E. and Chapman, K. M. (1963). *Am. J. Dig. Dis.* **8**, 54–102.
Datardina, S. P. and Linkens, D. A. (1977). *In* "Random Signal Analysis", IEE Conference Publication 159, IEE, London.
Davis, R. C. and Berry, F. (1963). *Psychol. Rep.* **12**, 135–137.
Davis, R. C., Garafolo, L. and Gault, F. P. (1957). *J. Comp. Physiol. Psychol.* **50**, 519–523.

Davis, R. C., Garafolo, L. and Kveim, K. (1959). *J. Comp. Physiol. Psychol.* **52**, 466–475.

Drieux, C., Garnier, D., Martin, A. and Moline, J. (1978). *J. Physiol. Paris* **74**, 703–707.

Duthie, H. L. (1974). *Gut* **15**, 669–681.

Duthie, H. L. (Ed.) (1978). "Gastrointestinal Motility in Health and Disease", MTP, Lancaster.

Duthie, H. L., Kwong, N. K., Brown, B. H. and Whittaker, G. E. (1971). *Gut* **12**, 250–256.

Engel, B. T. and McFall, R. A. (1959). *J. Appl. Physiol.* **14**, 1069–1070.

Fedor, J. H. and Russell, R. W. (1965). *Psychol. Rep.* **16**, 95–113.

Fukushima, K., Okada, I. and Nakasono, Y. (1951). *Med. J. Osaka Univ.* **2**, 83–95.

Furman, V. N. and Kazakov, V. N. (1970). British Library Translation RTS 7606 from Klinicheskaya Khirurgiya (1970) (6) (342) 16–21.

Gardner, F. M. (1979). "Phaselock Techniques", John Wiley and Sons, New York.

Geddes, L. A. (1972). "Electrodes and the Measurement of Bioelectric Events", Wiley Interscience, New York.

Golenhofen, K. (1970). *In* "Smooth Muscle" (Eds E. Bulbring, A. Brading, A. Jones and T. Tomita), pp. 316–342, Edward Arnold, London.

Goodman, E. N. (1943). *Surg. Gynecol. Obstet.* **75**, 583–592.

Goodman, E. N., Ginsberg, I. A. and Robinson, M. A. (1951). *Science* **113**, 682–683.

Goodman, E. N., Colcher, H., Katz, G. M. and Dangler, C. L. (1955). *Gastroenterology* **29**, 598–608.

Goodman, E. N., Colcher, H. and Schlaeger, R. (1959). *Bull. N.Y. Acad. Med.* **35**, 765–777.

Harris, F. J. (1978). *Proc. IEEE* **66**, 51–83.

Holzl, R. (1979). *In* "Psychosomatics and Biofeedback" (Eds W. H. G. Wolters and G. Sinnema), pp. 42–56, Martinus Nijhoff, The Hague.

Holzl, R., Schroder, G. and Kiefer, H. (1979). *Behav. Anal. Modif.* **3**, 77–97.

Ingram, P. W. and Richards, D. L. (1953). *Gastroenterology* **25**, 273–289.

Katzka, I., Lemon, H. M. and Jackson, F. (1955). *Gastroenterology* **28**, 717–730.

Kerremans, R. (1968). *Acta Gastro-Enterol. Belg.* **31**, 465–482.

Kohatsu, S. (1968). *Jap. J. Smooth Muscle Res.* **4**, 148–150.

Kohatsu, S. (1970). *Klin. Wochenschrift* **48**, 1315–1319.

Krasilnikov, L. G. and Fishzon-Russ, Yu I. (1963). *Ter. Arkh.* **35**, 56–62.

Kwok, H. L., Leung, C. T., Smallwood, R. H. and Linkens, D. A. (1980). *Automedica* **3**, 197–206.

Lilie, D. (1974). *In* "Bioelectric Recording Techniques. Part C: Receptor and Effector Processes" (Eds R. F. Thompson and M. M. Patterson), pp. 297–305, Academic Press, New York and London.

Linkens, D. A. (1978). *In* "Gastrointestinal Motility in Health and Disease" (Ed. H. L. Duthie), pp. 235–248, MTP, Lancaster.

Martin, W. S. and Morton, H. S. (1952). *Arch. Surg.* **65**, 382–397.

Martin, A. and Thillier, J.-L. (1971a). *C. R. Soc. Biol.* **165**, 1704-1710.

Martin, A. and Thillier, J.-L. (1971b). *Presse Med.* **79**, 1235-1237.

Martin, A. and Thillier, J.-L. (1972). *Nouv. Presse Med.* **1**, 453-456.

Martin, M., Thouvenot, J. and Touron, P. (1967). *C. R. Soc. Biol.* **161**, 2595-2600.

Martin, A., Thouvenot, J., Masson, J.-M. and Touron, P. (1970). *Ann. Medico-Psychol.*, Paris 1970 (1) 31–41.

Martin, A., Moline, J. and Murat, J. (1971). *Presse Med.* **79**, 1277–1278.
Martin, A., Thillier, J.-L. and Moline, J. (1972). *J. Physiol, Paris* **65**, 10.
McIntyre, J. A., Deitel, M., Baida, M. and Jalil, S. (1969). *Canad. J. Surg.* **12**, 275–284.
Monges, H., Salducci, J. and Roman, C. (1969). *Arch. Mal. Appar. Dig.* **58**, 517–530.
Morton, H. S. (1954). *Ann. R. Coll. Surg. Eng.* **15**, 351–373.
Morton, H. S. and Martin, W. S. (1953). *Rev. Gastroent.* **20**, 37–52.
Nelsen, T. S. (1967). Digest of the 7th International Conference on Medical and Biological Engineering, Stockholm. p. 337
Nelsen, T. S. and Kohatsu, S. (1968). *Am. J. Surg.* **116**, 215–222.
Nelsen, T. S. and Wallace, T. M. (1965). Digest of the 6th International Conference on Medical Electronics and Biological Engineering, Tokyo, pp. 369–370.
Nelsen, T. S., Eigenbrodt, E. H., Keoshian, L. A., Bunker, C. and Johnson, L. (1967). *Arch. Surg.* **94**, 821–835.
Oi, M., Tanaka, Y and Iwai, Y. (1967). Digest of the 7th International Conference on Medical and Biological Engineering, Stockholm, p. 335.
Oppenheim, A. V. and Schafer, R. W. (1975). "Digital Signal Processing", Prentice Hall, Englewood Cliffs.
Otterman, J. (1960). *IRE Trans.* **AC-5**, 11–17.
Puestow, C. B. (1932a). *Arch. Surg.* **24**, 565–573.
Puestow, C. B. (1932b). *Proc. Soc. Exp. Biol. Med.* **29**, 901–902.
Rabiner, L. R. and Gold, B. (1975). "Theory and Application of Digital Signal Processing", Prentice Hall, Englewood Cliffs.
Randall, R. B. (1977). "Application of B & K equipment to frequency analysis." B & K Naerum, Denmark.
Richter, C. P. (1924). *Am. J. Physiol.* **67**, 612–633.
Russell, R. W. and Stern, R. M. (1967). *In* "A Manual of Psychophysiological Methods" (Ed. P. H. Venables and I. Martin), pp. 219–243, North Holland, Amsterdam.
Schulz, J., Reitzig, P., Schulze, E., Barsch, J., Elzold, H. and Lisewski, G. (1973). *Dtsch Gesundh-Wesen* **28**, 2485–2488.
Schulz, J., Reitzig, P., Koblitz, F., Lisewski, G. and Schulze, E. (1975). *Dtsch Gesundh-Wesen* **30**, 956–958.
Sheiner, H. J. (1975). *Gut* **16**. 235–247.
Smallwood, R. H. (1976). Ph.D. Thesis, University of Sheffield.
Smallwood, R. H. (1978a). *Med. Biol. Engng Comp.* **16**, 507–512.
Smallwood, R. H. (1978b). *Med. Biol. Engng Comp.* **16**, 513–518.
Smallwood, R. H. (1978c). *Electr. Engng* **50**, (601), 27.
Smallwood, R. H. (1978d). *Electr. Engng* **50**, (612), 11.
Smallwood, R. H. (1978e). *Electr. Engng* **50**, (613), 15.
Smallwood, R. H. (1978f). *Analog Dial.* **13**, (2), 17.
Smallwood, R. H., Brown, B. H. and Duthie, H. L. (1975). *In* "Proceedings of the Fifth International Symposium on Gastrointestinal Motility" (Ed. G. Vantrappen), pp. 248–253, Typoff-Press, Herentals.
Smallwood, R. H., Linkens, D. A. and Stoddard, C. J. (1980a). *Clin. Phys. Physiol. Meas.* **1**, 47–58.
Smallwood, R. H., Linkens, D. A., Kwok, H. L. and Stoddard, C. J. (1980b). *Med. Biol. Engng Comp.* **18**, 591–600.
Smout, A. J. P. M. (1980). Thesis, Delft University Press.

Smout, A. J. P. M., van der Schee, E. J. and Grashuis, J. L. (1980a). *Dig. Dis. Sci.* **25**, 179–187.

Smout, A. J. P. M., van der Schee, E. J. and Grashuis, J. L. (1980b). *In* "Gastrointestinal Motility" (Ed. J. Christensen), pp. 187–194, Raven, New York.

Sobakin, M. A. and Privalov, I. A. (1976). *Bull. Exp. Biol. Med.* **81**, 636–638.

Sobakin, M. A., Smirnov, I. P. and Mishin, L. N. (1962). *IRE Trans. Bio-med. Electr.* **9**, 129–132.

Stern, R. M. (1964). *Psychol. Rep.* **14**, 799–802.

Stern, R. M. (1966). *Psychophysiology* **2**, 217–223.

Stevens, J. K. and Worrall, N. (1974). *Physiol. Psychol.* **2**, 175–180.

Stoddard, C. J., Brown, B. H. and Duthie, H. L. (1973). *Br. J. Surg.* **60**, 913.

Stoddard, C. J., Duthie, H. L., Smallwood, R. H. and Linkens, D. A. (1979). *Gut.* **20**, 476–483.

Szurszewski, J. H. (1969). *Am. J. Physiol.* **217**, 1757–1763.

Taylor, I., Smallwood, R. H. and Duthie, H. L. (1974a). *In* "Proceedings of the Fourth International Symposium on Gastrointestinal Motility" (Ed. E. E. Daniel), pp. 109–118, Mitchell, Vancouver.

Taylor, I., Duthie, H. L., Smallwood, R. H., Brown, B. H. and Linkens, D. A. (1974b). *Gut* **15**, 599–607.

Taylor, I., Duthie, H. L., Smallwood, R. H. and Linkens, D. A. (1975). *Gut* **16**, 808–814.

Thillier, J.-L. and Bertrand, J. (1975). *In* "Proceedings of the Fifth International Symposium on Gastrointestinal Motility" (Ed. G. Vantrappen), pp. 263–266, Typoff-Press, Herentals.

Thouvenot, J., Penaud, J., Arhan, P. and Faverdin, C. (1971). *J. Physiol., Paris* 102–103.

Thouvenot, J., Penaud, J. and Morelle, O. (1972). *J. Physiol., Paris* **65**, 170–171.

Thouvenot, J., Tonkovic, S. and Penaud, J. (1973). *Acta Med. Jug.* **27**, 227–247.

Tonkovic, S., Penaud, J., Thouvenot, J. and Mountafian, J.-P. (1975). *Med. Biol. Engng* **13**, 266–271.

Vantrappen, G. (Ed.) (1975). "Proceedings of the Fifth International Symposium on Gastrointestinal Motility", Typoff-Press, Herentals.

Walker, B. B. and Sandman, C. A. (1976). *Psychophysiology* **14**, 81.

Walker, B. B. and Sandman, C. A. (1977). *Psychophysiology* **14**, 393–400.

Walker, B. B., Lawton, C. A. and Sandman, C. A. (1978). *Psychosom. Med.* **40**, 610–619.

Wankling, W. J., Brown, B. H., Collins, C. D. and Duthie, H. L. (1968). *Gut* **9**, 457–460.

Waterfall, W. E., Duthie, H. L. and Brown, B. H. (1973). *Gut* **14**, 689–696.

Wenger, M. A., Engel, B. T. and Clemens, T. L. (1955). *Am. Psychol.* **10**, 452.

Wenger, M. A., Henderson, E. B. and Dinning, J. S. (1957). *Science* **125**, 990–991.

Wenger, M. A., Engel, B. T., Clemens, T. L. and Cullen, T. D. (1961). *Gastroenterology* **41**, 479–485.

Whitehead, W. E. and Drescher, V. M. (1980). *Psychophysiology* **17**, 552–558.

Wilson, A. J., Franks, C. I. and Freeston, I. L. (1982). *Med. Biol. Engng Comp.* **20**, 286–292.

6. LASER DOPPLER MEASUREMENT OF CUTANEOUS BLOOD FLOW

G. Allen Holloway

Center for Bioengineering, University of Washington, Seattle, Washington, USA

1.0 INTRODUCTION

Ever since the introduction of the plethysmographic method in the 1860s there has been great interest in a quantitative assessment of the level of blood perfusion of the skin. Because of this interest, a variety of techniques has been introduced in the hope of gaining a better understanding of the physiology and pathophysiology of skin circulation, both in terms of its nutritional function and in its role in the thermoregulation of the body. There has been interest in these methods by physicians from a number of clinical areas who have a need for this measurement both in the evaluation and in the treatment of patients whose skin is in some way compromised. Despite this interest, methods for evaluating skin circulation have proved less than ideal. They are either non-specific, or invasive, or slow or usable only on certain areas of the skin. One technique, of recent origin, which appears to offer the possibility to overcome most of these limitations is laser Doppler velocimetry, anemometry or spectroscopy as it is variously called. The method is non-invasive, instantaneous, and has the advantage that it can be used on any skin surface. It has been shown that it has definite advantages in the clinical assessment of cutaneous blood flow. It is the purpose of this chapter to describe this technique in some detail and to outline some of the ways in which it has been used to date.

NON-INVASIVE MEASUREMENTS: 2   *Copyright©1983 by Academic Press Inc. (London) Ltd.*
ISBN 0 12 593402 5   *All rights of reproduction in any form reserved*

2.0 BASIC LASER DOPPLER PRINCIPLES

Basic to the laser Doppler technique is the principle of the shift in frequency of waves of any type which are emitted or backscattered from a moving object. This is the Doppler principle described by Christian Doppler, a German physician, in 1842. Most people have experienced this phenomenon when a railroad train comes to a crossing and the pitch of the whistle decreases as it passes. Others have indirectly encountered the principle when stopped for speeding by a police radar system. This is also based on the Doppler shift as are a number of medical instruments which use ultrasound for the measurement of blood flow or other motion.

The principle states that sound which is emitted by an object in motion will be shifted in frequency relative to a stationary observer in proportion to the velocity of that object. The sound waves will be compressed as the object moves toward the observer, and will be closer spaced (i.e. have a shorter wavelength) and, therefore, higher in frequency when they reach the observer. If the object is moving away from the observer, the opposite is true, and the pitch heard by the observer is lower. Therefore, if one knows the frequency of the emitted sound, and the frequency of the sound when it reaches the observer, then the difference between the two will be the Doppler shift frequency. This can be expressed simply mathematically as

$$\Delta f = f_i - f_o, \tag{1}$$

where Δf is the Doppler frequency shift, f_i is the frequency emitted from the moving source, and f_o the observed frequency. If the sound is emitted from the observer and directed at a target moving toward the observer, the frequency is increased both on its way to and on its way from the target, producing a frequency shift twice as great as if the source of the sound was the target itself. This situation occurs in most practical uses of the Doppler principle as only rarely does the moving target act as the source of the sound.

Although the above examples have all used sound or ultrasound as the wave energy which is frequency shifted, the same effect will be produced for all types of energy in wave form. As light is a form of wave energy, it too will undergo a frequency shift when emitted or backscattered from a moving object. This phenomenon is made use of in the determination of velocities of stars in space. It is precisely this phenomenon which is employed in laser Doppler velocimetry systems. Laser light is used to illuminate a moving object or objects, and the light backscattered from these objects is frequency shifted due to their motion. The velocity of the objects can then be determined from the frequency shift that occurs.

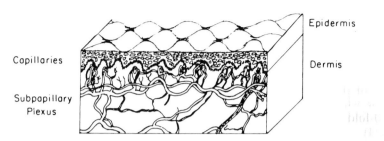

Epidermis

Capillaries

Dermis

Subpapillary
Plexus

FIG. 1   Cross-sectional diagram of the vascular anatomy of the upper dermis.

Although sources of light other than a laser might be considered, laser light is necessary due to both the single frequency or monochromatic light which it emits, and the coherent quality of the light. Since a difference in frequency is determined, it is important that the reference frequency be stable and of a single frequency. Laser light is monochromatic and by definition is a highly stable, single frequency radiation. It is also coherent which means that the light waves are in phase as they impinge upon the targets whose velocity will be measured. This coherent quality of light will be discussed in more detail later.

As the principle of laser Doppler velocimetry as used to measure cutaneous perfusion is dependent upon interfacing the instrument system with the biological system, most notably the skin, an understanding of the anatomy as well as physiological and physical properties of this organ is essential. The basic structure of the skin is pictured in Fig. 1, but it must be understood that this is quite variable in different areas of the body, with age and pathology contributing further variability. The outer layer, the epidermis, is a non-living keratin layer on the surface and progresses to a living cellular layer. This ranges from approximately 50–400 μm in thickness, the thickest areas being over the fingers, toes, palms of the hands and soles of the feet (Ryan, 1973). Beneath this lies the dermis in which all of the arterioles, capillaries and venules of the skin are contained. Arteries arise in the subdermis and penetrate vertically through the mid-dermis giving off arteriolar and capillary branches to the skin appendages on their way. They eventually terminate in the papillary dermis immediately below the epidermis where they form a rich vascular plexus, predominantly venular, from which the capillary loops arise. These loops are generally hairpin in shape, range from 200 to 400 μm in length and average 10 μm diam., each loop supplying about 0·1–0·2 mm² of skin surface area. Also found in the peripheral acral areas of the body including fingers, toes, palms, soles, and nose, are arteriovenous anastamoses (AVAs) which are short muscular vessels which allow shunting of blood past the capillaries and are felt to be most significant in the thermoregulation of the body. Large variations in flow rate can occur, particularly in areas where arteriovenous anastamoses are present, and may represent a 200-fold difference between the lowest to the highest flow rates (Rowell, 1974).

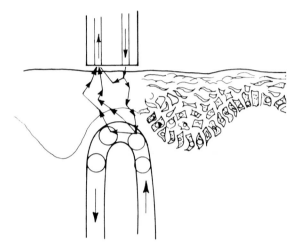

FIG. 2 Schematic representation of light interaction and scattering in the epidermis and dermis. Transmitted light is scattered both from non-moving structures in the epidermis and dermis, and from moving red blood cells and vessel walls in the vascular structures in the upper dermis.

The use of laser Doppler velocimetry to measure flow in biological systems utilizes the backscattering of laser light from moving red blood cells. However, when looking at blood flow in the skin as well as in other superficial capillary beds, the light must first penetrate through the non-vascular epidermis or other superficial layers. Light is incident upon the individual red blood cells at many different angles due both to the scattering as it penetrates the upper non-vascular layers, and the complex geometry of the capillaries in the dermis (Fig. 2). This is significant in that the Doppler equation includes a term for the angle which the incident energy makes with the moving targets, the red blood cells. Obviously this angle cannot be known because of the variable orientation of the capillaries with respect to the skin surface and the multiple scattering from the epidermis superimposed. The angle used, therefore, is a mean value for these various angles. As a statistical value the angle appears to remain the same between measurements both in a given individual and in different subjects. Thus, although the laser Doppler system can be calibrated absolutely in an *in vitro* set-up, the complex geometry of the skin precludes this in the *in vivo* situation (Stern, 1975). In fact, at present there is no system by which absolute measurements of cutaneous blood flow or blood flow velocity can be made, due mainly to the complexity of the skin microcirculation.

To determine the frequency of the light backscattered from the red blood cells, direct measurements of the frequency shifts cannot be made electronically, as with the ultrasonic Doppler, due to the extremely high frequencies of light; approximately 10^7 times higher than ultrasonic frequencies. Instead, the light which is scattered from the non-moving skin surface, and is the non-Doppler shifted reference signal, is mixed on the surface of a photodetector with the Doppler shifted light backscattered from the moving red blood cells. The mixing of these two components produces optical beating, or heterodyning, at a frequency which is the Doppler shifted frequency. As there are obviously many different frequency shifts in the backscattered light due to the different velocities of the many red blood cells, there are also many different frequency components to the Doppler shifted signal. The output of the photodetector is, therefore, a signal which is composed of a spectrum of frequencies, and the process of spectral broadening is said to have occurred.

As a spectrum cannot be simply characterized, the demonstration of the meaning of an individual flow spectrum or the comparison of several spectra becomes difficult to do in other than the non-real time research setting, particularly in the routine clinical situation where one would like to quantitate the flow state of an area of a patient's skin. For this reason, it has been suggested by Stern (1976), that the root mean square (RMS) bandwidth of the power spectrum of the Doppler shifted frequencies, a single number derived from a particular weighting of the spectrum, be used to represent flow. This value can be determined by analogue (or digital) electronic circuitry, and serves as a useful means of characterizing flow. It will be discussed in greater detail in Section 4.0. The values obtained with this algorithm have been compared with radiotracer clearance techniques in skin (Stern and Lappe, 1976; Holloway and Watkins, 1977) and microsphere deposition in kidney (Stern *et al.*, 1979) and have shown good correlation.

3.0 REVIEW OF THE LITERATURE

In 1958, Gould, and at almost the same time, Schawlow and Townes (1958) conceived and developed a source of coherent light of a single frequency, an optical maser, or as it came to be called, laser. Several years later Cummins *et al.* (1964) predicted that the rate of motion of macromolecules in solution could be detected by examining laser light backscattered from the solution and determining by what amount the frequency of the incident laser light had been shifted. The difference between the incident and backscattered frequencies was determined by mixing the two beams and observing the interference pattern, a technique used in radio called heterodyning. This was to be the cornerstone of present laser Doppler techniques. Later that year, Yeh and Cummins (1964) expanded their observations and described the use of this technique to measure fluid flow at velocities as low as 0.007 cm s^{-1} in a 10 cm diam. tube. They were able to measure flow velocity at several positions across the radius of the tube, and show that the observed flow profile corresponded almost exactly with the theoretically predicted profile. Progress continued, and an extensive review was written by Cummins and Swinney (1970) describing both theoretical and practical aspects of the method.

The first use of the laser Doppler method in a biological system was in 1972 by Riva et al. (1972). They first examined flow of 0.56 μm polystyrene beads and blood through 200 μm diameter capillary tubes and found that the results agreed with a measured pump flow rate. Subsequently, the system was used to measure blood flow in the retinal artery of a rabbit by aiming the incident laser light through the eye from the outside. No previously published data were available with which to compare their results, but values compared favourably with measured flow velocities in arteries of similar size in a dog. In their system, the reference (unshifted light) signal was light scattered from the glass wall of the capillary tube and was mixed with Doppler shifted light scattered from the moving red blood cells in the tube. Detection of the light was by a photomultiplier tube with a 1 mm aperture to limit the angular spread of the accepted light to 0.02 radians. In order to arrive at a value for flow from the spectrum of velocities contained in the photomultiplier output signal, the power spectrum of the signal was taken. In the case of the polystyrene beads, peak flow was determined from the sharp upper frequency cutoff of the power spectrum. However, using blood, the determination was more complex as the sharp cutoff was absent, presumably due to the multiple scattering of the light from red blood cells, but did result in values which appeared to agree with measured values for flow.

Tanaka et al. (1974) then applied an improved system to retinal arteries and veins which decreased retinal irradiance and shortened recording time in humans. Using the autocorrelation function flow rates arrived at were of the right order of magnitude although specific values for retinal vessel flow were, as mentioned, not available for comparison. More specific details of the system were later presented by Feke and Riva (1978).

This group continued to pioneer development in the area and next reported (Tanaka and Benedek, 1975) on the measurement of blood flow within the femoral vein of a rabbit using a fibre-optic catheter. They demonstrated that the laser light could be brought to and from the desired observation site through a small plastic or glass fibre with an overall diameter of 0·5 mm or less. Also presented was a more detailed theoretical consideration of the detection and signal analysis processes. They (Riva et al., 1979a) next presented a system for measuring absolute bi-directional blood velocity in the larger retinal vessels. Although this was theoretically an interesting approach, from a practical viewpoint they admitted the technique was both cumbersome to use and limiting since only steady state velocities could be reliably measured. It was felt, however, that it could be improved upon using a dual rather than single light collecting system. In a related paper (Riva et al., 1979b), an additional use of a laser Doppler system was described. The spectrum of laser light scattered from the cornea was used to measure the distribution of eye speeds in human subjects whose eyes were following moving targets, and correlated quite well with target speed save for the fact that their eyes moved intermittently versus the steady motion of the target.

Another application of the basic heterodyne laser Doppler system was presented by Einav et al. (1975a, b) whereby he and co-workers examined velocity profiles of red blood cells in small arterioles. To accomplish this, the laser light was brought to the microvessels through a special microscope system. By using a $60 \times$ water immersion lens the output was focused to a spot 22 μm in diameter, and could be positioned at different levels within the vessel to be examined. In this case the hamster cheek pouch was the experimental model used, and could be accurately positioned with a micrometer transport system. Collecting optics allowed mixing of the reference beam scattered forward at $0°$ with the shifted beam scattered at $4°$ on a photomultiplier tube. The output was analysed using both spectral analysis and correlation methods. Measurements were made at approximately 11 μm intervals across the vessel. A parabolic type flow profile was seen in the smaller (65 μm) arterioles which became progressively flatter as the vessel became larger (up to 95 μm), and correlated well with values obtained from a rotating prism particle velocity meter.

Le-Cong and Zweifach (1979) subsequently used a more sophisticated yet similar type of system to measure flow in individual rabbit mesentery microvessels ranging from 8 to 109 μm. They were able not only to measure accurately flow velocity and velocity profiles but also vessel dimensions, which allowed them to calculate volume flow. The velocity information was obtained here by processing the Doppler signal with a novel automatic frequency tracker, and velocity measurements agreed closely with those obtained from high-speed cinematographic recordings.

Other biological uses of laser Doppler velocimetry include those of Lee and Verdugo (1976, 1977) where they used this approach to study ciliary activity of cells both in the rabbit oviduct and in tissue culture, and Lee *et al.* (1977) where they evaluated the molecular structure of bovine cervical mucus by means of the molecular motion. Nossal and Chen (1972) had previously demonstrated the use of this method to measure bacterial motion.

The first use of laser Doppler velocimetry to assess flow and its physiology in the intact cutaneous microvascular system was in a short paper by Stern (1975). In his system, a 15 mW helium–neon laser was used to illuminate the surface of the skin of a fingertip. Backscattered light was detected by a photomultiplier tube with a pin-hole aperture, and the output passed through a variable bandwidth filter and then processed by a spectrum analyser. A clear difference could be seen between spectra obtained under normal resting conditions and those obtained when a blood pressure cuff was inflated on the arm, as could differences between before and after a vasodilator (ethanol) was administered. As the spectrum analyser could not be used in real-time, the signal power over a narrow band of frequencies was followed continuously and the blood pressure cuff occlusion studies repeated. The tracings from this showed cardiac pulsations as well as the expected decrease in output when arterial inflow was occluded. This clearly demonstrated that this methodology could be used to evaluate cutaneous blood flow continuously and non-invasively. However, Stern pointed out that the complexity of the multiple scattering of the light during its diffusion through the tissue

> precludes a quantitative theoretical calculation of the Doppler spectrum, which must instead be calibrated empirically against conventional physiological measures using the predictions of idealized theoretical models as a guide.

Stern *et al.* (1977) then elaborated on the initial report and presented both additional theoretical considerations and further experimental results. It was suggested that as the bandwidth of the Doppler spectrum should scale in proportion to red cell velocities, providing the geometry of the flow system did not change, a flow parameter could be calculated on an empirical basis, and the (RMS) bandwidth of the Doppler signal was used. Although it was stated that other weighted estimates of bandwidth could equally well be used, this one was chosen as it could be calculated in real-time using a relatively simple analogue circuit. Furthermore, this parameter could be easily normalized by division by the mean photomultiplier photocurrent to render the value independent of the laser power and tissue reflectivity. Comparison was made between flow values obtained in volunteer subjects using the [133]xenon radio-isotope clearance and laser Doppler methods under conditions of both normal resting flow and elevated flow secondary to ultraviolet irradiation, and this showed good correlation. The authors additionally examined flow in the renal cortex of the rat, and demonstrated that infusion of the vasoconstrictor nor-epinephrine, caused a decrease in flow which was dose related, and that infusion of a vasodilator caused an increase in flow, but to a lesser degree. Several arguments were also presented which suggested that the penetration depth of the laser Doppler method was in the neighbourhood of 1 mm, but would presumably vary with varying degrees of tissue pigmentation.

A subsequent study (Stern et al., 1979) also examined renal blood flow, both in the cortex and medullary ampulla, and compared flow measured by laser Doppler with that measured both by an electromagnetic flow probe placed on the renal artery, and by radioactive microsphere deposition under several different flow states. The laser Doppler flow parameter always changed directionally with the other two measures, but there was a tendency for it to decrease less than proportionally at low flows.

Although the system designed by Stern (1975; Stern et al., 1977, 1979) was functionally adequate, it was not easily utilized outside of the research laboratory and resulted in the fabrication by Watkins and Holloway (Holloway and Watkins, 1977; Watkins and Holloway, 1978) of a system which could be more easily used in the clinical setting. A photodiode rather than a photomultiplier tube and a fibre-optic system to carry the light to and from the surface to be examined were incorporated. Comparison with the [133]xenon clearance method yielded a correlation similar to that obtained by Stern et al. (1977).

Practical use of this system was then demonstrated by Holloway (1980) who used it to study the effect on local microvascular blood flow when inserting a 30-gauge needle into the skin with or without injection of fluid and/or vasoactive substances. An almost ten-fold increase in local flow was demonstrated with needle insertion alone. This instrument has subsequently been used by this group to evaluate microcirculatory flow simultaneously with the skin surface measurement of PO_2 in order to examine the flow dependence of the latter system (Piraino et al., 1979; Zick et al., 1980). Powers and Frayer (1978), in a very limited experiment, used a similar optical system and looked at flow in a glass tube, in a rat femoral vein and in a human finger. They examined the arithmetic difference between flow and no flow frequency spectra and showed that this difference varied depending upon the degree of flow present. Williams et al. (1977) also used Stern's system to map cortical strokes in Rhesus monkeys, and the same system has been used subsequently to examine flow in striated muscle and gingival oral mucosa (Bonner *et al.*, 1979; De Rijk *et al.*, 1980).

A slightly different RMS bandwidth processing algorithm has been proposed and used by Nilsson *et al.*, (1978, 1979, 1980a, b; Oberg, 1979) along with an interesting differential optical system to decrease noise, particularly laser noise. Using several receiving fibres surrounding the larger transmit fibre, light from half of the fibres is received by one photo-diode, and from the other half by a second. These are input into a differential amplifier where laser noise common to both fibres is common mode rejected whereas the Doppler signal, being a random scattering of light from red blood cells and tissues and not common to both detectors, is not. Subsequent processing originally examined a much narrower bandwidth (4–7 kHz) than the Stern (1977) algorithm (200 Hz to 20 kHz) but their later system utilized a larger bandwidth (100 Hz to 15 kHz). However, the signal-to-noise ratio and hence the useful signal level would appear to be similar in both. Nilsson *et al.* (1980a) have compared their system to a thermal clearance flow measurement system and have found a regression coefficient of 0·81. Oberg *et al.* (1979) have used it to evaluate skeletal muscle blood flow after experimental bullet wounding where comparison to both the [133]xenon radio-isotope clearance and microsphere deposition techniques have shown a good relationship in the small number of studies (five) performed.

4.0 DISCUSSION OF THE PRINCIPLES AND THE METHOD

4.1 Mathematical Principles

The full derivation of the mathematics involved in the analysis of the signal obtained from the photodetector is extensive, and well beyond the scope of this chapter. For the interested reader, however, this is discussed in some detail in the following references: Stern (1975), Nilsson *et al.* (1980b), Bonner *et al.* (1978), Cummins and Swinney (1970). The theory of how the photodetector output signal may be processed is pertinent, and will be presented and discussed here.

As has been stated, the Doppler shifted components of the backscattered light interfere both with the non-shifted light and with each other. This results in a spectrum of beat notes, which are for the most part within the auditory or audio range (although with high velocities, higher frequencies may be present), which is related to the velocity distribution of red blood cells flowing in the microcirculatory bed. Also as was stated earlier in reference to Eqn (1), the Doppler angle, θ, cannot be specifically defined due to the unknown capillary geometry and multiple scattering, and thus, the exact shape of this spectrum cannot be determined. However, Stern *et al.* (1977) have pointed out that the bandwidth of the spectrum of Doppler shifts, assuming a constant flow geometry, should scale in proportion to the red cell velocities, and also that the amplitude of the Doppler spectrum would be expected to increase with the number of red cells present in the sample volume although not necessarily linearly due to the effect of multiple scattering. On this basis they proposed a flow parameter, F which was the RMS bandwidth of the power spectrum of the Doppler signal:

$$F = \sqrt{\int_0^\infty \omega^2 P(\omega) d\omega},\qquad(2)$$

where ω is the angular frequency ($\omega = 2\pi f$, and f the frequency in Hz), $P(\omega)$ is the power spectrum of the Doppler signal. As was then stated, "In reality, changes in tissue perfusion and red cell content are generally accompanied by some change in the vascular geometry, red cell flow is not laminar, and red cells have a spinning motion as they flow. Therefore, the belief that F is proportional to flow must be based on empirical evidence." From the practical viewpoint, it is important that the flow parameter, F, in Eqn (2) can be arrived at in a more simple fashion not requiring a spectrum analyser. This can be done by fairly simple electronic means as outlined by Stern *et al.* (1977). The photocurrent output by the photo detector, i, is composed of two components, where

$$i = i_s + i_n\qquad(3)$$

where i_s is the Doppler shifted component and i_n is the photodetector shot noise component. If this signal is then passed through a bandpass filter with bandwidth just wider than the Doppler bandwidth, the noise is limited to reasonable levels, and the spectral power density $S(\omega)$ of this signal is simply the sum of those of the Doppler signal and the noise

$$S(\omega) = P(\omega) + N(\omega),\qquad(4)$$

where $N(\omega)$ is the noise spectrum when using a photodiode as a photodetector. By differentiating this signal, the power spectrum is effectively multiplied by ω^2, and by then taking the RMS value, the parameter R is obtained

$$R = \sqrt{\int_{B_L}^{B_H} \omega^2 S(\omega) d\omega} = \sqrt{\int_{B_L}^{B_H} \omega^2 P(\omega) d\omega + \int_{B_L}^{B_H} \omega^2 N(\omega) d\omega},\qquad(5)$$

where B_H and B_L are the high and low cut-off frequencies. The first term under the second square root sign is the flow parameter, defined earlier, squared, and the second term the noise spectrum processed similarly. This can be restated as

$$R = \sqrt{F^2 + a}, \tag{6}$$

where a is a constant

$$a = \sqrt{\int_{B_L}^{B_H} \omega^2 N(\omega) d\omega}, \tag{7}$$

and rearranged to give

$$F = \sqrt{R^2 - a}. \tag{8}$$

In practice, the noise term, a, can be eliminated by placing the probe on a non-moving surface with approximately the same optical scattering characteristics as skin. As no component due to flow, R, is present, the flow parameter, F, is then equal to the square root of a. If F then is electronically offset to be equal to zero, the noise term is removed:

$$F - (\text{offset}) = \sqrt{R^2 - a} = O; \quad R = O, (\text{offset}) = a. \tag{9}$$

Other measures of bandwidth or bandwidth weighting could have been, and have been, used. Bonner et al. (1978) suggested using an ω weighting resulting in Eqn (10) rather than the ω^2 weighting considered above in Eqn (2):

$$F = \sqrt{\int_{0}^{\infty} \omega P(\omega) d\omega}. \tag{10}$$

This was because of the sensitivity to the high frequency portion of the power spectrum with the ω^2 weighting which may correspond to multiple scattering within larger vessels. They showed that this parameter also scaled with flow both in vivo and with model systems, and that capillary flow information resided predominantly with low frequencies around 100 Hz which would be more significantly emphasised using the ω weighting.

It is also important to note that the value of the flow parameter is a function not only of the frequencies contained in the photodetector output, but also of the amplitude of the backscattered light at each frequency. If more light is backscattered, the flow parameter will be increased even though the velocites being measured remain constant. A decreased flow value will be obtained if the light level is decreased as for example from heavily pigmented skin where greater absorption occurs. It has, therefore, become quite standard to normalize the flow parameter to unit intensity by dividing it by the mean photo current, I,

$$F = \frac{\sqrt{R^2 - a}}{I}. \tag{11}$$

4.2 Physical Principles

4.2.1 Light–Tissue/Red Blood Cell Interaction

The interaction of light with tissue is a complex process and data concerning it, especially at the cellular level, are sparse. Light is transmitted to and through the epidermis, but with a considerable degree of scattering and absorption by the dermal hyaline and cellular components. A significant portion of the light is scattered back to the surface from these non-moving structures, and is not Doppler frequency shifted. It can, therefore, serve as a reference beam. The light which is not absorbed or backscattered by the non-moving structures penetrates to interact with the red blood cells in the capillaries and also most likely in the arteriovenous anastomoses and subpapillary plexus as well. Multiple scattering from more than a single red blood cell would seem to take place in a relatively small percentage of interactions. Nilsson et al. (1980b) have estimated that the whole-blood volume fraction in the skin capillary layer is only about 0·5% and, therefore, the red blood cell volume fraction in this tissue, assuming an haematocrit of 40%, would be only 0·2% of the total tissue volume. This may well be higher in the subpapillary plexus, but nevertheless, the red blood cells appear to be only a relatively small part of the scattering cross-sectional area of the cutaneous tissue.

Several investigators have examined the depth of penetration of light in human skin in vitro (Bachem and Reed, 1931; Jacquez and Kuppenheim, 1954; Jacquez et al. 1955a, b) and have noted an effective penetration depth of red light to be in the order of 0·6 mm. Stern et al. (1979) have studied this in the rat kidney with an estimated penetration depth of 1·0–1·5 mm. Nilsson et al. (1980b) have examined the problem in a non-biological model system and have also arrived at an estimate of effective penetration depth of 0·6 mm. From a practical viewpoint, no increase in flow is seen, for example, as one scans across the forearm skin and the visible subcutaneous veins where higher flow velocities would be expected if the light penetrated that far. These veins are generally more than 2 mm below the skin surface (Rushmer et al., 1966). It would thus appear that most, if not all, of the Doppler shifted signal is derived from flow in the cutaneous capillaries, subpapillary plexus and arteriovenous anastomoses.

4.2.2 *Optical Detection*

When using ultrasound the Doppler shift produced by moving red blood cells can be determined directly by electronically measuring the difference between the reference frequency and the frequency of the detected Doppler shifted signal. However, in the optical Doppler technique, the frequency of the light waves is too high to be measured directly and an indirect technique must be utilized. The method used takes advantage of the physical principle that when two waves of differing frequencies mix, the frequency of the resulting wave is the difference between the frequencies of the two mixed waves, or in this case the detected Doppler shift. In practice, the shifted and non-shifted waves mix on the surface of a photodetector as seen in Fig. 3. There they interfere both constructively and destructively producing an optical interference pattern which varies at the difference frequency between the two mixed waves. The photodetector then translates these interference frequencies into fluctuations in its output current which therefore also contains the Doppler shift information.

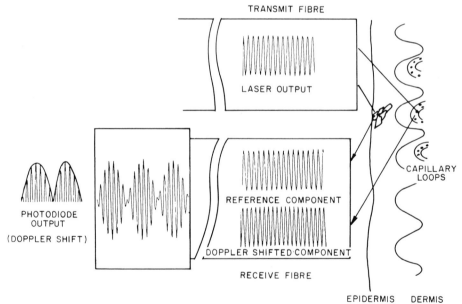

FIG. 3 Schematic diagram of the laser Doppler optical heterodyne process. Unshifted laser light backscattered from non-moving skin components serves as a reference and is mixed with Doppler shifted light from moving red blood cells. This mixing product impinges on the surface of the photodetector which converts the optical amplitude fluctuations into an electric current which in turn fluctuates at the Doppler shifted frequency.

However, in practice there are several differences from the ideal theoretical situation. In the Doppler equation as it applies to ultrasonic and laser Doppler systems, there is an angular term, θ, which represents the angle the incident wave energy, light or ultrasound, makes with the direction of motion of the red blood cells or other particles the velocity of which is being measured. The equation states that the Doppler shift frequency(ies), f

$$f = \frac{2vf_0 \cos\theta}{c} \qquad (12)$$

where v is the velocity of the moving target, f_0 the reference or transmitted frequency, c the speed of sound (ultrasound system) or light (laser Doppler system) in tissue, and θ = angle between the incident sound or light beam and the direction of red cell motion.

If θ is known, the true value of the Doppler shift can be calculated. In skin, because of the complex geometry of the capillary loops and sub-papillary plexus, this angle varies between 0° and 180° as the light enters from the surface and is scattered through many different angles. Thus the measured Doppler shift is from some mean or average angle, the specific value of which is unknown. As we shall see later, this is not a significant problem in the system as it now exists.

In the preceding, we have assumed that the photodetector output signal is formed by the mixing of the Doppler shifted backscattered light from the red blood cells with the light backscattered from non-moving tissues which represents the reference beam. This is considered to be what is known as the heterodyne type of interaction, and has generally been felt to be the predominant mechanism in the formation of the photodetector signal. However, it can be argued that a significant portion of the signal is composed of the interference of different frequency Doppler shifted signals interacting with each other rather than with the reference beam. This is referred to as the homodyne type of interaction. Nilsson *et al.* (1980b) have commented on this possibility but have assumed the hetero-dyne type of interaction where the power of the Doppler signal would be expected to be linearly related to the intensity of light scattered from the red blood cells, rather than a quadratic relationship which would hold for the homodyne model. *In vitro* studies by this group supported the fact that most of the power spectrum was derived from heterodyne interaction.

A variant differential method of optical detection has been used by Nilsson *et al.* (1980a, b). This variation consists of having from four to ten smaller receiving fibres arranged in a circle around the central trans-mitting fibre. One-half of the receiving fibres in one semi-circle are then coupled to one photodetector and the remainder in the other semicircle to a second photodetector. The output from each of these two photodetectors is then input to a differential amplifier where signals which are common to both photodetectors are rejected whereas those which are not common are passed. This allows common mode noise which is produced by the laser and seen in the backscattered non-shifted reference light to be rejected. On the other hand, red blood cell motion, as seen by each spatially separate receiving fibre, is a statistically independent realization of the same random process and produces a different signal in each photodetector. These signals, therefore, do not cancel out and are amplified and transmitted as the flow signal to the subsequent signal analysis circuits.

4.2.3 Signal Processing

In using laser Doppler velocimetry to measure blood flow in biological systems as implemented in current instruments, the light transmission and optical detection systems have all been essentially the same. This would include the system of Nilsson et al. (1980a) which, although using the differential optical technique for detection, still relies on the heterodyning of shifted and non-shifted light. The main difference between the systems has been in the methods of processing the signal from the optical detector.

Because this signal contains a broad band of frequencies ranging from near DC to in excess of 20–25 kHz, it was natural to employ frequency analysis techniques. Most of the original work from the group at MIT (Tanaka et al., 1974; Tanaka and Benedek, 1975) was carried out using standard frequency analysis to look at the range of frequencies contained in the signal. Although this method was and continues to be useful, it is difficult for an observer to actually compare the output frequency spectra and to appreciate either a flow level or change in flow. Additionally, frequency analysers are expensive and have generally not been practical for a clinically usable instrument. With new integrated circuit technology, however, this type of device will be available at lower cost and more suited to routine clinical use.

On the other hand, as Tanaka and Benedek (1975) have stated, the frequency spectra obtained from flowing blood are not direct representatives of the velocities present, as the light is incident at various angles as well as being multiply scattered from red blood cells so that the angle, θ, in the Doppler equation (Eqn 12) cannot be determined. The spectra, therefore, cannot be used to determine the mean flow value directly. This would indicate that the calculated mean velocity of blood flowing in the skin, as compared to measurements in controlled model systems using polystyrene spheres would possibly not be the same, and that the value for volumetric flow obtained when multiplying this velocity by the cross-sectional area of the vessel would therefore be different.

This same group has also used the autocorrelation function obtained from the photodetector output to examine flow. However, as the auto-correlation function is merely the inverse Fourier transform of the power spectrum, the considerations which hold for examination of the frequency spectrum hold for this as well. In order to obtain a single number as a representation of flow, they also measured the half-width of the autocor-relation function. Its reciprocal was then plotted against the mean flow velocity obtained in small glass tubes and a moderately good correlation was found between the two. Apparently this technique has not been used by other groups in investigations of biological systems.

The third and most commonly used signal processing method has been the use of weighted estimates of the bandwidth of the electrical power within the spectrum being examined. This was first suggested by Stern (1975) who proposed that the Doppler flow parameter be the unnormalized RMS bandwidth of the detected Doppler signal, the mathematical deriva-tion of which has been described in Section 4.1. As shown, this entails weighting the power spectrum by the square of the frequency. This derived parameter, although appearing dimensionally correct, requires such assumptions as the vascular geometry remaining unchanged during flow measurement, laminar red blood cell flow and red cell motion which does not include rotation. These all can be shown not to hold under normal conditions of capillary flow and to accept this flow parameter as being proportional to flow at present must be based on empirical evidence, not on theoretical arguments. From a practical point of view, however, this flow parameter or variants of it have been shown to correlate quite closely with other more traditional methods of measuring micro-vascular flow such as radio-isotope clearance, microsphere deposition and volume plethysmographic methods. Generally, the difference between the several frequency weighted bandwidth algorithms has been the method of frequency weighting. Stern (1975), and our group (Piraino *et al.*, 1979) have weighted the spectrum by the square of the frequency, and have shown good correlation with other methods. However, Nilsson *et al.* (1980b) as well as unpublished data from others, have pointed out that weighting by the non-squared frequency is more sensitive at low flows. As there is no appropriate model system against which to compare these two different algorithms, the question of whether either is linearly related to flow, particularly at low flow states, is at present unanswerable. The *in vitro* model developed by Nilsson *et al.* (1980b), however, did show that the product of red cell volume and velocity showed a linear correlation with their unsquared frequency weighted signal. Moreover, as low flow states are frequently encountered in the clinical situation, it may not be unreasonable to use the algorithm which appears to give better sensitivity at low flows until the true accuracy of the method can be better defined.

5.0 SYSTEM DESIGN

5.1 Light Source

The criteria that have been used in the selection of a light source first include the need for monochromatic light which is spatially coherent at least over small areas. Power output must be adequate to allow penetration through areas with varying degrees of pigmentation, and if the instrument is to be portable, size is an additional consideration. The helium–neon laser has been universally used in these systems primarily because of its availability and cost. It provides monochromatic coherent red light at a wavelength of 632·8 nm which theoretically gives the greatest back-scattering from red blood cells. Practical experiments, however, would indicate that this latter consideration is of minimal importance as lasers of other wavelengths which have been tested have given similar results. The power used has ranged from approximately 2 to 15 mW in different systems and is determined primarily by the amount of backscattered Doppler shifted light which can be collected and brought to the photo-detector surface. In systems where relatively small optic fibres have been used, power from 5 to 15 mW has been necessary due to the small collecting area of the fibres. In the system described by Nilsson *et al.* (1980a) they have been able to use a 2 mW laser due to their large optical fibre size. A 5 mW helium–neon laser is currently the largest size which can be incorporated into a portable system. New solid state power supplies have also been developed which decrease the size and weight. Additionally, the helium–neon laser is still the least expensive, the most available and most widely used of any laser.

5.2 Fibre-optic System

Light leaving the laser must be coupled into one of the optical fibres of the probe. The laser output beam is approximately 1 mm in diameter and must be focused onto the fibre which has an effective diameter of 62 μm. This has been accomplished by using both a lens and a spatial translation system. The laser output is focused by means of a microscope objective lens onto the polished or cleaved tip of the fibre which is held within the spatial translation system. The latter is necessary as the alignment of the small fibre optic aperture with respect to the focal point of the lens is highly critical. Small deviations in either the vertical or transverse axis or from the laser beam axis will cause significant loss of transmitted power. It is generally expected that power transmission will be in excess of 60% of the laser output if the alignment is correct.

Coupling of the receiving fibre to the photodetector is less critical in terms of alignment. In this case, the fibre is merely brought into contact with the front glass surface of the photodiode. With the optical fibre connectors which are now available and which will be discussed subsequently, centering of the fibre with respect to the photodetector is easily accomplished.

The optical fibres which have been used with the laser Doppler systems have generally consisted of two types. In our system, as has been indicated above, we have used small fibres with a core diameter of 62 μm. The most effective fibres in our experience have been of quartz glass in which the core has a graded optical index. A quartz glass cladding surrounds the core and in turn has a 13 μm lacquer coat to provide the excellent flexibility and moderate strength against breakage. We have usually found it necessary to add external jacketing to each pair of fibres to protect them further from stress and breakage. These fibres give excellent optical transmission with a loss of $<0.5\%$ m^{-1} of length (6 dB km^{-1}), resulting in essentially no loss of light within the fibres themselves over the short lengths of several metres which are used in the probes.

Nilsson *et al.* (1980a) have used larger fibres ranging from 500 to 1000 μm in core diameter to increase the light coupled into and backscattered from the skin. They have not been concerned with the problem of laser mode competition as their differential optical processing technique has compensated for this. It has been found, however, that the larger fibres are much less flexible and are less advantageous in terms of probe design. At the present time, a wide variety of optical fibres in terms of size, strength, coating, transmission, and cost are manufactured and are easily available.

Of importance in the practical design of laser Doppler systems are mechanisms to hold the fibres in correct alignment relative to both the spatial translation system in the laser output and the photodiode. Over the past several years a significant advance has taken place in optical fibre connectors and a number of different types are now commercially available. These are not only capable of placing and holding the optical fibre in the correct alignment, but can also be connected and disconnected rapidly and repeatedly, permitting interchange of probes either in case of breakage or the need to use different probes for different measurements. Although these were expensive initially, their cost is falling quite rapidly and the concept of truly disposable probes is certainly not far off.

Although the transmission loss is very low within the fibres, significant loss can and does occur where light is either coupled into or out of the fibres. To achieve the lowest possible loss, entrance and exit surfaces are of great importance at the termination of the fibres and should be as flat as possible. All of the good quality new connectors have been designed in relation to a method for producing these optically smooth surfaces. Basically two techniques are available for this. The first is achieved by holding the fibre to be cleaved at a tension and bend radius specific to that fibre type and size and then lightly nicking the fibre with a sharp instrument. This can be done quickly and easily, and can produce an excellent optical surface. Several of the connector manufacturers provide tools to accomplish this.

Another method is to produce a smooth optical surface by polishing. Several manufacturers have utilized such a polishing system for use with their connectors. Very good optical surfaces can be produced with both methods and result in low interface losses.

In the earliest laser Doppler systems detection of the light back-scattered from the skin surface was accomplished by using a photo-multiplier tube. Although it does have some advantages in a photodetection system, there are also significant disadvantages which have made the photodiode preferable as the light detector. The problem of photo-multiplier tube size, the necessity for a high voltage power supply, and its quite high cost have made it impractical to consider its use in systems in which portability is either desired or necessary. The basic requirements for the photodetector method used consist of a detector which has low electrical noise characteristics and which can deliver both adequate gain and bandwidth. Some of the photodiodes available now are combined with an integrated circuit containing the preamplifier which gives improved noise characteristics as well as decreased size. Although there is currently no general agreement as to what the system bandwidth should be, minimum requirements would appear to extend from about 50 Hz to in excess of 15–20 kHz to encompass the frequencies encountered at both low and high flows.

5.3 Signal Processor

The signal processing configuration described here is the configuration which we are currently using. However, as stated previously, variations on this system have been proposed and are being used. The basic system is that outlined in block diagram in Fig. 4. The voltage signal coming from the photodetector and preamplifier is divided. Part of this signal passes through a low pass filter allowing only DC to pass and to serve as the

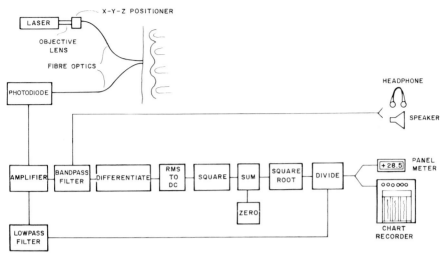

Fɪɢ. 4   Block diagram of the laser Doppler system.

normalizing factor in the normalization section of the circuit. The remainder of the signal is bandpass filtered, differentiated and passed through
an RMS to DC converter. The output of the bandpass filter is also fed to
either a headphone or speaker which permits listening to the Doppler
shifted signal which is within the audio range.

The output of the RMS converter containing both flow signal and noise
is then squared and an offset, representing the noise signal, subtracted
from the square of the flow signal in the summing device. Finding the
noise value is accomplished with the probe placed on a still surface. In
this condition with no motion, there is no Doppler shifted or flow component to the signal, only the noise component which can be removed by
subtracting an offset from the signal until the meter reads zero. The
square root of this output which represents only the flow signal is then
taken and led to a dividing circuit which normalizes the output by the
total intensity of the backscattered light obtained from the first part of
the circuit. This then permits comparison of perfusion levels obtained
on skin surfaces with different backscattering characteristics. The value
may then be displayed on a panel meter and/or strip chart recorder.

The fact that other investigators have used slightly different configurations of the system (Nilsson et al., 1980a) was mentioned previously. The
basic difference has been in the frequency weighting utilized. Whereas
the differentiator circuit weights by the square of the frequency, the most
common alternative has been weighting by the non-squared frequency
which can be accomplished by a resistor–capacitor lattice network or
other relatively simple circuits.

6.0 PRACTICAL APPLICATION OF THE METHOD

6.1 Factors Affecting the Probe Sample Volume

As discussed above, the volume of tissue sampled by the laser Doppler system is approximately 1 mm³. Stern *et al.* (1977) identified an effective penetration depth of 1–1·5 mm when using rat renal cortex whereas Nilsson *et al.* (1980b) identified the depth of maximum signal as being 0·6 mm in a synthetic model for skin using a polymer, Delrin, as a scatterer. These two values obtained by different methods would appear to be quite similar. Two factors which can affect the probe sample volume in living skin are light intensity and its distribution and pigmentation. Although neither the depth of penetration nor the percentage of light which is backscattered to the collecting fibre is determined by the intensity of the light impinging on the skin surface, when incident light levels become low, the photodetector has progressively less light returned to its surface until its output can no longer be distinguished from noise. Optimization of the signal-to-noise ratio and, therefore, optimal resolution of low flows are best obtained with higher levels of incident laser light. However, as indicated, practical considerations limit the power of the laser which can be included in this type of instrument.

Important in addition to the level of incident light is the amount of backscattered light collected by the receiving fibre and transmitted to the photodetector. Quite naturally this is improved by the use of larger fibres with larger acceptance angles. In this case, fibre choice is determined by the limiting size of the aperture necessary to avoid the laser mode competition. It is expected that these compromises can be better optimized in the future.

It would also seem likely that the position of the end of the transmitting fibre with respect to the receiving fibre would be of importance. Experience has led to the empirical conclusion that separation of the small fibres used in our system is optimized at approximately 1 mm. Teleologically this would appear to be the condition in which there was lesser coupling of the non-shifted laser light between the two fibres and greater collection of light transmitted deeper and backscattered from moving red blood cells as determined by overlapping areas of the transmitting and receiving fibres (Fig. 5). It is of interest that the signal remains unchanged as one moves the probe away from the skin surface up to a distance of approximately 1 mm at which point the signal begins to degrade. We have also noted that there appears to be relatively little effect of angulating one fibre with respect to the other through an angle of up to 35°.

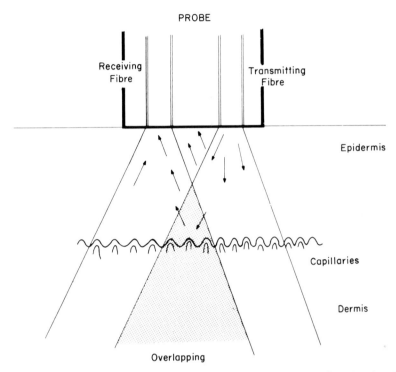

PROBE

Receiving
Fibre

Transmitting
Fibre

Epidermis

Capillaries

Dermis

Overlapping

FIG. 5   Schematic diagram of the sampled volume of blood flow in the skin. Light from the laser and the transmitting optical fibre enters the skin in the form of a cone. Light backscattered from the stippled area is within the receiving area of the receiving fibre and is transmitted back to the photodetector. Light outside the stippled area does not participate in the flow measurement process.

The effective depth of the laser Doppler measurement would also appear to be more related to depth of light penetration in blood containing tissue, namely the dermis, than to the thickness of the avascular epidermis. Flow measurements seem to be unaffected by epidermal thickness and are similar even when comparing values obtained in areas of highly thickened epidermis, such as the soles of the feet, with those of lesser epidermal thickness. It seems likely that the epidermis functions as a scatterer but does not attenuate the incident light to anywhere near the degree that the dermis does.

Pigmentation is the second important factor in determining the amount of incident light which reaches the red blood cells in the upper dermis. Jacquez and co-workers (Jacquez and Kuppenheim, 1954; Jacquez et al., 1955a, b) have shown that reflectance from the forearm of a fair complexioned white subject using a wavelength the same as that of the helium–neon laser, 632·8 nm, was approximately 70% whereas this value in a heavily pigmented black subject fell to approximately 35%. Thus, with the same power of incident laser light, the heavily pigmented black subject will have a significant decrease in signal-to-noise ratio as compared to the lesser pigmented white individual although, except in very low flow states, the flow values should be the same due to the normalization. From a practical viewpoint we have been able to measure flow in subjects with any type of pigmentation although the mean backscattered light level which becomes the denominator in the normalization process may be so low in the most heavily pigmented individuals as to cause electronic errors. There are still reservations, however, as to whether identical flow levels would produce the same flow parameter values in both lightly and heavily pigmented skin, and because of technical difficulties this concern has not been fully tested.

6.2 Calibration

Laser Doppler velocimetry is a technique which can be calibrated against a flow standard in controlled systems and because of this has been used in industry to measure flow both invasively and non-invasively in various situations. The problem arises, however, as to how to calibrate this system for absolute flow in *in vivo* biological situations. The calibration standards for these laser Doppler systems used for microcirculatory flow measurements are divided into *in vitro* and *in vivo* systems.

Two types of systems have been used for *in vitro* calibration. The first is to calibrate against a true volume flow and to compare the laser Doppler output to measured flow through small translucent tubes. The laser Doppler has been shown to scale with flow under these conditions. The second type of calibration standard involves comparing the laser Doppler output with standards which are moving at known velocities. These have consisted of rotating cylinders or discs or a vertical shaker the displacement of which is modulated by a sine wave input. When the output is compared with the known velocities in these two test set-ups, the laser Doppler has shown linear scaling. Good repeatability has been achieved in comparison with each of these methods.

Calibration against *in vivo* capillary blood flow, however, is a more difficult problem. Although a variety of methods have been used to measure capillary blood flow, none of these can serve as a "gold standard" at present. The laser Doppler measurement has been compared with xenon clearance, microsphere deposition, and photoplethysmography (Stern, 1977; Holloway, and Watkins, 1977; Nilsson *et al.*, 1980a) and has shown good correlation. However, each of these alternative techniques measures a different parameter in a different volume and each is known to have a number of shortcomings. To be noted is the fact that the first two of the above methods are time averaging techniques and will not measure the instantaneous changes seen with a laser Doppler velocimeter.

It is also important to observe that it is not uncommon to see marked local differences in flow over separation distances of 5–20 mm and thus to calibrate the system accurately, the "gold-standard" method must be examining exactly the same area as the laser Doppler.

6.3 Methods of Use

From a clinical and practical viewpoint, the laser Doppler instrument is easily used. The probe is placed on the area of skin where flow is to be determined and values read directly from either the panel meter or the strip chart recorder. It is frequently very handy to listen to the audio output from the speaker, particularly when scanning across an area, as differences in flow are easily appreciated by the change in overall pitch of the output. It is then easy to proceed back over the areas where significant differences exist and obtain numerical values for perfusion at those specific sites.

One must be aware, however, that as the instrument measures motion, any motion of the surface being measured relative to the probe will produce a signal. It is important, therefore, to avoid motion of the probe. We have found the simplest way to do this is to attach the probe lightly to the skin surface with double-sided tape. This is particularly convenient when a specific skin area is being examined before and after a perturbation, and avoids the problem of pressure loading of the skin which can occur when the probe is held by hand on the skin surface. Very small amounts of pressure loading can cause significant decreases in blood flow in the probe area and it is almost impossible to maintain a light, constant pressure with a hand-held probe. It is also easy to just set the probe on the skin surface in areas where use of tape is to be avoided, as in disease states. Although less stability is obtained, this is adequate for rapid, non-continuous flow measurements.

It is further important to maintain the environment under controlled conditions as would be indicated in any measurement of skin blood flow. Changes in either environmental or skin temperature will cause marked changes in perfusion. In fact, measurement of environmental temperature, skin temperature, and core temperature, although worthwhile, probably do not provide adequate sensitivity to evaluate the thermal status of the patient. In this respect, very slight increases in core temperature which may not be distinguishable from normal by our standard measurement techniques, may cause some degree of peripheral vasodilatation whereas a slight decrease will cause a similar decrease in flow. Additionally, changes in the emotional state of the patient are capable of causing transient flow changes and as neutral an emotional state as possible should be maintained. However, the mere thought of, for example, a missed appointment or perhaps a difficult decision, can be shown to be capable of causing significant changes in flow. For this there are obviously no preventative measures except for several repetitions of the measurement.

In making clinical measurements, one can use either the actual numerical output of the system or the relative measurements made by comparing the values obtained from the area of interest with values obtained from a "normal" or reference area. As an absolute non-biological zero flow is obtained prior to making a measurement by zeroing the instrument on a non-moving surface, all measurements are referenced to this value. With this baseline, we found we were able to make serial measurements at any time interval and be able to compare directly the values obtained. Therefore, a higher value obtained during a subsequent measurement is indicative of the fact that an increase in flow has occurred and that a similar value at a subsequent time indicates the same level of perfusion. In our experience, comparison of perfusion values made in the same location in a given individual on a number of occasions and under baseline conditions showed them to be quite constant. Comparison of these values between different individuals also shows similar readings although because of the complex factors involved they do differ to some degree. Because of the ease of making measurements with this system, relative measurements comparing a reference site with a test site or a test site with values measured at an earlier time are easily obtained.

Because of the wide normal variations which are present in cutaneous blood flow particularly in the hands and feet under conditions which we define as "baseline" or "normal", we have found it most useful to combine the laser Doppler velocimetry measurement with a stress or perturbation in order to evaluate the cutaneous flow reserve. The stresses which we have found most useful have been thermal using either local or more general heating or cooling, the production of ischaemia and subsequent hyperaemic response in peripheral areas by means of arterial occlusion, and finally the application of graded external mechanical pressure to reduce flow. It will await the results of future studies to determine which, if any of these, is the most effective in determining flow reserve in specific clinical situations.

As a final note, it should also be mentioned that laser Doppler velocimetry measurements of superficial microvascular blood flow need not be limited to the skin alone. Any tissue with capillaries near the surface will give perfusion measurements which appear to vary as one would expect with stimuli which cause changes in flow. Results in comparison with other techniques have been reported in the kidney. Perfusion has also been measured in the heart, brain, gastro-intestinal tract, oral mucosa and various internal and external surfaces within the genital tract. Although it is necessary to await further studies to document the early results, laser Doppler velocimetry would appear to be an easily used tool which can be applied to measurements of perfusion in these organ systems.

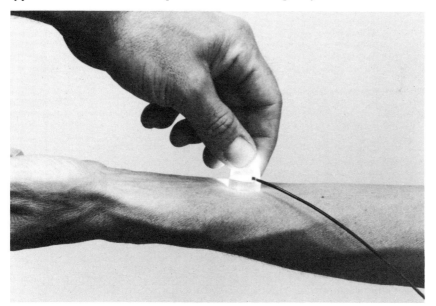

FIG. 6   Laser Doppler probe in position on the forearm of a subject.

6.4 Sample Results of Clinical Use

Of the variety of applications for the laser Doppler system just alluded to, it seems worthwhile to present one as an example to demonstrate the output and use of the system. Figure 6 is a photograph of a probe in position on the forearm of a subject. The optical fibres are encased in the black sheath leading to the instrument. Tracings of the system output from two different anatomical locations are shown in Fig. 7. The subject is a patient with moderately severe lower extremity ischaemic vascular disease. In the upper tracing taken from the dorsum of the foot, a baseline level of flow with superimposed fluctuations of spontaneous vasomotion can be seen. A heater which surrounds the probe was then turned on to 44°C

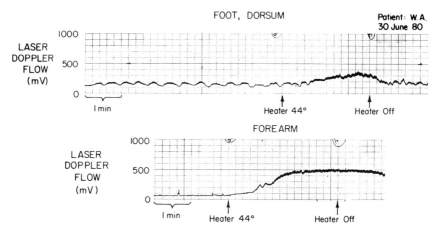

FIG. 7 Skin perfusion in the ischaemic leg and non-ischaemic forearm of a patient as measured by the laser Doppler system. A full description of these data is presented in the text.

and an increase in flow occurred over the next several minutes until after the heater was turned off. In the lower tracing, the same protocol was followed with the probe placed on the forearm. It should be noted that although the baseline resting flow in the forearm was lower, the response to heating was much greater in this location where there was no evidence of vascular disease. Also, despite the fact that the tracing was not continued for an extended period, the increased forearm flow persisted for a longer period of time. It is additionally worthy of comment that the spontaneous vasomotion was not seen in the forearm tracing. In our experience, vasomotion which we can demonstrate in this manner occurs only intermittently and would appear to be associated with conditions of a "neutral" state of the microvasculature; i.e. without thermal, emotional or other stimuli which would tend to cause either vasoconstriction or vasodilatation.

This example thus serves to demonstrate one application of the instrument and some of the changes in perfusion which may be seen. Other aspects, e.g. the difference between peak systolic and peak diastolic flow levels and the change in this value with changes in physiological state, are not presented due to space limitation but are significant and will be investigated and more fully described in future work.

7.0 SUMMARY

Laser Doppler velocimetry is an instrument system which has only recently been applied to the evaluation and quantitation of perfusion in the microvascular bed. The instrument is based on the Doppler principle, but uses low power laser light rather than the more commonly used ultrasound, and has a sample volume of approximately 1 mm³. As it is non-invasive, it can be used on any skin surface or exposed microvascular bed and provides a continuous semi-quantitative measure of microcirculatory perfusion, it has a number of advantages as compared to other cutaneous blood flow measurement techniques.

Initial studies have shown that it is easily used, and it has demonstrated good correlation with both xenon radio-isotope clearance and microsphere deposition techniques. Areas of current evaluation and utilization are in most major areas of medicine and surgery and include plastic, vascular and orthopaedic surgery, dermatology, gastro-enterology, rheumatology, burns and anaesthesiology.

It appears that this technique has a useful place in both physiology and in clinical areas, and should prove to be valuable in the assessment of perfusion in the microvasculature.

Acknowledgements

This work was supported under National Institutes of Health grants HL23711 and HL00800. My thanks to Professor Greg Zick and especially to Daniel Piraino, Ph.D. for their help in the preparation and review of the manuscript.

References

Bachem, H. and Reed, C. I. (1931). The penetration of light through human skin. *Am. J. Physiol.* **97**, 86–91.

Bonner, R. F., Bowen, P., Bowman, R. L. and Nossal, R. (1978). Realtime monitoring of tissue blood flow by laser Doppler velocimetry. *Proc. of the Technical Program, Electro-Optics Laser* pp. 539–550.

Cummins, H. Z. and Swinney, H. L. (1970). Light beating spectroscopy. *In* "Progress in Optics" (Ed. E. Wolf), Vol. VIII, p. 130, North-Holland Publ. Co., Amsterdam.

Cummins, H. Z., Knable, N. and Yeh, Y. (1964). Observation of diffusion broadening of Rayleigh scattered light. *Phys. Rev. Letters* **12**, 150–153.

De Rijk, W. G., Bowen, P. D. and Bonner, R. F. (1980). Preliminary results with laser Doppler velocimetry. *Am. Assoc. Dent. Res. Abs.* 325.

Einav, S., Berman, H. J., Fuhro, R. L., DiGiovanni, P. R., Fridman, J. D. and Fine, S. (1975a). Measurement of blood flow *in vivo* by laser Doppler anemometry through a microscope. *Biorheology* **12**, 203–205.

Einav, S., Berman, H. J., Fuhro, R. L., DiGiovanni, P. R., Fine, S. and Fridman, J. D. (1975b). Measurements of velocity profiles of red blood cells in the microcirculation by laser Doppler anemometry (LDA). *Biorheology* **12**, 207–210.

Feke, G. T. and Riva, C. E. (1978). Laser Doppler measurements of blood velocity in human retinal vessels. *J. Optical Soc. Am.* **68**, 526–531.

Holloway, G. A. (1980). Cutaneous blood flow responses to injection trauma measured by laser Doppler velocimetry. *J. Invest. Derm.* **74**, 1–4.

Holloway, Jr, G. A. and Watkins, D. W. (1977). Laser Doppler measurement of cutaneous blood flow. *J. Invest. Derm.* **69**, 306–309.

Jacquez, J. A. and Kuppenheim, H. F. (1954). Spectral reflectance of the human skin in the region 235–1000 μm. *J. Appl. Physiol.* **7**, 523–528.

Jacquez, J. A., Kuppenheim, H. F., Dimitroff, J. M., McKeehan, W. and Huss, J. (1955a). Spectral reflectance of human skin in the region 235–700 μm. *J. Appl. Physiol.* **8**, 212–214.

Jacquez, J. A., Dimitroff, J. M. and Kuppenheim, H. F. (1955b). Spectral reflectance of the human skin in the region 0·7–2·6 mm. *J. Appl. Physiol.* **8**, 297–299.

Le-Cong, P. and Zweifach, B. W. (1979). *In vivo* and *in vitro* velocity measurements in microvasculature with a laser. *Microvasc. Res.* **17**, 131–141.

Lee, W. I. and Verdugo, P. (1976). Laser light-scattering spectroscopy: A new application in the study of ciliary activity. *Biophys. J.* **16**, 1115–1119.

Lee, W. I. and Verdugo, P. (1977). Ciliary activity by laser light-scattering spectroscopy. *Ann. Biomed. Engng* **5**, 248–259.

Lee, W. I., Verdugo, P., Blandau, R. J. and Gaddum-Rosse, P. (1977). Molecular arrangement of cervical mucus: A reevaluation based on laser light-scattering spectroscopy. *Gynecol. Invest.* **8**, 254–266.

Nilsson, G. E., Tenland, T. and Oberg, P. A. (1978). Continuous measurement of capillary blood flow by light beating spectroscopy. *Proc. Conf. Transducers and Measurements*, Madrid, Spain, Oct. 10–14.

Nilsson, G. E., Tenland, T. and Oberg, P. A. (1979). Laser Doppler flowmetry—A non-invasive method for microvascular studies. *XIII International Conference on Med. and Bio. Engng*, Jerusalem, Israel, Aug. 19–24.

Nilsson, G. E., Tenland, T. and Oberg, P. A. (1980a). A new instrument for continuous measurement of tissue blood flow by light beating spectroscopy. *IEEE Trans. Bio-Med. Engng* BME-27, 12–19.

Nilsson, G. E., Tenland, T. and Oberg, P. A. (1980b). Evaluation of a laser Doppler flowmeter for measurement of tissue blood flow. *IEEE Trans. Bio-Med. Engng* BME-27, 597–604.

Nossal, R. and Chen, S. H. (1972). Light scattering from motile bacteria. *J. Phys. (Paris) Colloq.* C-1 **33**, 171–176.

Oberg, P. A., Nilsson, G. E., Tenland, T., Holmstrom, A. and Lewis, D. H. (1979). Use of a new laser Doppler flowmeter for measurement of capillary blood flow in skeletal muscle after capillary bullet wounding. *Acta Chir. Scand. (Suppl.)* **489**, 145–150.

Piraino, D. W., Zick, G. L. and Holloway, G. A. (1979). An instrumentation system for the simultaneous measurement of transcutaneous Oxygen and skin blood flow. "Frontiers of Engineering in Health Care", IEEE Press Catalog No. 79CH1440-7, pp. 55–58.

Powers, E. W. and Frayer, W. W. (1978). Laser Doppler measurements of the microcirculation. *Plast. Reconstr. Surg.* **61** (2), 250–255.

Riva, C., Ross, B. and Benedek, G. B. (1972). Laser Doppler measurements of blood flow in capillary tubes and retinal arteries. *Invest. Ophth.* **11** (ii), 936–944.

Riva, C. E., Feke, G. T., Eberli, B. and Benary, V. (1979a). Bi-directional LDV system for absolute measurement of blood speed in retinal vessels. *Appl. Optics* **18** (13), 2301–2306.

Riva, C. E., Timberlake, G. T. and Feke, G. T. (1979b). Laser Doppler techniques for measurement of eye movement. *Appl. Optics* **18**, 2486–2490.

Rowell, L. R. (1974). The cutaneous circulation. *In* "Physiology and Biophysics" (Eds T. C. Ruch and H. D. Patton), Vol. II, pp. 185–195, W. B. Saunders Co., Philadelphia.

Rushmer, R. F., Buettner, K. J. K., Short, J. M. and Odland, G. F. (1966). The skin. *Science* **154**, 343–348.

Ryan, T. J. (1973). Structure and shape of blood vessels of skin. *In* "The Physiology and Pathophysiology of the Skin" (Ed. A. Jarrett), Vol. 2, pp. 619–625, Academic Press, London and New York.

Schawlow, A. L. and Townes, C. H. (1958). Infrared and optic lasers. *Phys. Rev.* **112**, 1940.

Stern, M. D. (1975). *In vivo* evaluation of microcirculation by coherent light scattering. *Nature* **524**, 56–58.

Stern, M. D. and Lappe, D. L. (1976). Measurement of local tissue blood flow by laser Doppler spectroscopy. *Fedn. Proc.* **35**, 234.

Stern, M. D., Lappe, D., Bowen, P. D., Chimosky, J. E., Holloway, Jr, G. A., Keiser, H. R. and Bowman, R. L. (1977). Continuous measurement of tissue blood flow by laser Doppler spectroscopy. *Am. J. Physiol.* **232** (4), H441–H448.

Stern, M. D., Bowen, P. D., Parma, R., Osgood, R. W., Bowman, R. L. and Stein, J. H. (1979). Studies of the measurement of renal cortical and medullary blood flow by laser Doppler spectroscopy in the rat. *Am. J. Physiol.* **236** (1), F80–F87.

Tanaka, T. and Benedek, G. B. (1975). Measurement of the velocity of blood flow (*in vivo*) using a fiber optic catheter and optical mixing spectroscopy. *Appl. Optics* **14** (1), 189–196.

Tanaka, T., Riva, C. and Ben-Sira, I. (1974). Blood velocity measurements in human retinal vessels. *Science* **186**, 830–831.

Watkins, D. W. and Holloway, Jr, G. A. (1978). An instrument to measure cutaneous blood flow using the Doppler shift of laser light. *IEEE Trans. Bio-Med. Engng* BME-25, 28–33.

Williams, P. C., Stern, M. D., Bowen, P. D., Brooks, R. A., Hammock, M. K., Bowman, R. L. and Di Chiro, G. (1977). Mapping of cerebral cortical strokes in rhesus monkeys by laser Doppler spectroscopy. *Med. Res. Engng* **13**, 3–5.

Yeh, Y. and Cummins, H. Z. (1964). Localized fluid flow measurements with an He–Ne laser spectrometer. *Appl. Phys. Letters* **4** (10), 176–178.

Zick, G. L., Holloway, G. A., Jr and Piraino, D. W. (1980). Simultaneous measurement of tcPO$_2$ and capillary blood flow. *Proc. Int. Conf. Memb. for Blood Gases, Blood Ion Conc. and Resp. Gas Exch.*

7. NON-INVASIVE SPECTRO-PHOTOMETRIC ESTIMATION OF ARTERIAL OXYGEN SATURATION

I. Yoshiya and Y. Shimada

Department of Anesthesiology, Osaka University Medical School, Japan

Intensive Care Unit, Osaka University Hospital, Japan

1.0 INTRODUCTION WITH HISTORICAL PERSPECTIVE

The non-invasive evaluation of arterial oxygenation has been a most difficult problem in the management of patients with respiratory insufficiency. Current advances in medical bio-electronics have provided us with two ingenious devices which are capable of bed-side estimation of arterial oxygenation quantitatively and non-invasively. One of them is the non-invasive transcutaneous oxygen partial pressure (PO_2) electrode which is described in the preceding volume of this monograph (Huch *et al.*, 1979). The other is the *in vivo* oximeter which estimates the oxygen saturation of arterial blood (arterial SO_2) by the spectrophotometric analysis of the transmitted light through the pinna of the ear or the fingertip. It is non-invasive and gives us quantitative information on the status of arterial oxygenation.

Arterial PO_2 and SO_2, both of which are very good indicators of arterial oxygenation, have different physiological implications. Arterial PO_2 represents the status of lung function, since it is a determinant of alveolar ventilation, diffusion across the alveolar membrane, venous admixture, and alterations in ventilation–perfusion ratios. Arterial SO_2, on the other hand, is a direct measure of how much oxygen per unit of blood is carried to the peripheral organs. PO_2 and SO_2 are related to each other by means of the oxygen dissociation curve for haemoglobin which is not linear but is represented by an S-shaped curve (Fig. 1). As shown in Fig. 1, PO_2 is a more sensitive indicator of changes in arterial oxygenation in the right upper portion of the dissociation curve where the slope is not steep. In contrast, SO_2 is more sensitive in the steeper portion of the dissociation curve, and so is a better indicator of change during impaired arterial oxygenation in severely hypoxaemic patients.

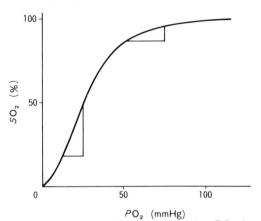

FIG. 1   Oxygen dissociation curve for haemoglobin. PO_2 is a more sensitive indicator of changes in arterial oxygenation in the right upper portion of the curve, whereas SO_2 is more sensitive in the steeper portion of the curve.

In the present chapter, we are to summarize the present status of oximetry, which is defined in terms of the spectrophotometric devices which estimate arterial SO_2 transcutaneously and non-invasively.

Historically, there have been three important steps in the development of non-invasive oximetry. The first step was the determination of the absorption spectra of oxyhaemoglobin (HbO_2) and reduced haemoglobin (Hb) (Horecker, 1943), and the application of the Lambert–Beer law for the absorption of light to solutions of haemoglobin derivatives. Hüfner (1900) described the analysis of two-compartment systems, measuring the extinction of light by haemoglobin derivatives at two different wavelengths, one at an isobestic point where the absorption coefficients for HbO_2 and Hb are equal, and the other at a wavelength at which there is a significant difference in the extinction of light by the individual components, as shown in Fig. 2. The two-wavelength method has been extensively developed for the determination of SO_2 of haemoglobin solutions and haemolysed blood. Many of the commercially available haemoximeters are based on this principle.

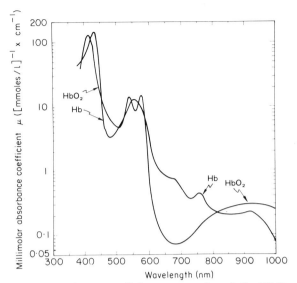

FIG. 2   Millimolar absorbance coefficient vs. wavelength for HbO_2 and Hb. The millimolar absorbance coefficient, μ, is defined by the equation: $D = \mu cl$, where D is the optical density of the haemoglobin solution, c is the concentration of haemoglobin moieties in millimoles per litre, and l is the light path in the solution. (Redrawn from Laing *et al.* (1975), with permission from the author and the publisher.)

The second step was the application of haemoximetry to non-haemolysed, whole blood or red cell suspensions. Kramer *et al.* (1951) found that red blood cell suspensions failed to follow the Lambert–Beer law but showed the non-linear behaviour of light absorption. The non-linearity has been claimed to result from the scattering, reflection and refraction of light by the red blood cells and the suspending medium (saline or plasma). As with haemoglobin solutions, haematocrit and the depth of the cuvette also affect the linearity of the light absorption by the whole blood and red blood cell suspensions.

The third step is ascribed to Wood and Geraci (1949) who developed a new method of obtaining the absolute value for arterial SO_2 *in vivo* by compressing the ear pinna, thus obtaining the baseline absorption of light by the bloodless tissues.

Many important developments of *in vivo* oximeters were made during the first half of the twentieth century by Kramer (1935), Matthes (1935), Millikan (1942), Wood and Geraci (1949), Squire (1940), Zijlstra (1953), Nilsson (1956), and others. An extensive review of the development of oximetry until 1960 is available elsewhere (Nilsson, 1960).

Recently, two unique oximeters which seem able to measure the absolute values for arterial SO_2 have been developed; a multiple-wavelength ear oximeter utilizing computer simulation (Merrick and Hayes, 1976), and a pulse-wave type oximeter which analyses the increase in extinction of light by the inflow of arterial blood to the fingertip (Liappis, 1979; Yoshiya et al., 1980).

In this review, we will mention briefly the basic principles of oximetry, and describe in detail some of the recent advances in non-invasive in vivo oximetry and its applications to the clinical settings.

2.0 BASIC PRINCIPLES

The absorption spectra of HbO_2 and Hb significantly differ from each other, as shown in Fig. 2. The ratio of HbO_2 to total haemoglobin in haemoglobin solutions, haemolysed blood, red blood cell suspensions and whole blood has been successfully estimated by analysing the optical densities of transmitted light (Drabkin, 1950; van Assendelft, 1970). The absorption of trans-illuminated light by haemoglobin solution and haemolysed blood has been known to follow the Lambert–Beer law which is described as:

$$\log_{10}\frac{I}{I_0} = -\alpha cL, \tag{1}$$

where I_0 = intensity of incident light,
I = intensity of transmitted light,
c = concentration of absorbing material,
α = specific absorption coefficient,
L = length of light path.
When the absorption coefficients for HbO_2 and Hb at a particular wavelength and the concentration of total haemoglobin are known, Eqn (1) can be re-written as:

$$\log_{10}\frac{I}{I_0} = -[\alpha_1 c_1 + \alpha_2(c - c_1)]L, \tag{2}$$

where α_1 and α_2 = specific absorption coefficients for HbO_2 and Hb,
c_1 and c = concentrations of HbO_2 and total haemoglobin.
Therefore, SO_2 can be derived from the ratio of transmitted to incident light, total haemoglobin concentration and the specific absorption coefficients for HbO_2 and Hb as:

$$SO_2 = \frac{c_1}{c} = \frac{-1}{(\alpha_1 - \alpha_2)cL}\log\frac{I}{I_0} - \frac{\alpha_2}{\alpha_1 - \alpha_2}. \tag{3}$$

Equation (3) shows that SO_2 is dependent upon the haemoglobin concentration and the length of the light path. The effects of the haemoglobin concentration and the length of light path are successfully cancelled by using the two-wavelength method, preferably one at an isobestic point and the other at which absorption coefficients of HbO_2 and Hb differ significantly. The equation for the two-wavelength method is written as:

$$SO_2 = \frac{\alpha_2 \log(I/I_0)\lambda_1 - \beta_2 \log(I/I_0)\lambda_2}{(\alpha_2 - \alpha_1)\log(I/I_0)\lambda_1 - (\beta_2 - \beta_1)\log(I/I_0)\lambda_2}$$

$$= \frac{\alpha_2 Q - \beta_2}{(\alpha_2 - \alpha_1)Q - (\beta_2 - \beta_1)}, \tag{4}$$

where α_1 and α_2 = absorption coefficients for HbO_2 and Hb at wavelength λ_2,

β_1 and β_2 = absorption coefficients for HbO_2 and Hb at wavelength λ_1,

$Q = \log(I/I_0)\lambda_1/\log(I/I_0)\lambda_2$.

When an isobestic wavelength (λ_2) is employed ($\alpha_1 = \alpha_2 = \alpha$), the equation becomes:

$$SO_2 = \frac{| -\alpha |}{\beta_2 - \beta_1} Q + \frac{\beta_2}{\beta_2 - \beta_1}. \tag{5}$$

Equation (5) indicates that by using two wavelengths, one at an isobestic point, SO_2 can be derived from the absorption coefficients for HbO_2 and Hb at each wavelength and the ratio of the transmittance of light at both wavelengths. SO_2 is independent of the concentration of haemoglobin and the depth of light path provided the absorption coefficients are not influenced by these parameters. The two-wavelength technique has also been applied to reflection oximetry (Kramer *et al.*, 1956; Cohen and Wadsworth, 1972; Laing *et al.*, 1975). A detailed description of the oximeter of this type is available in a later section.

3.0 PROBLEMS ASSOCIATED WITH *IN VIVO* ESTIMATION OF ARTERIAL SO_2

3.1 Multiple Scattering, Reflection and Refraction of Light in Whole Blood

It has been well documented that the Lambert–Beer law is not valid for non-haemolysed red cell suspensions and whole blood because of the non-linear increase in the absorption coefficients of oxygenated and reduced blood with increasing haematocrit (Kramer *et al.*, 1951; Nilsson, 1960; Loewinger *et al.*, 1964). The increase is due to the scattering, reflection and refraction of light by the red cells and plasma. Practically, SO_2 of whole blood can be estimated if adequate compensation for the effects of scattering is made. Twersky (1970) has developed a theory for the transmission of light through a suspension of particles in terms of the absorption and scattering properties of the particles. Twersky's equation can be adopted for the light transmission properties of whole blood in a simple notation including the scattering term B as:

$$\log I/I_0 = \alpha c L + B. \tag{6}$$

With the two-wavelength method, the effect of scattering is only partially compensated for, because the scattering term B has a wavelength dependence. Pittman and Duling (1975a, b) tackled the problem by using three wavelengths, two isobestic (546 and 520 nm) and one non-isobestic (555 nm), cancelling the scattering term B with the assumption that B, at the isobestic points, are equal. The compensation by Pittman and Duling, however, has not been applied to the estimation of *in vivo* arterial SO_2, although they postulated the application of the method using wavelengths in the red and near infrared regions.

With currently available *in vivo* oximeters, another approach of compensation for the scattering effect is undertaken as described later.

3.2 Choice of Wavelengths

As shown in Fig. 2, there are several isobestic points for HbO_2 and Hb absorption spectra. *In vivo* spectrophotometric determination of SO_2 requires suitable wavelengths at which the absorption coefficients for HbO_2 and Hb are great in comparison with that of tissue, yet not so great as to make measurement difficult. Accordingly, wavelengths at the red and infrared regions have been extensively used. At the wavelength of 660 nm, the sensitivity of the saturation measurement is maximum (Mook *et al.*, 1969), being three times greater than at wavelengths near 550 nm, which are frequently used with haemoglobin solutions and whole blood. At 805 nm, in the near infrared region, the absorption coefficients for HbO_2 and Hb are isobestic. A combination of wavelengths at 660 and 805 nm has therefore been used for *in vivo* estimation of arterial SO_2. However, the reflection (and scattering) characteristics of HbO_2 and Hb differ significantly from each other at these wavelengths (Mook *et al.*, 1968; Cheung *et al.*, 1977).

Wavelengths shorter than 600 nm are not suitable for *in vivo* oximetry because the human ear, as well as the fingertip, transmit practically no light at wavelengths in these spectral regions.

3.3 Arterialization

Most ear oximeters need the blood in the ear to be arterialized. The venous blood returning from the ear must be at least 95% saturated in order to get an accurate estimation of arterial SO_2 (Merrick and Hayes, 1976). The arterialization can be accomplished either by the application of vasodilators or by increasing the ear perfusion by heating. A popularly employed method for arterialization is the initial rubbing of the ear followed by the application of radiant heat so as to maintain the temperature of the ear at 41°C. Topical application of vasodilators such as histamine ointment or a rubefacient ointment (0·4% noylic acid vanyllylamide and 2·5% of the α-butoxyethyl ester of nicotinic acid) is also recommended (Schneider *et al.*, 1978). An adequate arterialization of the ear is particularly important in patients with impaired local or systemic perfusion such as arterial hypotension, hypovolaemia, or intense peripheral vasoconstriction.

3.4 Sites of Measurement

The ear pinna, fingertip, interdigit web and, in infants, intermetacarpal space have been chosen for transmission oximetry. The ear pinna, which has been most popularly chosen, is readily arterialized and is almost entirely an inactive organ. Therefore, it is suitable for the measurement of arterial SO_2 during exercise as well as during the resting state.

For the continuous measurement of arterial SO_2, the ear-probe must be secured to the pinna of the ear by means of a head mount. The fingertip is the best site for the pulse-wave type oximeter since the pulse wave is easily detected here.

3.5 Emission, Transmission and Detection of Light

Current advances in optical and electronics technology have provided us with suitable light sources, narrow band filters and sensitive photodetectors. In the past, the light source, filters and the photodetector were chosen on the basis of their selectivity in the red and near infrared spectra and the size and weight suitable for mounting them in the ear-probe. Since glass-fibre optics have been adopted for conveying light from the source to the site of measurement and thence to the photodetector, the

choice for the light source and the detector has become relatively unrestricted.

3.5.1 *Light Source*

Tungsten filament lamps were frequently used in the past since they can
be made in a miniature size. The spectra of output depends upon the
current through the filament. Light emitting diodes (LEDs), which have
been recently developed, can be made very compact and conveniently
have fairly narrow spectral outputs. Currently available *in vivo* oximeters
which utilize glass-fibre optics for the light guides incorporate the light
source in the chassis of the amplifier. Halogen lamps or tungsten–iodine
lamps are used with these instruments.

3.5.2. *Optical Filters*

Narrow band filters which selectively transmit light in the red and near
infrared regions are required to be sensitive in the measurement of optical
density. Light guides, collimating lenses and optical splitters also have
their own spectral characteristics. When a reference light path in addition
to the measuring light path is used to obtain the optical density of the
incident light (I_0), a standardization procedure is required to cancel the
difference in the attenuation of light by the two light paths. Optical
filters are also important in minimizing the possibility of thermal injury
to tissues.

3.5.3 *Optical Transmitter*

Glass-fibre optics have been extensively used in currently available *in vivo*
oximeters such as the multiple-wavelength oximeter (Merrick, 1975) or
the pulse-wave type oximeter (Yoshiya *et al.*, 1980), as well as in the photo-
electric plethysmograph (Challoner and Ramsay, 1974). One of the
advantages to the use of glass-fibre optics is the unrestricted choice of a
light source, a photodetector, narrow band filters and optical beam splitters.
The possibility of electrical hazards can also be minimized. Thermal
injury to the tissue can be minimized because the intensity and the optical
spectra of incident light are readily controlled. One of the disadvantages
in using glass-fibre optics is that the light guides are much heavier than
the ear-probe of the Wood-type oximeter. This is a disadvantage when
using it for small infants. Another disadvantage of glass-fibre optics is
the fragility. Careless handling of the light guides may break some of the
glass fibres and decrease the light path to the photodetector.

3.5.4 *Photo-electric Detector*

Many sensitive and compact photo-electric detectors are now available due to the recent advances in semiconductor technology. There are several types of photo-electric detectors. Photo-emissive tubes, which emit electrons from a surface upon which sufficiently energetic photons fall, have not been used recently because they are delicate and bulky. A very sensitive photomultiplier tube has been used in *in vivo* oximetry as well as in photo-electric plethysmography, but Challoner (1979) states that it is so sensitive that the design current is easily exceeded by the background illumination.

There are two types of semiconductors which are currently being used as photodetectors; a photoconductive type which changes its resistance directly with photon absorption (e.g. selenium barrier-layer photocell, cadmium–sulphur photocell, lead–sulphur photocell), and a photovoltaic type which generates a voltage directly as a result of the absorption of a photon (e.g. silicon photodiode, gallium–arsenic–phosphor photodiode).

Of these photo-electric elements, those which are sufficiently sensitive to the red and near infrared spectra are chosen for the spectrophotometric estimation of arterial SO_2. The linearity of the output of the photo-electric cell is not a significant requirement for accurate measurement, because the non-linearity can be compensated for by means of an adequate linearizer. The temperature coefficient of the photo-electric cell is a problem because each cell has its own temperature coefficient and the sensitivity changes with the temperature of the cell. The temperature of the photo-electric cell must be maintained constant or compensation for the temperature change must be performed. With currently available *in vivo* oximeters, the effect of the temperature coefficient of the photo-electric detector can be successfully cancelled, as described in a later section.

The effect of ambient light entering the photo-electric cell either directly or through the fibre-optic light guides will be such that it increases the background optical density incident to the photo-electric cell leading to considerable errors with most oximeters. To combat this problem, a combination of a photodetector and narrow band filters which have selective sensitivities at the wavelengths to be measured is used. In addition, it is recommended to carry out the test with a dark covering on the test area and probe.

Choice of the light source and photodetector in photo-electric plethysmography are extensively discussed by Challoner (1979) in the preceding volume.

3.6 Determination of the Absolute Value for Arterial SO_2

Since the first application of the ear oximeter in clinical medicine by Matthes (1935), efforts have been directed towards the *in vivo* determination of the absolute value for arterial SO_2 because both the ear and the fingertip are imperfect cuvettes. The transilluminated ear or fingertip contains various tissues other than red blood cells; e.g. skin, muscle, fat, bone, cartilage, blood vessels, plasma, leukocytes and so forth. All of these attenuate the incident light by absorption, multiple scattering, reflection and refraction according to their spectral characteristics. The final goal at which transmittance and reflectance oximeters are aiming is the determination of the ratio of HbO_2 and Hb that are present in the arterial blood. For this purpose, attempts have been made to take out only the amount of light absorbed by the haemoglobin derivatives specifically from the total attenuation of light by the other above mentioned tissues. The first important step in achieving this goal has been ascribed to Wood and Geraci (1949) who compensated for the effect of tissues by subtracting the attenuation of light by the bloodless tissues from the total attenuation of light by the blood-containing ear. Until recently, the Wood-type ear oximeters have been extensively used although they have limited accuracies and reliabilites for use in clinical fields. Recently, two types of *in vivo* oximeters have been developed with considerable success. In the following section, the design and principal characteristics of the Wood-type and the new types of oximeters are described.

4.0 CONSTRUCTION AND OPERATION OF THREE TYPES OF *IN VIVO* TRANSMITTANCE OXIMETERS, AND A REFLECTION OXIMETER

4.1 Wood-type Ear Oximeter

The ear-piece of the original oximeter described by Wood and Geraci (1949) consists of a tungsten filament lamp, two barrier-layer iron–selenium photocells one covered with a green Wratten 87N filter and the other a red Wratten 29F filter, and a rubber diaphragm of pressure capsule which is interposed between the lamp and the pinna of the ear. A combination of the photocell and the filters allows the selective measurement of optical density of the transmitted light in the red and near infrared spectral regions. The outputs of the photocells are recorded by means of two galvanometers which are adjusted to produce equal deflection. After the arterialization of the ear is complete the pressure capsule is inflated to 200 mmHg rendering the pinna of the ear bloodless. The galvanometer

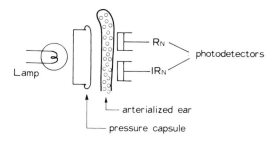

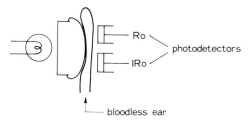

FIG. 3   Schematic representation of the Wood-type ear oximeter. See text for explanations.

deflections for the red and infrared cells (R_0 and IR_0, respectively) are recorded with the bloodless ear. Then, the pressure capsule is deflated and the galvanometer deflection for the blood-containing ear (R_N and IR_N, respectively) are obtained (Fig. 3). The arterial SO_2 is derived from the following equation:

$$SO_2 = A\frac{\log(R_0/R_N)}{\log(IR_0/IR_N)} - B, \tag{7}$$

where A and B are constants related to the absorption coefficients of HbO_2 and Hb at the measurement wavelengths:

$$A = \frac{\beta}{\alpha_2 - \alpha_1} \text{ and } B = \frac{\alpha_2}{\alpha_1 - \alpha_2},$$

where α_1 and α_2 are absorption coefficients for HbO_2 and Hb at the wavelengths of λ_2, β is the absorption coefficient for HbO_2 and Hb at an isobestic wavelength of λ_1.

The improvements attained by the Wood-type oximeter are: (a) estimation of the absolute values for arterial SO_2 can be performed without calibrating the instrument by blood sampling; (b) the measurement can be made with similar accuracy in deeply pigmented ear as well as in less pigmented ear; (c) the accuracy is not affected by the haemoglobin concentration.

Although the Wood-type oximeter was an important step towards the absolute determination of arterial SO_2, it still has a number of considerable limitations which prevent it being widely used as an *in vivo* oximeter, making arterial puncture unnecessary. The problems associated with the Wood-type oximeter are: (a) the ear cannot be made completely bloodless by compression with a rubber balloon; (b) the tissue being compressed changes its volume and composition and accordingly changes its absorbance characteristics; (c) any movement of the ear-probe relative to the pinna requires remeasurement of the bloodless ear because the absorbance characteristics change with the change in the measuring site; (d) arterialization of the ear may not always be complete when there is a peripheral circulatory disturbance. The latter is also a problem with most of the currently available oximeters which require arterialization. Despite the problems listed above, the Wood-type oximeters have been known to give good accuracies as compared to currently developed *in vivo* oximeters, provided that the characteristics of the oximeter are fully known and the measurement is carefully performed. However, the Wood-type oximeter has a serious limitation when used in severely hypoxaemic patients, firstly because it requires frequent standardization calibrations, and secondly because great care must be paid to the constancy of the ear-piece position and exclusion of extraneous light (Saunders *et al.*, 1976).

The Wood-type oximeter has undoubtedly been a milestone in the development of non-invasive spectrophotometric arterial SO_2 estimation. It has offered a considerable amount of information for the development of the more recent novel *in vivo* oximeters which are to be described in the following section.

4.2 Multiple-wavelength Ear Oximeter

A spectrophotometric technique which analyses the optical density of light-absorbing substances using more than two wavelengths has been widely used by analytical chemists. It has also been adopted for transmittance oximetry of whole blood (Pittman and Duling, 1975a, b) as well as for *in vivo* reflectance oximetry (Cheung *et al.* 1977; Lübbers, 1973). Merrick and Hayes (1976) successfully applied the technique to *in vivo* estimation of arterial SO_2 by using eight selective wavelengths between 650 and 1050 nm. The principle, design and operation of a multiple-wavelength ear oximeter are described briefly in the following section.

4.2.1 *Principle*

The multiple-wavelength ear oximeter calculates the absolute value for the arterial SO_2 from the optical densities measured at eight selective wavelengths by using a digital processor on an empirical basis. The blood-containing ear is composed of a number of tissues other than blood which are light absorbers. Assuming that the light absorbers act independently and that no combination of a pair of absorbers has the same absorbance spectrum as HbO_2 or Hb, a set of simultaneous equations can be written for the total absorbance of light by m substances at each of n wavelengths:

$$A_1 = d(K_{11}C_1 + K_{12}C_2 + K_{13}C_3 \; - \; - \; - \; - \; - \; K_{1m}C_m)$$
$$A_2 = d(K_{21}C_1 + K_{22}C_2 + K_{23}C_3 \; - \; - \; - \; - \; - \; K_{2m}C_m$$

$$A_n = d(K_{n1}C_1 + K_{n2}C_2 + K_{n3}C_3 \; - \; - \; - \; - \; - \; K_{nm}C_m), \tag{8}$$

where K_{nm} is the absorption coefficient of an absorber m at the wavelength n, d is the thickness of the light path, and C_m is the concentration of a substance m. Merrick and Hayes (1976) claimed that, when measurements are made of many patients with different skin pigmentation, haemoglobin, and so forth, the analysis at eight wavelengths suffices for the estimation of arterial SO_2 with an acceptable accuracy. By representing the concentrations of HbO_2 and Hb by C_1 and C_2, respectively, the above matrix can be solved for C_1 and C_2 according to Kramer's law:

$$C_1 = \frac{\begin{bmatrix} \dfrac{A_1}{d}K_{12}K_{13}K_{14} & - & - & - & K_{18} \\ \dfrac{A_2}{d}K_{22}K_{23}K_{24} & - & - & - & K_{28} \\ - & - & - & - & - \\ - & - & - & - & - \\ \dfrac{A_8}{d}K_{82}K_{83}K_{84} & - & - & - & K_{88} \end{bmatrix}}{\begin{bmatrix} K_{11}K_{12}K_{13} & - & - & - & K_{18} \\ K_{21}K_{22}K_{23} & - & - & - & K_{28} \\ - & - & - & - & - \\ K_{81}K_{82}K_{83} & - & - & - & K_{88} \end{bmatrix}} = \frac{\sum\limits_{n=1}^{8} AnFn}{\det K} \tag{9}$$

$$C_2 = \cfrac{\begin{bmatrix} K_{11}\cfrac{A_1}{d}K_{13}K_{14} & - & - & - & K_{18} \\ K_{21}\cfrac{A_2}{d}K_{23}K_{24} & - & - & - & K_{28} \\ - & - & - & - & \\ - & - & - & - & \\ K_{81}\cfrac{A_8}{d}K_{83}K_{84} & - & - & - & K_{8b} \end{bmatrix}}{\det K} = \cfrac{\sum\limits_{n=1}^{8} AnGn}{\det K}, \qquad (10)$$

where

$$\det K = \begin{bmatrix} K_{11}K_{12}K_{13} & - & - & - & K_{18} \\ K_{21}K_{22}K_{23} & - & - & - & K_{28} \\ - & - & - & - & \\ K_{81}K_{82}K_{83} & - & - & - & K_{88} \end{bmatrix}$$

Then,

$$SO_2 = 100 \times \frac{C_1}{C_1 + C_2} = 100 \times \frac{\sum\limits_{n=1}^{8} AnFn}{\sum\limits_{n=1}^{8} An(Fn + Gn)}. \qquad (11)$$

By relating the absorbance, A, to the transmittance, T; $An = -\log Tn$. The equation can be simplified as:

$$SO_2 = 100 \times \frac{\sum\limits_{n=1}^{8} Kn \log Tn}{\sum\limits_{n=1}^{8} Ln \log Tn}, \qquad (12)$$

where $Kn = Fn$, $Ln = Fn + Gn$. K_0 and L_0 are introduced to account for the absorption of non-varying components. Thus

$$SO_2 = 100 \times \frac{K_0 + \sum\limits_{n=1}^{8} Kn \log Tn}{L_0 + \sum\limits_{n=1}^{8} Ln \log Tn}. \qquad (13)$$

To determine the constants K_0 to K_8 and L_0 to L_8, 750 data points were collected on 22 volunteers of varying races and smoking habits. Various SO_2 values from 65 to 100% were obtained while subjects were breathing varying oxygen concentrations. The transmittance values for each of the eight wavelengths were processed by a digital computer. The eighteen constants, K_0 to K_8 and L_0 to L_8, were adjusted to produce a best fit to the actual values for SO_2 according to the method for optimization by Kowalick and Osborne (1968). This method of estimating the arterial SO_2 is to compensate for the effects of tissue pigmentation, haemoglobin content, depth of the light path and multiple scattering of light by tissues and blood constituents on an empirical basis. Current developments in optical as well as computer technology have rendered the method practical.

4.2.2 Design and Operation

The optical section of the multiple-wavelength ear oximeter (HP 47201A) (Hewlett-Packard Co., Ltd., Waltham, Massachusetts, USA) is composed of a light source, a beam splitter, eight narrow band filters mounted on a rotating filter wheel, a fibre-optic path and a silicon photodetector as illustrated in Fig. 4. The filter wheel rotates at a speed of 1300 r min^{-1}, feeding a set of selective wavelengths sequentially to the pinna of the ear through a fibre-optic guide. The transmitted light then travels through another light guide to reach the photodetector. To determine the intensity

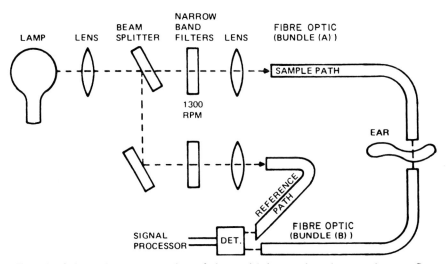

FIG. 4   Schematic representation of the multiple-wavelength ear oximeter. See text for explanations. (From Flick and Block (1977), with permission from the author and the publisher.)

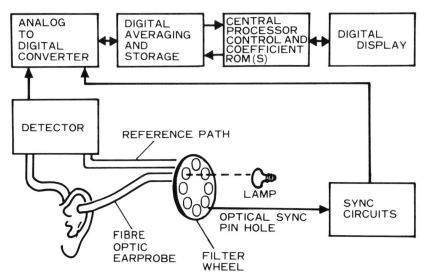

Fig. 5 Simplified block diagram of the multiple-wavelength ear oximeter. (Courtesy of Hewlett-Packard.)

of light incident to the ear, a second light beam which is separated from the collimated light by a beam splitter is applied to the filter wheel. The filters are so located as to pass only one beam at a time. As the wheel rotates, the light beam at a particular wavelength is first fed to the photo-detector through the ear-probe, and then through the reference path to the photodetector. The reference path provides a reference light intensity which can be used to correct for short-term changes in light intensity due to source or detector changes (Fig. 5). The ratio of the transmitted light intensity to the incident light intensity is obtained by a data processor and the saturation values are calculated using the equation described above. The standardization procedure before the measurement is required to store reference light intensities for each of the eight wave lengths. It is accomplished simply by placing the probe into a standardization cavity of the instrument. The front view of the instrument is illustrated in Fig. 6. Calibration of the instrument against SO_2 values obtained by the blood sampling analysis is practically unnecessary with the multiple-wavelength oximeter. The SO_2 values are independent of skin pigmentation and insensitive to the movement of the ear-probe relative to the pinna of the ear. One of the problems in constructing the instrument has been the choice of a set of narrow band filters which have the accurate filter-bandwidth because the centre-wavelength deviation considerably worsens the accuracy of SO_2 measurement. Sensitivity to ambient light is very low. The light shield is provided to protect from intense light. In ordinary

FIG. 6 The front view of the multiple-wavelength ear oximeter. (Courtesy of Hewlett-Packard.)

room light, there should be no significant effect with the shield on or off. Arterialization of the pinna of the ear is another major problem for the multiple-wavelength oximeter because it is sometimes difficult in patients with circulatory insufficiency, as described later.

4.3 Pulse-wave Type *In Vivo* Oximeter

Recently, an ingenious method for obtaining the absolute values for arterial SO_2 has been developed (Nakajima *et al.*, 1975; Liappis, 1979; Yoshiya *et al.*, 1980). In contrast to the empirical method employed in the multiple-wavelength oximeter, the calculations made by the pulse-wave type oximeter are based on an absolute relationship. Although the pulse-wave type method still has theoretical sources of error, it seems that it has many possibilities for the future development of oximetry. In the preceding volume of this series, Challoner (1979) described photo-electric plethysmography which provides indications of cutaneous blood flow by analysing the change in optical density caused by the arterial pulse. It detects the volume change in blood components due to the pumping action of the heart. It necessarily should eliminate the effect of changes in blood oxygen saturation, and an isobestic point at 805 nm has been chosen for the spectrophotometric analysis to be independent of blood oxygen saturation. Pulse-wave type oximetry deals with the same phenomenon as the photo-electric plethysmography in a different manner: the decrease in optical density due to the inflow of arterial blood into the site of measurement.

We are to describe the principle, design and operation of a currently available oximeter (Minolta-Oximeter 101, Minolta Camera Co., Ltd., Osaka, Japan) which analyses arterial SO_2 in the fingertip.

4.3.1 *Principle*

The pulse-wave type *in vivo* oximeter performs a spectrophotometric analysis of light transmitted through the fingertip, considering only the change in absorbance due to the inflow of arterial blood into the fingertip. The fingertip is composed of "blood" and "non-blood" components. The "blood" component consists of both arterial and venous blood, and the "non-blood" component consists of skin, muscle, bone and other tissues. The attenuation of light by the "non-blood" component hardly changes with pulsation. The "blood" component, on the other hand, changes its volume as the blood flows into and out of the fingertip with pulsation. Accordingly, the attenuation of light pulsates as shown in Fig. 7. Assuming that the pulsating portion of the attenuation of light is caused solely by the arterial blood which flows into the fingertip during the inflow phase, arterial SO_2 can be obtained by spectrophotometric analysis of the pulsating AC component. The AC component is obtained by subtracting the DC component from the total attenuation of light as described below.

Assuming that the Lambert–Beer law can be used for the whole blood that is present in the fingertip, the optical density of the transmitted light I is written as:

$$I = I_0 F_T 10^{-\alpha' d} 10^{-\alpha \iota}, \tag{14}$$

where I_0 = optical density of light incident to the fingertip,
 F_T = absorption of light by the "non-blood" component,
 d and α' = quantity of blood that is present at the end of the outflow phase and its absorption coefficient,
 ι and α = quantity of blood that flows into the fingertip and its absorption coefficient.

The total output of the photo-electric element is given by:

$$E_{DC + AC} = AI = AI_0 F_T 10^{-\alpha' \gamma d} 10^{-\alpha \gamma \iota}, \tag{15}$$

where A and γ are constants specific to the photodetector.
The AC component of the photo-electric output is given by:

$$E_{DC} = AI_0 F_T 10^{-\alpha' \gamma d}. \tag{16}$$

The logarithmic difference of $E_{DC + AC}$ and E_{DC} is given by:

$$Y = \log (E_{DC + AC}/E_{DC}) = -\alpha \gamma \iota. \tag{17}$$

The Ys at the wavelengths of 650 and 805 nm (Y_{650} and Y_{805}, respectively) are calculated as:

$$
\left.\begin{array}{l}
Y_{605} = - \alpha_{650}\gamma^{\iota} \\
Y_{805} = - \alpha_{805}\gamma^{\iota}
\end{array}\right\} \tag{18}
$$

therefore,

$$
\alpha_{650}/\alpha_{805} = Y_{650}/Y_{805}. \tag{19}
$$

Accordingly, the ratio of the absorption coefficients at wavelengths of 650 and 805 nm is determined by measuring the ratio of Ys at the respective wavelengths. In terms of Y_{650}/Y_{805}, Eqn (3) can be rewritten as:

$$
SO_2 = A - B(Y_{650}/Y_{805}). \tag{20}
$$

In this manner, by measuring the logarithmic difference of the total attenuation of light ($E_{DC + AC}$) and the DC component of the attenuation (E_{DC}) at wavelengths of 650 and 805 nm, the absolute saturation value for the arterial blood can be obtained.

The present method uses several assumptions in determining arterial SO_2 in the fingertip. The increase in the attenuation of light during the inflow phase is assumed to be caused solely by the arterial blood. The optical density of the transmitted light may fluctuate also by the changing saturation of the blood in the fingertip due to the consumption of oxygen by the tissues. The effect of this fluctuation is neglected with the present method, although it might produce unknown amounts of error provided the blood flow is very low compared to the oxygen uptake of the tissue at the site of measurement. Any possible change in the depth of tissue with pulsation is also neglected. Another assumption is that the amount of light attenuated by the scattering, reflection and refraction of light by the blood and tissues is independent of the differences in the wavelengths, as well as the depth of light path which changes with pulsation.

The present method is unique in that it measures the absolute values for arterial SO_2 on a theoretical basis. It requires neither arterialization nor compression of the measuring site. Skin pigmentation does not affect the readings theoretically.

4.3.2 Design and Operation

Light emitted by a halogen lamp is filtered by a broad band filter which attenuates the infrared portion of the emitted light according to the spectral characteristics illustrated in Fig. 8. The light is then carried by a glass fibre-optic to a finger-probe. The transmitted light travels through another glass fibre-optic and is bisected with a beam splitter, as shown in Fig. 9.

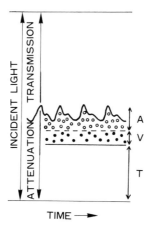

FIG. 7 Schematic representation of the pulsatile photo-electric output. The attenuation of the incident light is caused by A, arterial blood; V, venous blood and T, tissues. (From Yoshiya *et al.* (1980), with permission from the publisher.)

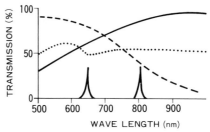

FIG. 8 Spectral transmittance curve of the broad band filter (— — — — —), the glass fibres (. . . . .), the narrow band filters (——— at 650 and 805 nm), and a spectral emittance curve of the halogen lamp (———). (From Yoshiya *et al.* (1980), with permission from the publisher.)

Each light beam is then filtered by narrow band filters which selectively transmit light at wavelengths of 650 and 805 nm (Fig. 8). At each wavelength, the photo-electric conversion is carried out with a pair of silicon-photodiodes. The saturation of oxygen is electronically calculated according to the equation described above. The calculated value for oxygen saturation is held and up-dated every 5 s by a sample-and-hold circuit, and displayed on a digital meter. The pulsatile output, Y_{805}, is displayed on a galvanometer which represents the amplitude of the AC signal. With the galvanometer we can determine if the photo-electric output is adequate for analysis. There is also a control circuit which makes the same decision and indicates if the displayed value becomes unreliable by means of an

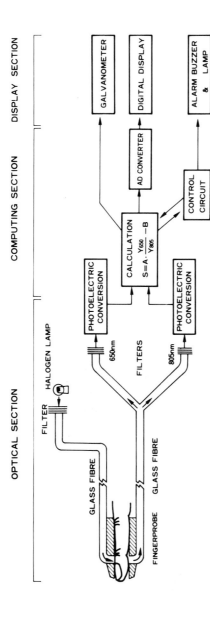

FIG. 9 Block diagram of the pulse-wave oximeter. See text for explanations. (From Yoshiya *et al.* (1980), with permission from the publisher.)

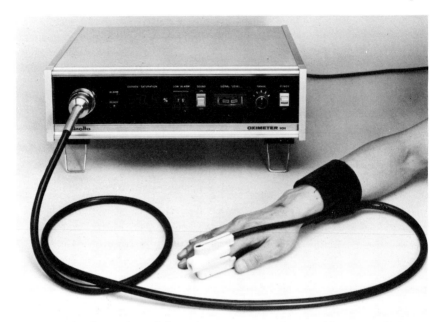

FIG. 10   The front view of the pulse-wave type oximeter. (Courtesy of Minolta Camera.)

alarm signal. Neither standardization nor calibration procedures are necessary with the present instrument because the logarithmic differences of total and pulsatile signals are calculated at the wavelengths of 650 and 805 nm, as described above. Since the absolute value for optical density of the transmitted light is not necessary, the oxygen saturation value is uninfluenced by both the intensity of the light source and the intensity of light incident on the photodetectors. The contamination of the incident light on the photodetectors with ambient light does not seriously affect the measurement, which is in contrast with the Wood-type oximeter, and this is because of the subtraction method. Similarly, neither the difference in sensitivity of the photo-electric elements nor the uneven transmittance of the optical paths become critical problems in measuring oxygen saturation.

The actual measurement of arterial SO_2 can easily be made by introducing a fingertip of a patient into the finger-probe (Fig. 10). In small children, intermetacarpal space can be used with a small modification of the finger-probe. So long as the pulsatile photo-electric output is stable, as shown in Fig. 7, the oxygen saturation values are reliable irrespective of the measuring site. However, the measurement is interrupted when the fingertip changes its position relative to the axis of the light beam. This is a problem in clinical situations when, for example, the patient is uncooperative or moving involuntarily. The measurement is automatically resumed soon after the cessation of the movement.

The accuracy and clinical limitations of operation of the three types of *in vivo* transmittance oximeters will be discussed later in this section.

5.0 REFLECTION OXIMETER

The oxygen saturation of non-haemolysed whole blood can be estimated by the spectrophotometric analysis of the reflectance of light by blood. The reflectance has been assumed to be exponentially related to oxygen saturation by several investigators (Brinkman and Zijlstra, 1949; Kramer *et al.*, 1956; Refsum and Hisdal, 1958). Anderson and Sekelj (1967) have demonstrated that the reflection and transmission are linearly related and so the former is exponentially related to the absorption coefficient of haemoglobin. Accordingly, the oxygen saturation of whole blood can be estimated with a method similar to transmission oximetry by analysing the optical density of the reflected light utilizing the two-wavelength method (Rodrigo, 1953; Falholt, 1965). The two-wavelength method has also been successfully applied to the direct intravascular measurement of SO_2 using a catheter fibre-optic reflection oximeter (Polanyi and Hehir, 1964; Enson *et al.*, 1962; Kapany and Silbertrust, 1964; Mook *et al.*, 1968; Wilkinson *et al.*, 1979).

There have been several attempts to make *in vivo* estimations of SO_2 of blood present in intact organs (Kramer *et al.*, 1956; Cheung *et al.*, 1977), the skin (Zijlstra and Mook, 1962; Cohen and Wadsworth, 1972), and the choroidal tissue (Laing *et al.*, 1975). However, at present, no reflection oximeters are available which measure the arterial SO_2 quantitatively and non-invasively. The subtraction method of Wood and Geraci (1949) which derives the absolute values for arterial SO_2 by subtracting the optical density of the bloodless ear from that of the blood-containing ear could not be applied to reflection oximetry. The reflectance characteristics of the compressed, bloodless ear would differ considerably from the blood-containing tissue because the reflectance of light is known to be a non-linear function of the sample path (Kramer *et al.*, 1956; Anderson and Sekelj, 1967). Cheung *et al.* (1977) have constructed a reflection oximeter by using five different wavelengths in the red and near infrared regions. Although they failed to obtain absolute values for arterial SO_2, the multiple-wavelength method is one possibility for the future development of non-invasive, *in vivo* reflection oximetry. Another possibility for the reflection method would be to utilize the pulse-wave type method described earlier, since sufficiently large arterial volume pulses are known to be obtained with reflection photoplethysmography (Weinman *et al.*, 1977; Challoner and Ramsay, 1974; Challoner, 1979).

6.0 CLINICAL APPLICATION OF *IN VIVO* OXIMETRY

For the measurement of arterial SO_2 in clinical fields, oximetry provides a continuous record of the amount of oxygen carried by the blood to the peripheral tissues. It is non-invasive and provides us with a rapid assessment of the status of arterial oxygenation.

We will review in some detail several of the clinical applications of oximetry which might be of help in a variety of clinical settings.

6.1 Anaesthesia and Intensive Care

Non-invasive oximetry is suitable for continuous monitoring of arterial oxygenation during anaesthesia and intensive care. The rapidity of the response indicates to us the severity of arterial hypoxaemia during endotracheal intubation, tracheal suctioning and toiletting (Fig. 11). It is also suitable for the recognition of hypoxaemia during one-lung ventilation in thoracic surgery. Positive end-expiratory pressure (PEEP) is often used in

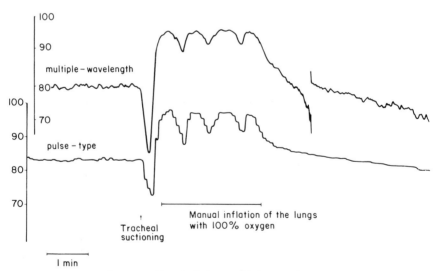

FIG. 11   Tracing from readings of the multiple-wavelength oximeter and the pulse-wave type oximeter in a 3-month-old boy with respiratory insufficiency secondary to interstitial pneumonia. Both oximeter readings abruptly dropped during tracheal suctioning as indicated by the arrow. Note that manual inflation of the lungs with 100% oxygen markedly increased arterial SO_2.

intensive care for the management of patients with acute respiratory insufficiency. Optimal adjustment of PEEP is mandatory and PEEP will sometimes cause an abrupt drop in arterial PO_2 and SO_2, especially in patients with hypovolaemia (Qvist et al., 1975), cardiac failure (Powers et al., 1973), and exaggerated ventilation–perfusion inequalities (Hammon et al., 1976). Oximetry provides quick recognition of the detrimental adjustment of PEEP. The other situation where oximetry is of value in monitoring is the prevention of oxygen toxicity in patients with severe respiratory insufficiency. Since high inspiratory oxygen concentration brings about pulmonary oxygen toxicity (Frank and Massaro, 1979), it should be reduced below 50% so long as the arterial SO_2 remains above 85–90%. Continuous monitoring of SO_2 with alarm systems is suitable in those patients.

6.2 Bronchoscopy

Fibre-optic bronchoscopy is widely used both as a diagnostic and thera-peutic procedure in clinical settings, and has proved to be safe. However, it can cause severe hypoxaemia in patients who already have damaged lungs (Berman and Jones, 1976). Instantaneous knowledge of arterial SO_2 during fibre-optic bronchoscopy will allow hypoxaemia to be reversed rapidly by using respiratory manoeuvres and will allow suctioning to be carried out safely and effectively.

6.3 Pulmonary Function Test

Non-invasive oximetry combined with mild physiological loading such as excercise, hypoxia, or pharmacological agents provides us with invaluable information relating to pulmonary function at rest and during exercise in patients with lung disease.

Al-Dulymi and Hainsworth (1975) used a Wood-type ear oximeter for the assessment of pulmonary function in patients with lung disease. They estimated arterial SO_2 while subjects breathed 15% oxygen in nitrogen, and found that arterial SO_2 values obtained from patients with lung disease were clearly separated from those obtained from normal subjects, and the extent of the lowering of saturation was related to the clinical severity of the disease. They inferred that the pulmonary function test should clearly separate the results from patients with lung disease from normal subjects, and that there should be no normal subject with abnormal results and vice versa. Thus ear oximetry combined with the breathing of mildly hypoxic gas mixture is shown to be valuable in detecting impaired pulmonary function. Ishikawa et al. (1978) measured the change in arterial

SO_2 during physiological exertion in patients with chronic obstructive lung disease. They showed a marked drop in arterial SO_2 during cough and bowel movement as well as eating in those patients.

For the assessment of the effects of physiotherapy, pharmacological agents and oxygen therapy in patients with lung disease, oximetry provides more physiological data without disturbance in respiratory physiology such as breath-holding or hyperventilation, that painful invasive techniques frequently produce (Trask and Cree, 1962; Ishikawa et al., 1978; Girsch and Girsch, 1979).

6.4 Diagnosis and Follow-up of Cardiac Patients

Oximetry has often been used for the diagnosis and follow-up of cardiac patients. In pre-operative patients, oximetry may aid in the diagnosis of congenital heart disease or the evaluation of its severity (McIlroy, 1959; Keane et al., 1975). Repeated evaluation of arterial SO_2 at rest and during exercise will help to determine the severity of the patient's illness and the need for surgery. In the immediate post-operative critical period, oximetry is not a reliable substitute for direct blood gas measurement because of inaccuracies produced by disturbances in peripheral perfusion, as will be discussed later. However, for patients in the follow-up period, oximetry can provide a wealth of information relating to residual intracardiac shunt, ventilation perfusion inequalities associated with long-standing pulmonary arterial hypertension, effect of pulmonary banding procedure and palliative procedures, such as the shunting operations of Waterstone, Blalock and so forth.

6.5 Physiological Studies during Extreme Conditions

Oximetry has been utilized for the evaluation of arterial SO_2 at high altitude. West et al. (1962) measured arterial SO_2 with the Wood-type oximeter at 5800 m on six members of the Himalayan Scientific and Mountaineering Expedition and found significant arterial desaturation both at rest and during work. Powles et al. (1978) studied the changes in arterial SO_2 during sleep by means of the multiple-wavelength oximeter at 5369 m, and found a significant drop in arterial SO_2. The severity of hypoxaemia was reported not to differ whether the subject had been acclimatized to the high altitude or not. Arterial cannulation at high altitude is unjustified because of the thrombotic complication in the presence of polycythaemia. Oximetry has the great advantage in that it is non-invasive and it can repeatedly measure arterial SO_2 while the subject is sleeping or working at high altitude.

The other situation where oximetry is of value is in aerospace medicine. Human subjects in high performance aircraft are routinely exposed to high acceleration forces for prolonged periods. Nolan *et al.* (1963) described arterial desaturation under such conditions using the Wood-type oximeter. However, the oximeter is not without the influence of changes in blood content in the trans-illuminated ear tissue and haemoglobin content due to acceleration. Besch *et al.* (1978) used a multiple-wavelength ear oximeter and compared the observed arterial SO_2 values with those calculated indirectly from arterial PO_2 measured from sampled arterial blood with a blood gas analyser, and corrected for pH and base excess at different levels of acceleration. Regression analysis of the data showed that the ear oximeter is accurate, with a correlation coefficient of 0·95. The result showed that the changes in blood content and blood haemoglobin concentration in the trans-illuminated ear due to acceleration forces do not affect the operation of the multiple-wavelength oximeter. Oximetry becomes one of the essential measuring procedures in terms of a pilot's mental alertness and safety of operation in the dynamic environment of high sustained acceleration forces.

6.6 Accuracy and Limitations in the Clinical Settings

There have been many reports concerning the problems encountered in *in vivo* non-invasive oximetry in clinical settings. Among them, the accuracy and stability of the measurements are of prime importance.

Wood and Geraci (1949) assessed the accuracy of their modified ear oximeter and found that 95% of the 275 photo-electric determinations were within ±5% of values obtained by the Van Slyke method. Saunders *et al.* (1976) evaluated the accuracies of two types of oximeters, the Wood-type and the multiple-wavelength type, in healthy subjects during acute and progressive hypoxia induced by breathing hypoxic gas mixtures. They found that both types of oximeters have similar linearities and accuracies as compared with directly measured arterial blood SO_2 (Fig. 12). Several studies with similar results have been reported on the accuracies and linearities of the multiple-wavelength oximeter both in normal subjects and patients with varying degrees of pulmonary disease during rest or exercise (Flick and Block, 1977; Poppius and Viljanen, 1977; Scoggin *et al.*, 1977; Krauss *et al.*, 1978).

Liappis (1979) and Yoshiya *et al.* (1980) evaluated the linearity and accuracy of the pulse-wave type oximeter in patients with pulmonary disease and found similar results to those obtained with the multiple-wavelength oximeter.

However, the new methods are not without limitations. Chaudhary and Burki (1978) observed the relation between the multiple-wavelength oximeter SO_2 and the SO_2 calculated from blood gas analysis, and found that there is a much wider variation in the readings with lower arterial PO_2 (PO_2 lower than 61 mmHg) compared to those with normoxia. Douglas *et al.* (1979) observed that the oximeter constantly underestimates arterial SO_2 compared to the *in vitro* oximeter at arterial SO_2 lower than 65%.

Recently, Sarnquist *et al.* (1980) observed that the pulse-wave type oximeter over-estimates arterial SO_2 derived from an *in vitro* oximeter at arterial SO_2 lower than 70% in healthy subjects. We have also noticed a similar tendency in patients with pulmonary disease and in healthy subjects.

Although both oximeters either over- or under-estimate arterial SO_2 derived from *in vitro* oximetry, the deviations are linear and may, therefore, be corrected.

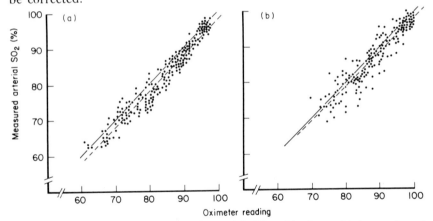

FIG. 12(a)   Comparison of measured arterial SO_2 with the multiple-wavelength ear oximeter (HP 47201A) readings. ———, line of identity; - - - - - -, line of regression. $y = 0.99x - 1.52$; SEE = 2.52; $r = 0.97$. (From Saunders *et al.* (1976), with permission from the author and the publisher.)

FIG. 12(b)   Comparison of measured arterial SO_2 with the Wood-type ear oximeter (Waters 0-1100) readings. ———, line of identity; - - - - - -, line of regression. $y = 0.95x + 2.59$; SEE = 2.98; $r = 0.93$. (From Saunders *et al.* (1976), with permission from the author and the publisher.)

It should be recognized that significant errors can be introduced in the presence of carboxyhaemoglobin (COHb) with the Wood-type oximeter (Gordy and Drabkin, 1957). Douglas *et al.* (1979) have indicated that the multiple-wavelength oximeter is also sensitive to COHb, and that it over-estimates arterial SO_2 compared to the *in vitro* oximeter (Instrumentation Laboratory co-oximeter 182). This error would be important when studying hypoxaemic patients with chronic bronchitis or emphysema, many of whom are heavy cigarette smokers. All other oximeters, including the pulse-wave type, could have the same error in the presence of COHb. The effect of COHb on *in vivo* oximetry needs further investigation.

Both oximeters, multiple-wavelength type and pulse-wave type, have limitations of operation in the presence of poor peripheral perfusion. Chaudhary and Burki (1978) observed no systemic difference between the multiple-wavelength oximeter SO_2 and SO_2 derived from direct blood gas analysis in patients with low cardiac output state. When an insufficient amount of blood perfuses the ear-piece field of view, and hence the quantity of total haemoglobin decreases, relatively little light is absorbed by the haemoglobin. If the blood volume is less than the smallest volume in the calibration series of the oximeter, then the "off-ear" light comes on. The multiple-wavelength oximeter is equipped with a perfusion indicator which measures the adequacy of pulsatile blood flow at the site of measurement by using photo-electric plethysmography. It calibrates pulsatile blood flow in terms of the expected oximeter error.

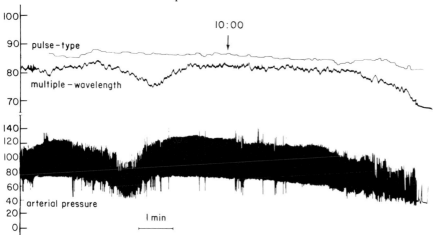

FIG. 13   Tracing from readings of the pulse-wave type oximeter, multiple-wavelength oximeter, and radial arterial pressure in a 10-year-old boy with single atrium and single ventricle measured during corrective surgery (bilateral Glenn's procedure). See text for explanations.

We have assessed the influence of impaired circulatory states upon the digital read-outs of the multiple-wavelength ear oximeter and the pulse-wave type finger oximeter. Figure 13 shows that the multiple-wavelength oximeter SO_2 drops concomitant with a short-term drop in arterial blood pressure, whereas the pulse-wave type oximeter SO_2 is relatively uninfluenced. The following are possible explanations for the difference between the two oximeter readings: (a) when the arterial blood pressure drops, less blood volume would be present in the ear-piece field of view and the multiple-wavelength oximeter reading needs correction by means of the perfusion indicator, as described above; (b) a genuine fall in arterial SO_2 might have been picked up by the faster response of the multiple-wavelength oximeter; or (c) the pulse-wave type oximeter could have detected SO_2 in the larger arterial vessel, whereas the multiple-wavelength oximeter could have detected SO_2 in arterialized capillary blood which may have been more influenced by a drop in arterial blood pressure. The relative stability of the pulse-wave type oximeter is confirmed by applying varying levels of pressure with a sphygmomanometric cuff proximal to the finger of a healthy volunteer (Fig. 14). The figure shows that until the cuff

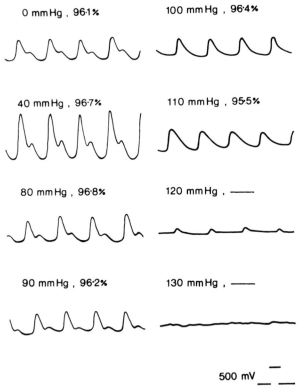

FIG. 14   Photo-electric output of a healthy subject when a sphygmomanometric cuff applied proximal to the finger was inflated in a stepwise fashion. The figures on each tracing are the cuff pressure and digital readout of the instrument, respectively. (From Yoshiya et al. (1980), with permission from the publisher.)

pressure is increased to 90 mmHg there are no gross changes in either the photo-electric output or the SO_2 reading. When the cuff pressure is increased to 100 and 110 mmHg, the photo-electric output becomes damped whilst reasonable SO_2 values are shown. When the cuff pressure is increased to 120 mmHg, the photo-electric output is decreased less than 500 mV in amplitude, and the arterial SO_2 cannot be measured.

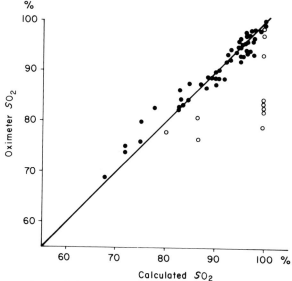

FIG. 15   Comparison of the calculated arterial SO_2 with the pulse-wave type oximeter readings. ●, Data from patients without circulatory disturbances; ○, data from patients after cardiopulmonary bypass and low cardiac output.

In spite of this finding the pulse-wave type oximeter is not without the influence of low cardiac output, general and/or local cooling, or poor peripheral perfusion. Figure 15 illustrates the correlation between the arterial SO_2 measured with the pulse-wave type oximeter and the SO_2 calculated from blood gas analysis. Although there is a good correlation in patients without circulatory disturbances (solid circle), we cannot find any correlation in patients after cardiopulmonary bypass and low cardiac output (open circle). The reasons are still not clear, but may be due to: (a) the changing saturation of the blood in the fingertip secondary to the consumption of oxygen by the tissues in the presence of poor perfusion; (b) the exaggerated effects of the changes in haemoglobin concentration, scattering, reflection and refraction of light in the presence or sludging and stasis of red blood cells; and (c) the possible pulsatile (AC) fluctuation of the venous component in the fingertip.

In summary, both oximeters have limitations of operation in the presence of poor peripheral perfusion in the site of measurement. The multiple-wavelength ear oximeter is equipped with a perfusion indicator which measures the adequacy of the pulsatile blood flow at the site of measurement by using photo-electric plethysmography as described above, and the correction of arterial SO_2 values by means of the different levels of blood flow seems feasible. Regarding the pulse-wave type oximeter, it theoretically measures arterial SO_2 as long as pulsatile blood flow is detected. However, further studies are required to ameliorate the accuracy of the instrument in the presence of poor peripheral perfusion.

7.0 SUMMARY AND CONCLUSIONS

Spectrophotometric *in vivo* oximetry is an accurate, reproducible, rapid and totally non-invasive way of measuring the absolute value of arterial SO_2 in the clinical field. More than 30 years have passed since the first application of spectrophotometric oximetry to the *in vivo* estimation of arterial SO_2. Several problems primarily associated with *in vivo* oximetry, namely, multiple scattering, reflection and refraction of light, arterialization at the site of measurement and so forth, have been partially resolved by the current advances in biomedical engineering. The multiple-wavelength ear oximeter has attempted to solve the problems by using eight selective wavelengths instead of two and by the optimization method of the empirically derived data by means of computer technology. The pulse-wave type finger oximeter has ameliorated the problems by combining photo-electric plethysmographic analysis with spectrophotometry for the detection of the absolute value of arterial SO_2 in the fingertip.

However, the current methods still have significant sources of error when they are applied in certain clinical situations, such as low cardiac output state, local and/or general body cooling and disturbed peripheral perfusion. We expect that the majority of the problems will be resolved either empirically by using computer technology or theoretically by using photo-electric technology.

Note Added in Proof

The theoretical basis of the pulse-type finger oximeter described here has been evolved on the assumption that effects of multiple scattering are sufficiently small to be considered negligible. Recent studies in the authors' laboratories have shown that this assumption is not valid for the accurate estimation of SO_2 below 90%. The results of these studies are in accord with the Kubelka–Munk equation (Kubelka, P. (1948) *J. Opt. Soc. Amer.*

38, 448–457), which describes the scattering properties of intensely light-scattering materials. Therefore, the accuracy of the pulse-type oximeter could be improved significantly by incorporating compensation for multiple-scattering effects.

References

Al-Dulymi, R. and Hainsworth, R. (1975). Use of an ear oximeter to assess lung function. *Thorax* **30**, 566–568.
Anderson, N. M. and Sekelj, P. (1967). Reflection and transmission of light by thin films of nonhaemolysed blood. *Phys. Med. Biol.* **12**, 185–192.
Berman, L. and Jones, N. L. (1976). Fibroptic bronchoscopy; monitoring oxygenation with ear oximetry. *Am. Rev. Respir. Dis.* **114**, 651.
Besch, E. L., Baumgarden, F. W., Burton, R. R., Gillingham, K. K., McPherson, R. F. and Leverett, S. D., Jr (1978). Calibration of a new ear oximeter in humans during exposure to centrifugation. *J. Appl. Physiol.* **44**, 483–487.
Brinkman, R. and Zijlstra, W. G. (1949). Determination and continuous registration of the percentage oxygen saturation in small amounts of blood. *Arch. Chir. Neerl.* **1**, 177–183.
Challoner, A. V. J. (1979). Photoelectric plethysmography for estimating cutaneous blood flow. *In* "Non-invasive Physiological Measurements" (Ed. P. Rolfe), Vol. I, pp. 125–151. Academic Press, London and New York.
Challoner, A. V. J. and Ramsay, C. A. (1974). A photoelectric plethysmograph for the measurement of cutaneous blood flow. *Phys. Med. Biol.* **19**, 317–328.
Chaudhary, B. A. and Burki, N. K. (1978). Ear oximetry in clinical practice. *Am. Rev. Respir. Dis.* **117**, 173–175.
Cheung, P. W., Takatani, S. and Ernst, E. A. (1977). Multiple wavelength reflectance oximetry in peripheral tissues. *Adv. Exp. Med. Biol.* **94**, 69–75.
Cohen, A. and Wadsworth, W. (1972). A light emitting diode skin reflectance oximeter. *Med. Biol. Engng* **10**, 385–391.
Douglas, N. J., Brash, H. M., Wratth, P. K., Calverley, P. M. A., Leggett, R. J. E., McElderry, L. and Flenley, D. C. (1979). Accuracy, sensitivity to carboxy-hemoglobin, and speed of response of the Hewlett-Packard 47201A ear oximeter. *Am. Rev. Respir. Dis.* **119**, 311–313.
Drabkin, D. L. (1950). Measurement of O_2 saturation of blood by direct spectro-photometric determination. *In* "Methods in Medical Research" (Ed. J. H. Comroe), Vol. II, pp. 159–161, Year Book Medical Publisher, Chicago.
Enson, Y., Briscoe, W. A., Polanyi, N. L. and Coumand, A. (1962). *In vivo* studies with an intravascular and intracardiac reflection oximeter. *J. Appl. Physiol.* **15**, 552–558.
Falholt, W. (1963). Blood oxygen saturation determined spectrophotometrically. *Scand. J. Clin. Lab. Invest.* **15**, 67–72.
Flick, M. R. and Block, A. J. (1977). Continuous *in vivo* measurement of arterial oxygen saturation by oximetry. *Heart Lung* **6**, 990–993.
Frank, L. and Massaro, D. (1979). The lung and oxygen toxicity. *Arch. Intern. Med.* **139**, 347–359.
Girsch, L. S. and Girsch, B. J. (1979). Arterial oxygen saturation by non-invasive oximetry technique: an office test for determining jeopardy of status asthmaticus. *Ann. Allerg.* **42**, 14–18.

Gordy, E. and Drabkin, D. L. (1957). Spectrophotometric studies. XVI. Determination of the oxygen saturation of blood by a simplified technique, applicable to standard equipment. *J. Biol. Chem.* **227**, 285–296.

Hammon, J. W., Jr, Wolfe, W. G., Moran, J. F., Jones, R. H. and Sabiston, D. C., Jr (1976). The effect of positive end-expiratory pressure on regional ventilation and perfusion in the normal and injured primate lung. *J. Thorac. Cardiovasc. Surg.* **72**, 680–689.

Horecker, B. L. (1943). The absorption spectra of hemoglobin and its derivatives in the visible and near infrared regions. *J. Biol. Chem.* **148**, 173–183.

Huch, R., Huch, A. and Rolfe, P. (1979). Transcutaneous measurement of PO_2 using electrochemical analysis. *In* "Non-invasive Physiological Measurements" (Ed. P. Rolfe), Vol. I, pp. 313–329, Academic Press, London and New York.

Hüfner, G. (1900). Ueber die gleichzeitige quantitative Bestimmung zweier Farbstoffe im Blute mit Hülfe des Spectrophotometers. *Arch. Anat. Physiol., Physiol. Abt.* 39–48.

Ishikawa, S., Linzmayer, I. and Segal, M. S. (1978). Clinical use of a non-invasive method for determining oxygen saturation: ear oximetry. *Ann. Allerg.* **41**, 18–20.

Kapany, N. S. and Silbertrust, N. (1964). Fibre optics spectrophotometer for *in vivo* oximetry. *Nature* **204**, 138–142.

Keane, J. F., Williams, R., Treves, S. and Rosenthal, A. (1975). Assessment of the postoperative patients by noninvasive technique. *Prog. Cardiovasc. Dis.* **18**, 57–74.

Kowalick, J. and Osborne, M. R. (1968). "Methods for Unconstrained Optimization Problems", American Elsevier Publishing, New York.

Kramer, K. (1935). Ein Verfahren zur fortlaufenden Messung des Sauerstoffgehaltes im strömenden Blute an uneröffneten *Gefässen. Zeitschr. Biol.* **96**, 61–75.

Kramer, K., Elam, Z. O., Saxton, G. A. and Elam, W. N., Jr (1951). Influence of oxygen saturation, erythrocyte concentration and optical depth upon the red and near-infrared light transmittance of whole blood. *Am. J. Physiol.* **165**, 229–246.

Kramer, K., Graft, R. and Overbeck, W. (1956). Photoelektrische Reflektometrie an tierischen Organen. *Pflügers Arch.* **262**, 285–300.

Krauss, A. N., Waldman, S., Frayer, W. W. and Auld, P. A. M. (1978). Noninvasive estimation of arterial oxygenation in newborn infants. *J. Pediatr.* **93**, 275–278.

Laing, R. A., Danisch, L. A. and Young, L. R. (1975). The choroidal eye oximeter: an instrument for measuring oxygen saturation of choroidal blood *in vivo*. *IEEE Trans. Biomed. Engng* BME 22, 183–195.

Liappis, N. (1979). Nichtinvasive Messung der Sauerstoffsättigung mit dem Oxygenmet-Oximeter an Fingern, Mittelhand und Handgelenk von Säuglingen. Vergleich mit der berechneten Sauerstoffsättigung aus pH und PO_2 der Blutgasanalyse. *Klin. Pädiat.* **191**, 467–471.

Loewinger, E., Gordon, A., Weinreb, A. and Gross, J. (1964). Analysis of a micromethod for transmission oximetry of whole blood. *J. Appl. Physiol.* **19**, 1179–1184.

Lübbers, D. W. (1973). "Oxygen Transport to Tissue" (Eds H. I. Bicher and D. Brunley), Plenum Press, New York.

Matthes, K. (1935). Untersuchungen über die Sauerstoffsättigung des Arterienblutes. *Arch. Exp. Path. Pharmacol.* **179**, 698–711.

McIlroy, M. B. (1959). The clinical use of oximeter. *Br. Heart J.* **21**, 293–314.

Merrick, E. B. (1975). A multi-wavelength ear oximeter. Biocept 75, International Conference on Biomedical Transducers, Paris.

Merrick, E. B. and Hayes, T. J. (1976). Continuous, non-invasive measurements of arterial oxygen levels. *Hewlett-Packard J.* pp. 2–9.

Millikan, G. A. (1942). The oximeter, an instrument for measuring continuously oxygen saturation of arterial blood. *Rev. Sci. Instrum.* **13**, 434–444.

Mook, G. A., Osypka, P., Sturm, R. E. and Wood, E. H. (1968). Fibre-optic reflection photometry on blood. *Cardiovasc. Res.* **2**, 199–209.

Mook, G. A., van Assendelft, O. W. and Zijlstra, W. G. (1969). Wavelength dependency of the spectrophotometric determination of blood oxygen saturation. *Clin. Chim. Acta* **26**, 170–173.

Nakajima, S., Hirai, Y., Takase, H., Kuse, A., Aoyagi, S., Kishi, M. and Yamaguchi, K. (1975). New pulse-type ear oximeter. *Kokyu to Junkan* **23**, 709–713. (Japanese).

Nilsson, N. J. (1956). Ein vereinfachtes Spektrophotometer zur Bestimmung des Hämoglobin- und Sauerstoffgehaltes im *Blut. Arch. Ges. Physiol.* **262**, 595–615.

Nilsson, N. J. (1960). Oximetry. *Physiol. Rev.* **40**, 1–26.

Nolan, A. C., Marshall, H. W., Cronin, L., Sutterer, W. F. and Wood, E. H. (1963). Decreases in arterial oxygen saturation and associated changes in pressures and roentgenographic appearance of the thorax during forward ($+$Gx) acceleration. *Aerospace Med.* **34**, 797–813.

Pittman, R. N. and Duling, B. R. (1975a). A new method for the measurement of percent oxyhemoglobin. *J. Appl. Physiol.* **38**, 315–320.

Pittman, R. N. and Duling, B. R. (1975b). Measurement of percent oxyhemoglobin in the microvasculature. *J. Appl. Physiol.* **38**, 321–327.

Polanyi, M. L. and Hehir, R. M. (1962). *In vivo* oximeter with fast dynamic response. *Rev. Sci. Instrum.* **33**, 1050–1054.

Poppius H. and Viljanen, A. A. (1977). A new ear oximeter for assessment of exercise-induced arterial desaturation in patients with pulmonary diseases. *Scand. J. Resp. Dis.* **58**, 279–283.

Powers, S. R., Mannal, R., Neclerio, M., English, M., Man, C., Leather, R., Ueda, H., Williams, G., Custead, W. and Dutton, R. (1973). Physiologic consequences of positive end-expiratory pressure (PEEP) ventilation. *Ann. Surg.* **178**, 265–272.

Powles, A. C. P., Sutton, J. R., Gray, G. W., Mansell, A. L., McFaden, M. and Houston, C. S. (1978). Sleep hypoxaemia at altitude: its relationship to acute mountain sickness and ventilatory responsiveness to hypoxia and hypercapnia. *In* "Environmental Stress, Individual Human Adaptations" (Eds L. J. Filinsbee, J. A. Wagner, J. F. Bergia, B. L. Drinkwater, J. A. Gliner and J. F. Bedi), pp. 373–381, Academic Press, New York and London.

Qvist, J., Pontoppidan, H., Wilson, R. S., Lowenstein, E. and Laver, M. B. (1975). Hemodynamic responses to mechanical ventilation with PEEP: the effect of hypervolemia. *Anesthesiology* **42**, 45–55.

Refsum, H. E. and Hisdal, B. (1958). Construction and use of a simple reflectometer for determination of hemoglobin oxygen saturation in blood. *Scand. J. Clin. Lab. Invest.* **10**, 439–443.

Rodrigo, F. A. (1953). The determination of the oxygenation of blood *in vitro* by using reflected light. *Am. Heart J.* **45**, 809–822.

Sarnquist, F. H., Todd, C. and Whitcher, C. (1980). Accuracy of a new non-invasive oxygen saturation monitor. *Anesthesiology* **53**, S163.

Saunders, N. A., Powles, A. C. P. and Rebuck, A. S. (1976). Ear oximetry: accuracy and practicability in the assessment of arterial oxygenation. *Am. Rev. Respir. Dis.* **113**, 745–749.

Schneider, A. J. L., Miller, R. and Baetz, W. R. (1978). Improved ear oximeter with a rubefacient ointment. *Crit. Care Med.* **6**, 384–386.

Scoggin, C., Nett, L. and Petty, T. L. (1977). Clinical evaluation of a new oximeter. *Heart Lung* **6**, 121–126.

Squire, J. R. (1940). Instrument for measuring quantity of blood and its degree of oxygenation in web of the hand. *Clin. Sci.* **4**, 331–337.

Trask, C. H. and Cree, E. M. (1962). Oximeter studies on patients with chronic obstructive emphysema, awake and during sleep. *N. Engl. J. Med.* **266**, 639–642.

Twersky, V. (1970). Interface effects in multiple scattering by large, low-refracting absorbing particles. *J. Opt. Soc. Am.* **60**, 908–914.

van Assendelft, O. W. (1970). "Spectrophotometry of Hemoglobin Derivatives", pp. 119–130, Springfield, Illinois.

Weinman, J., Hayat, A. and Raviv, G. (1977). Reflection photoplethysmography of arterial-blood-volume pulses. *Med. Biol. Engng Comput.* **15**, 22–31.

West, J. B., Lahiri, S., Gill, M. B., Milledge, J. S., Pugh, L. G. C. E. and Ward, M. P. (1962). Arterial oxygen saturation during exercise at high altitude. *J. Appl. Physiol.* **17**, 617–621.

Wilkinson, A. R., Phibbs, R. H. and Gregory, G. A. (1979). Continuous *in vivo* oxygen saturation in newborn infants with pulmonary disease: a new fiberoptic catheter oximeter. *Crit. Care Med.* **7**, 232–236.

Wood, E. H. and Geraci, J. E. (1949). Photoelectric determination of arterial oxygen saturation in man. *J. Lab. Clin. Med.* **34**, 387–401.

Yoshiya, I., Shimada, Y. and Tanaka, K. (1980). Spectrophotometric monitoring of arterial oxygen saturation in the fingertip. *Med. Biol. Engng Comput.* **18**, 27–32.

Zijlstra, W. G. (1953). "Fundamentals and Applications of Clinical Oximetry", 2nd edn, van Gorcum, Assen, Netherlands.

Zijlstra, W. G. and Mook, G. A. (1962). "Medical Reflection Photometry", van Gorcum, The Hague, Netherlands.

8. TRANSCUTANEOUS BILIRUBIN MEASUREMENT

I. Yamanouchi and A. Yamanishi

Children's Medical Centre, Okayama, Japan
Minolta Camera Co. Ltd., Osaka, Japan

1.0 INTRODUCTION

Jaundice, often termed icterus, means simply a yellowish staining of the integument and deeper tissues due to the presence of excessive quantities of a bile pigment "bilirubin". In general, this excess of bilirubin can result from three conditions: firstly, underdevelopment of the enzyme system necessary for the excretion of bilirubin; secondly, excessive breakdown of red cells; thirdly, any process which blocks bile flow to the duodenum. Because of the limited ability of newborn infants to clear bilirubin from blood plasma, and the increased load of bilirubin on the liver cells, a large proportion of all newborn infants develop jaundice, and this reaches a peak on about the fourth to fifth day of life. This type of jaundice in the newborn is called "physiological jaundice". Premature and low birth-weight infants may be even less able than full-term infants to handle bilirubin, so jaundice may be more marked and may stay longer. In Caucasian infants bilirubin concentrations exceeding 12 mg dl^{-1} (full-term) or 15 mg dl^{-1} (pre-term) suggest the diagnosis of "physiological jaundice".

The neonatal period is the only time at which "hyperbilirubinaemia", an elevated plasma bilirubin concentration, may be life threatening, when it produces bilirubin-encephalopathy, or kernicterus. Hyperbilirubinaemia becomes important in terms of the incidence of kernicterus with serum bilirubin levels over 18–20 mg dl^{-1}. It is therefore important in the care of the newborn to have convenient and reliable methods with which to detect the presence and severity of jaundice.

NON-INVASIVE MEASUREMENTS: 2   *Copyright©1983 by Academic Press Inc. (London) Ltd.*
ISBN 0 12 593402 5   *All rights of reproduction in any form reserved*

It was almost a quarter of a century ago when Gosset (1960) originally devised an "icterometer". It was a simple device for estimating quickly the depth of jaundice in newborn babies at the cot-side without taking a blood specimen. It was made of a strip of transparent Perspex, 1/8 in. thick, 1 1/4 in. wide and 7 in. long. On its convex surface were painted five transverse yellow stripes in slightly different shades. When the icterometer is used, the painted convex side is pressed against the tip of the baby's nose until the skin is blanched. Skin colour can then be matched with the yellow stripes on the scale. The reading is recorded according to the number of the best match. The device was highly successful, for it showed at once whether the baby's skin was more or less jaundiced than it had been the day before; this is often all one needs to know. But the simplicity of the icterometer had led to problems which have been described by several investigators (Gosset, 1960; Culley *et al.*, 1960; Morrison and Wilkinson, 1962). For example, the yellow stripes do not always give a perfect match with the baby's skin; the device may not be reliable in artificial light; the colour shades of icterometers made from different batches of paint may not be identical.

To date, repeated blood sampling and *in vitro* analysis has been the only acceptable methodology available for monitoring the jaundiced newborn infant. In an effort to reduce the number, and the trauma, of invasive procedures in the neonatal nursery, an automatic photo-electric icterometer, which allows serum bilirubin concentration to be estimated by measurements made through the skin of the newborn, was developed. (Yamanouchi *et al.*, 1980).

2.0 BACKGROUND

Measurements of the spectral reflectance of the skin were carried out by Ballowitz and Avery (1970) in order to determine the lighting for nurseries that would be optimal to detect jaundice and cyanosis in newborn babies, and also in order to learn more of the mechanisms of phototherapy in jaundice. The authors found that in icteric newborns there was reduced reflectance between about 430–520 nm. This was also confirmed in icteric rats.

Hannemann *et al.* (1978) investigated a non-invasive method for serum bilirubin estimation from neonatal skin reflectance. Reflectance measurements in the spectral region 400–750 nm were carried out as follows. Monochromatic light, derived from a tungsten light and grating monochromater, was applied to the infant's skin through one branch of a bifurcated fibre-optic bundle. The second branch of the bundle carried the reflected light to an optical detector the amplified output of which was recorded on one channel of an FM tape recorder with a wavelength reference signal recorded on a second channel. The recorded analogue signals were digitized and processed by a computer. Estimates of serum bilirubin levels were then obtained from the reflectance spectra by statistical regression analysis. The analyses of samples from 30 Caucasian newborn infants were compared with serum bilirubin measurements based on a diazo reaction (Jendrassik and Cleghorn, 1936). Using a double natural logarithm method with five wavelengths (425, 460, 525, 535 and 545 nm) the best correlation coefficient was found to be 0·931. The 95% prediction limit for this empirical relationship was shown to be ±2 mg dl^{-1}.

Hannemann and co-workers made a further study in 1979 in order to analyse statistically the relationship for predicting serum bilirubin level from various combinations of spectral data, and to establish a physical basis for the relationship between skin reflectance and serum bilirubin. Reflectance spectra (380–800 nm) and concurrent serum bilirubin measurements were taken on a sample population of 58 white and 45 black, full-term infants (1–3 days of age).

For the first purpose of the study, the correlations considered were based upon: reflectance in a single wavelength band; ratio of the reflectance in two wavelength bands; and multiple linear regressions. For the white group, a correlation based upon the ratio of reflectance at 450 nm to the reflectance at 520 nm, yielded $r = 0·81$. For the black infant group, the ratio of reflectance at 480 nm to the reflectance at 500 nm proved to be the best combination with $r = 0·83$. Multiple linear regression analysis, comprised of six wavelengths (410, 420, 440, 490, 520 and 550 nm) gave a correlation coefficient, $r = 0·831$ with a standard error of estimate ±1·45 mg dl^{-1} for the white group. For the black infant group, a four wavelength (400, 490, 520 and 590 nm) analysis produced $r = 0·877$ with a standard error of estimate ±1.46 mg dl^{-1}.

For the second purpose of the study, they assumed the skin to be a one-dimensional, homogeneous medium having a certain absorption coefficient and a certain scattering coefficient. They then applied the Kubelka–Munk formulae, substituting the skin surface reflectance for that of a layer so thick that further increase in thickness fails to change the reflectance. From the relation given by three wavelengths (420, 460, 520 nm), they obtained the results, in which, for the white infant group (58 infants, 124 observations) $r = 0·778$ and the standard error of estimate was ±1·4 mg dl^{-1}, and for the black infant group (45 infants, 96 observations) $r = 0·865$ and the standard error of estimate was ±1·4 mg dl^{-1}.

Independently of Hannemann et al., Peevy et al. (1978) made a study to estimate the serum bilirubin level by the light reflectance of skin at five wavelengths (424, 465, 511, 556 and 629 nm). Thirty black and 14 Caucasian infants were studied with simultaneous spectral reflectance and serum bilirubin level measurements. Serum bilirubin levels ranged from 1 to 12 mg dl^{-1} and correlated with spectral reflectance estimates with a coefficient of 0·94. Agreement was within 2 mg dl^{-1} in all cases. They concluded that the measurement of spectral reflectance of skin is an accurate and non-invasive screening method for the estimation of serum bilirubin.

These techniques are based on the use of large computers, and it is expected that they will continue to provide useful information for fundamental studies. However, the complete system is too complicated for use by clinicians. In an effort to simplify the instrumentation, independently of these preceding studies, a portable bilirubin meter which allows serum bilirubin concentration to be estimated transcutaneously was developed (Yamanouchi *et al.*, 1980).

3.0 BASIC PRINCIPLE

3.1 Direct Spectrophotometry of Total Serum Bilirubin

In order to measure total serum bilirubin by spectrophotometry, Beer's law is applied either by diluting the serum or by shortening the path length of the beam.

Consider two wavelengths, λ_1 and λ_2, such that the spectral absorption of bilirubin is greater at λ_1 than at λ_2. The respective absorptivities, $A(\lambda_1)$ and $A(\lambda_2)$, are then given by:

$$A(\lambda_1) = \varepsilon_{Hb}(\lambda_1) \cdot C_{Hb} + \varepsilon_b(\lambda_1) \cdot C_b \tag{1}$$

$$A(\lambda_2) = \varepsilon_{Hb}(\lambda_2) \cdot C_{Hb} + \varepsilon_b(\lambda_2) \cdot C_b \tag{2}$$

where $\varepsilon_{Hb}(\lambda)$ is an absorption constant; C, concentration; suffix Hb, haemoglobin; and suffix b, bilirubin.

Accordingly, by eliminating C_{Hb} from (1) and (2) we obtain:

$$
\begin{aligned}
C_b &= \frac{1}{\varepsilon_{Hb}(\lambda_2) \cdot \varepsilon_b(\lambda_1) - \varepsilon_{Hb}(\lambda_1) \cdot \varepsilon_b(\lambda_2)} \times \\
&\quad (\varepsilon_{Hb}(\lambda_2) A(\lambda_1) - \varepsilon_{Hb}(\lambda_1) A(\lambda_2)). \\
&= \alpha_A(\lambda_1) - \beta_A(\lambda_2),
\end{aligned} \tag{3}
$$

where α and β are constants. This means that the concentration of bilirubin is obtained from the absorptivities $A\,(\lambda_1)$ and $A\,(\lambda_2)$. This implies that the effect of haemolysis is eliminated.

Furthermore, if we choose wavelengths λ_1 and λ_2 such that:

$$\varepsilon_b\,(\lambda_1) \neq \varepsilon_b\,(\lambda_2)$$
$$\varepsilon_{Hb}\,(\lambda_1) = \varepsilon_{Hb}\,(\lambda_2)$$

then,

$$C_b = \frac{1}{\varepsilon_b\,(\lambda_1) - \varepsilon_b\,(\lambda_2)}\left(A\,(\lambda_1) - A\,(\lambda_2) \right) \tag{4}$$

thus the concentration of bilirubin is simply proportional to the difference between the absorptivities $A\,(\lambda_1)$ and $A\,(\lambda_2)$.

3.2 Determination of Serum Bilirubin by Skin Reflectance

Hanley and DeWitt (1978) assumed that the skin was a one-dimensional, homogeneous medium, and tried to apply the Kubelka–Munk Analysis. Kubelka and Munk (Kubelka *et al.*, 1931; Wyszecki and Stiles, 1967) assumed, firstly, a medium having no border with the air, and secondly, a layer of particles distributed uniformly and randomly. They showed that when perfectly diffused light is incident on the layer, the optical characteristics of the layer of particles are determined by two constants, i.e. the absorption coefficient, K, and the scattering coefficient, S.

As shown in Fig. 1, they assumed a layer of particles and a unit layer of thickness dx at depth x. If a luminous flux incident upon the unit layer is called i and the flux passing the unit layer in the direction of reflection is called j, the following differential equations apply:

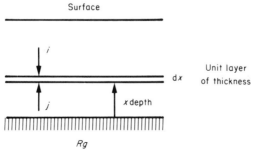

FIG. 1   Representation of a particle layer.

$$\left.\begin{array}{l} d\dot{j} = -(S + K)\,\dot{j}dx + Si dx \\ -d\dot{i} = -(S + K)\,\dot{i}dx + Sj dx \end{array}\right\}. \tag{5}$$

Solutions of these equations are obtained under various boundary conditions. They considered $R\infty$ to be the reflectance of a layer so thick that further increase in thickness fails to change the reflectance, and they produced the following relation between $R\infty$ and K/S:

$$R\infty = 1 + \frac{K}{S} - \left(\frac{K^2}{S^2} + \frac{2K}{S}\right)^{1/2} \tag{6}$$

or

$$\frac{K}{S} = \frac{R\infty}{2} + \frac{1}{2R\infty} - 1. \tag{7}$$

3.3 Principle of Minolta Transcutaneous Bilirubin Meter

In the Minolta transcutaneous bilirubin meter, light is carried through a glass fibre bundle to the skin surface. The light penetrates the skin without passing through the air. The light coming back out of the skin is then returned through the glass fibre bundle to the spectrophotometric module, once again without passing through air. The measurement is made on blanched skin, produced by a pre-set pressure-activated switch attached to the glass light guide (see Section 4.0). Under this condition, the measured light intensity is related mainly to the absorption by bilirubin and/or melanin, and the absorption of bilirubin is considered to conform to Beer's law. Accordingly, the detected light intensity, $I(\lambda)$, according to Lambert–Beer's law, is given by

$$I(\lambda) = I_0(\lambda) \cdot F(\lambda) \cdot 10^{-\varepsilon_b(\lambda) \cdot C \cdot t} \tag{8}$$

where I_0 is the intensity of the incident light, $F(\lambda)$ is the rate of absorption of light by the "non-bilirubin" compartment, $\varepsilon_b(\lambda)$ is the absorption coefficient of bilirubin, C is the concentration of bilirubin, t is the effective optical path-length of bilirubin. As described above, two wavelengths, λ_1 and λ_2, are chosen in order to minimize the residual effect of haemoglobin. Each wavelength has a different bilirubin absorption coefficient, and both have approximately equal haemoglobin absorption coefficients. Then the difference between respective optical densities, O.D., is given by

$$\text{O.D.}\,(\lambda_1) - \text{O.D.}\,(\lambda_2) = \log\,(F(\lambda_2)/F(\lambda_1)) + (\varepsilon_b\,(\lambda_1) - \varepsilon_b(\lambda_2))C \cdot t. \tag{9}$$

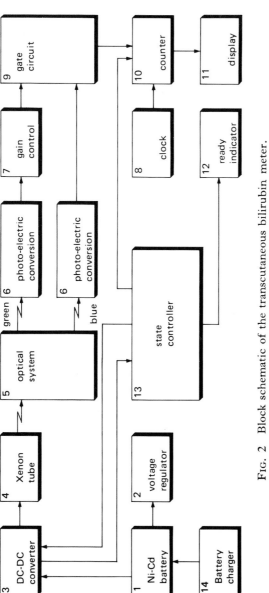

FIG. 2   Block schematic of the transcutaneous bilirubin meter.

Thus, the terms other than those related to bilirubin absorption are offset, and the output is proportional to the bilirubin concentration.

4.0 TECHNICAL DESCRIPTION OF
MINOLTA TRANSCUTANEOUS BILIRUBIN METER

The instrument is operated by built-in rechargeable batteries, is 16 cm long × 7 cm wide × 3 cm deep, and weighs *c*. 300 g. It is hand held, and can be operated easily, even in an incubator.

4.1 Description of the Instrument

Figure 2 is a block schematic of the instrument. The Ni–Cd battery, 1, is rechargeable and has stable discharge characteristics even at heavy discharge, as the internal resistance is very low. The voltage regulator, 2, supplies a constant voltage to the various parts of the circuit, in the face of variations with battery discharge. The DC–DC converter, 3, raises the voltage of the Ni–Cd battery to that necessary for firing the xenon tube, and it also charges a capacitor.

The xenon tube, 4, is similar to those commonly used in photography. The energy emitted for one flash is 2 J, that is about 1/10 of the energy used in photography. The intensity of the tube output used here is made to follow an exponential decay, and this is used in the calculation of optical density.

The optical system, 5, consisting of the bundle of glass fibre and the spectrophotometric module, is shown in more detail in Fig. 3. The

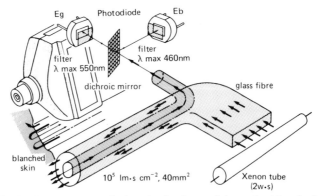

FIG. 3   Light from a xenon flash-tube is directed into the blanched skin via a concentric optical fibre arrangement. The reflected and backscattered light is split by a dichroic mirror and passed through filters (550 and 460 nm) before being detected by photodiodes (Eg and Eb). (Copyright © 1980 American Academy of Pediatrics. Reproduced from *Pediatrics* **65**(2), 195–202, with permission.)

light intensity is measured by means of a GaAsP photocell, the output current of which is converted into a voltage by the I–V converter. As the gain of the I–V converter is very high, the output will be saturated until the light intensity decays below a threshold value. The photocurrent during the saturation period is compensated automatically in order to maintain the ideal state of the I–V converter.

The gain control, 7, adjusts the instrument display to zero when a standard white reflector is used.

The clock, 8, generates pulses with a fixed time interval, and these are fed to a counter. A gate circuit allows clock pulses, occurring between time points set when the "blue" and "green" voltages cross pre-set thresholds, to be counted. The "count numbers" obtained represent the difference in optical density of the blue and green wavelengths. The counter, 10, is a 2-digit decade counter, the output of which is interfaced to two LED 7-segment displays, 11. A ready indicator, 12, is lit when the capacitor of the DC–DC converter has stored sufficient energy to fire the flash tube. The state controller, 13, is a logic circuit used to ensure that a constant voltage on the capacitor of the DC–DC converter is attained, the ready indicator is lit before letting the xenon tube trigger.

When the force applied between the skin and the glass fibre of the instrument reaches approximately 200 g the xenon tube is fired. The intense light (2J) travels through the outer part of the concentric optical glass fibre bundle to a flattened skin contact area, or probe, as indicated in Fig. 3.

The light from the probe area penetrates the blanched skin and trans-illuminates the cutaneous and subcutaneous tissue. The scattered light is subject to absorption by the bilirubin, and a component is returned, through the inner part of the concentric optical glass fibre bundle, to the spectrophotometric module.

Inside the module, light is divided by a dichroic mirror into two components, one of which passes through a blue filter with a maximum transmittance at 460 nm and the other through a green filter with maximum transmittance at 550 nm. The intensity of the yellow colour is obtained as the difference between the optical densities of blue and green (such difference corrected to zero for white). This intensity is then converted to an arbitrary scale, shown on the digital display, linearly related to bilirubin.

The use of the transcutaneous bilirubin meter on the forehead of an infant is shown in Fig. 4.

4.2 Operating Procedure

The operating procedure of the instrument is simple. Once the power switch has been activated, the reset button should be depressed; this will charge the photoflash circuit and subsequently cause the "ready" lamp to

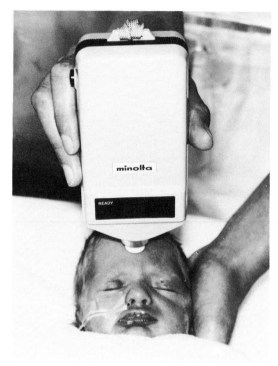

FIG. 4   Bilirubin meter being used on the forehead of a newborn infant.

be illuminated. The photoprobe may then be positioned on the skin of the
neonate. The instrument is then pressed steadily against the skin surface
until an audible click occurs, at which time the reading should be noted.
When the measurement has been completed, the power should be switched
off.

5.0 CLINICAL PERFORMANCE DATA

Clinical trials to evaluate the performance of the transcutaneous bilirubin
meter have been focused primarily on precision, accuracy and technique
dependence. The following sections present data relating to the coefficients
of variation for each operational aspect studied.

5.1 Precision of Transcutaneous Bilirubin Meter

The transcutaneous bilirubin meter was tested on five different neonatal
patients with serum bilirubin concentration levels ranging from 2·5 to

TABLE I

Precision of transcutaneous bilirubin measurement

Serum bilirubin concentration (mg $(100\,ml)^{-1}$)	Sequence of reading					Mean	Coefficient of variation (%)
	1st	2nd	3rd	4th	5th		
2·5	9	9	9	9	10	9·2	4·86
5·0	12	12	12	12	12	12·0	0
10·5	18	16	17	17	18	17·2	4·90
14·5	21	21	22	22	20	21·2	3·95
19·0	26	27	28	27	27	27·0	2·62

19·0 mg dl^{-1}. In each instance, the same test was performed on the same patient, at the same body site, by the same operator. Results, shown in Table I, indicate that the 25-test sampling for precision produced a mean coefficient of variation of $<5\cdot0\%$. It can be assumed that the marginal percentage of variation is predictable, and that the transcutaneous bilirubin meter is sufficiently precise to provide a reliable indication of bilirubin concentration levels in neonates with a clinical diagnosis of icterus.

5.2 Technique-dependence in Transcutaneous Bilirubin Measurement

In testing the transcutaneous bilirubin meter for technique-dependence, four neonates, each with different, previously established levels of total serum bilirubin concentration (ranging from 1·5 to 13·5 mg dl^{-1}), were tested by four different individuals: one neonatologist, two resident physicians and one registered nurse. Each individual was asked to use the instrument in each infant, at the same body site, five times in succession. The coefficients of variation, (Table II) ranged from 2·13 to 4·98%. Whilst the coefficient of variation did not exceed 5·0% in any case, it should be noted that the magnitude of the variation did not appear to increase proportionately with the level of total serum bilirubin concentration.

Although no tests have been performed to evaluate the effects of varying technique among non-professional personnel, the simplicity of operation suggests that technique-dependence should only be of marginal significance to the overall performance.

TABLE II

Operator difference in transcutaneous bilirubin measurement

Serum bilirubin concentration (mg $(100 \text{ ml})^{-1}$)	Test administrator				Mean	Coefficient of variation (%)
	MD[a]	MD[b]	MD[b]	RN		
1·5	9·0	9·0	10·0	9·8	9·45	2·13
7·5	15·2	16·2	15·4	17·0	15·95	3·69
11·0	21·0	20·4	18·8	21·6	20·45	4·09
13·5	22·6	23·4	22·8	22·8	22·93	4·98

[a]Neonatologist.
[b]Resident.
(Copyright © 1980 American Academy of Pediatrics. Reproduced from *Pediatrics* **65**(2), 195–202, with permission.)

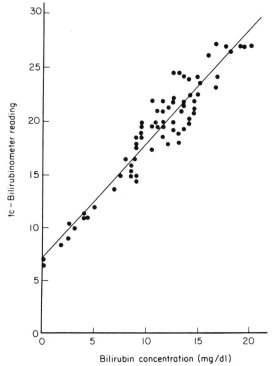

FIG. 5 Correlation between transcutaneous bilirubin meter reading and serum total bilirubin concentration measured with the American Optical bilirubinometer. Triplicate measurements were made, on the forehead, by one neonatologist. $y = 1·08x + 7·22$; $r = 0·951$; $P < 0·001$; $n = 73$. (Copyright © 1980 American Academy of Pediatrics. Reproduced from *Pediatrics* **65**(2), 195–202, with permission.)

5.3 Accuracy of Transcutaneous Bilirubin Measurement

The transcutaneous bilirubin meter was tested for accuracy in both full-term and low birth weight neonates.

5.3.1. Correlation between Transcutaneous and Serum Bilirubin Measurement

The transcutaneous bilirubin meter was tested for accuracy by comparison with measurements on withdrawn blood samples using the American Optical Corporation bilirubinometer. The precision, expressed by coefficient of variation, of this measurement has been reported by Levkoff *et al.* (1970) to be 2·4% at a level of 20·3 mg dl⁻¹, which is the important region as regards detection of hyperbilirubinaemia. Blood sampling for the invasive procedure was accomplished simultaneously with the transcutaneous measurement. It should be noted, however, that no tests were conducted on any infants diagnosed to have medical complications other than physiological jaundice. Triplicated transcutaneous measurements on the forehead of the infant were carried out by one neonatologist throughout this series of tests. As illustrated in Fig. 5, a linear relationship was found

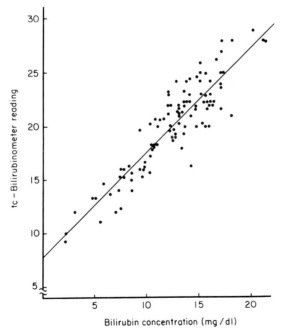

FIG. 6   A similar comparison between transcutaneous and serum total bilirubin concentration to that shown in Fig. 5. A separate group of infants was used and the measurements made by a different operator (resident physician). $y = 0\cdot989x + 7\cdot812$; $r = 0\cdot896$; $P < 0\cdot001$; $n = 105$.

to exist between the two procedures, expressed as $y = 1·08x + 7·22$; $r = 0·951$; $P < 0·001$; $n = 73$.

In order to check the individual characteristics in transcutaneous measurements, a correlation study was performed by another individual: a resident physician on another set of normal term neonates. As illustrated in Fig. 6, a similar linear relationship was found to exist: $y = 0·989x + 7·812$; $r = 0·896$; $P < 0·001$; $n = 105$.

At very high levels of hyperbilirubinaemia, ranging between 26 and 30 mg dl^{-1}, the transcutaneous method lost its accuracy. As illustrated in Fig. 7, there is obviously a plateau in the reading of the transcutaneous bilirubin meter at these high levels. However, this saturation effect has nothing to do with the characteristics of the instrument itself. It is clear to say that this plateau effect is a limitation to transcutaneous bilirubin measurement.

It is well known that the skin colour of obstructive jaundice is slightly different from that of haemolytic jaundice. For this reason, the transcuta-

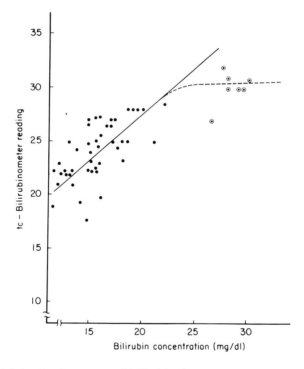

FIG. 7   At high levels of serum total bilirubin the transcutaneous measurements are proportionately low.

neous bilirubin meter has been tested for accuracy in seven older infants with obstructive jaundice due to congenital bile duct atresia or neonatal hepatitis. Figure 8 illustrates clearly that the bilirubin meter readings in older infants in these conditions vary considerably from those of normal term newborns.

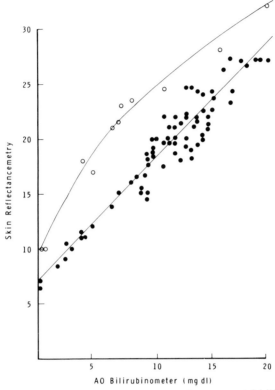

FIG. 8   Relationships between transcutaneous and serum total bilirubin in older infants with indirect and direct reacting hyperbilirubinaemia. ● Indirect reacting hyperbilirubinaemia; ○ direct reacting hyperbilirubinaemia.

The transcutaneous bilirubin meter was also tested by comparison with invasive serum bilirubin measurements using Michaelsson's colorimetric alkali azobilirubin method. Once again, blood sampling for the invasive procedure was carried out simultaneously with the transcutaneous measurement. As shown in Fig. 9, a linear relationship was found to exist between the measurements made by the two procedures: $y = 1{\cdot}11x + 7{\cdot}96$; $r = 0{\cdot}933$; $P < 0{\cdot}001$; $n = 66$.

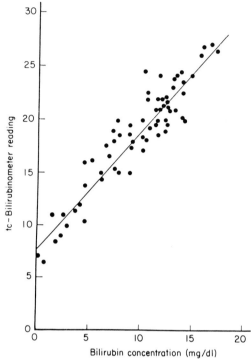

Fig. 9 Correlation between transcutaneous readings and serum bilirubin concentration measured by the alkali azobilirubin method. $y = 1\cdot11x + 7\cdot96$; $r = 0\cdot933$; $P < 0\cdot001$; $n = 66$. (Copyright © 1980 American Academy of Pediatrics. Reproduced from *Pediatrics* **65**(2), 195–202, with permission.)

TABLE III

Coefficient of correlation between serum bilirubin concentration and reading on transcutaneous bilirubin measurements at different sites of body surface

	Fore-head	Right upper chest	Middle of sternum	Right upper abdo-men	Upper back	Lower back	Sole	Heel
No. of measurements	40	40	40	39	39	39	40	40
Coefficient of correlation	0·877	0·780	0·864	0·887	0·903	0·887	0·792	0·559
$P <$	0·001	0·001	0·001	0·001	0·001	0·001	0·001	0·001

(Copyright © 1980 American Academy of Pediatrics. Reproduced from *Pediatrics* **65**(2), 195–202, with permission.)

5.3.2. *Accuracy of Transcutaneous Bilirubin Measurement at Different Body-sites*

The forehead is recommended as the site of choice for transcutaneous bilirubin measurement because: (1) it is usually accessible; (2) the frontal bone structure offers the necessary resistance to activate the xenon tube mechanism; and (3) the skin surface of the forehead is taut and firm. It must be assumed, however, that instances will arise when the forehead may be contra-indicated as a measurement site. For this reason, the transcutaneous bilirubin meter has been tested for accuracy at eight different body sites: the forehead, the right upper chest, the middle sternum, the right upper abdomen, the upper back, the lower back, the right sole and the right heel. Each body site was tested three times on 40 different neonates representing a random sampling of term infants.

As shown in Table III, the coefficients of correlation between total serum bilirubin concentration levels and the transcutaneous measurements were high with the exceptions of the sole and heel areas. As there are no definitive data available at this time, it is reasonable to assume that these lower coefficients of correlation are attributable to the differences in subcutaneous and epidermal composition at these sites, e.g. the contents of keratin and elastin.

5.3.3. *Correlation between Transcutaneous and Invasive Bilirubin Measurement in Infants Diagnosed to have Medical and Surgical Complications*

The transcutaneous bilirubin meter was tested for accuracy in full-term neonates admitted to our Neonatal Intensive Care Unit having medical and/or surgical complications including haemolytic hyperbilirubinaemia. Blood sampling for the invasive procedure was carried out essentially simultaneously with the triplicated transcutaneous measurement on the forehead. The neonates tested were divided into three groups by post-natal age: group one, neonates aged less than 7 days; group two, neonates

TABLE IV

Correlation between transcutaneous and serum bilirubin measurement in infants diagnosed to have medical and surgical complications

Age (days)	n	r	$y = ax + b$	P
7	61	0·830	$y = 0·86x + 10·31$	0·001
7–14	67	0·883	$y = 0·89x + 9·59$	0·001
14–28	27	0·908	$y = 0·98x + 8·92$	0·001

aged from 7 to 14 days; group three, neonates aged from 14 to 28 days. Neonates treated by exchange transfusion and/or phototherapy were excluded. As can be seen in Table IV, good correlation was found between the transcutaneous and serum measurements in these neonates having complications.

5.3.4. *Relationship between Serum Bilirubin Concentration and Transcutaneous Bilirubin Meter Measurements during the Course of Hyperbilirubinaemia*

When neonatal jaundice develops rapidly the serum bilirubin can rise so fast that the skin colour lags behind. Similarly, when serum levels are falling rapidly the fading of the skin colour is delayed. In order to see the discrepancies between the level of serum bilirubin and the degree of skin jaundice, observations were made during the course of hyperbilirubinaemia in four term neonates. Figure 10 illustrates that in the infants with a decrease in serum bilirubin concentrations there is a moderate delay in the fading of skin jaundice compared to the decrease in serum bilirubin. Skin colour lags behind a decreasing serum bilirubin concentration, thus introducing a bias such that the transcutaneous bilirubin meter indicates a higher value than is actually present.

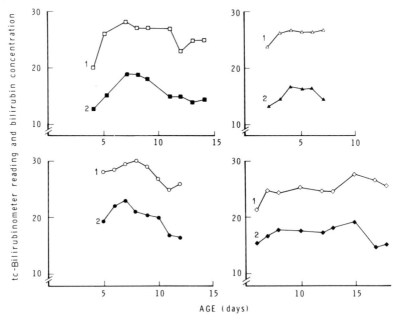

FIG. 10 The long-term relationships between transcutaneous readings and serum bilirubin concentrations in four subjects. 1: Transcutaneous bilirubin meter reading; 2: serum bilirubin concentration.

5.3.5. *Transcutaneous Bilirubin Measurement in Premature and Low Birth-weight Infants*

Naturally, skin colour cannot be overlooked as a possible influence on the accuracy of the transcutaneous bilirubin meter. This is important since premature and low birth-weight neonates commonly exhibit darker skin than do term neonates. Post *et al.* (1976) and Krauss *et al.* (1976) studied infant groups of various gestational ages and found decreasing skin reflectance with gestational age.

When testing instrument accuracy exclusively in premature or low birth-weight infants, the subjects were segregated into four groups according to weight. Serum bilirubin measurement (American Optical Bilirubinometer) and transcutaneous bilirubin measurement were compared, the latter being made on both the forehead and the inner side of the forearm of each baby.

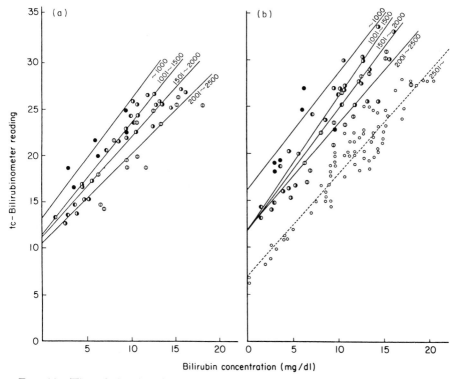

FIG. 11   The relationships between transcutaneous and serum bilirubin measurements in pre-term and low birth-weight infants. (a) Forearm. (b) Forehead. ○ 2501 ~, ⬓ 2001 ~ 2500 g, ◗ 1501 ~ 2000, ◐ 1001 ~ 1500, ● ~ 1000. (Copyright © 1980 American Academy of Pediatrics. Reproduced from *Pediatrics* **65**(2), 195–202, with permission.)

Figures 11(a) and (b) illustrate clearly that the regression lines vary considerably from group to group. It is of particular interest that the regression line for the lowest birth-weight group is the most positive. It is entirely conceivable, therefore, that the differences in the coefficients of correlation and P values in the tests performed in term and low birth-weight neonates could be the result of the instrument's sensitivity to changes in pigmentation and subcutaneous bilirubin (Tables V and VI).

TABLE V

Accuracy of transcutaneous bilirubin measurement in different birth-weight groups[a]

Birth-weight group	n	r	$y = ax + b$	P
2001–2500	14	0·848	$y = 0·95x + 10·63$	<0·001
1501–2000	23	0·929	$y = 1·08x + 11·30$	<0·001
1001–1500	7	0·975	$y = 1·22x + 11·41$	<0·001
< 1000	6	0·949	$y = 1·30x + 13·29$	<0·005

[a] Correlation between transcutaneous bilirubin measurement on forearm and serum total bilirubin concentration measured by American Optical bilirubino-meter.
(Copyright © 1980 American Academy of Pediatrics. Reproduced from *Pediatrics* 65(2), 195–202, with permission.)

TABLE VI

Accuracy of transcutaneous bilirubin measurement in different birth-weight groups[a]

Birth-weight group	n	r	$y = ax + b$	P
2501 <	73	0·951	$y = 1·08x + 7·22$	<0·001
2001–2500	14	0·888	$y = 1·13x + 11·92$	<0·001
1501–2000	23	0·922	$y = 1·32x + 11·76$	<0·001
1001–1500	8	0·977	$y = 1·46x + 11·65$	<0·001
< 1000	6	0·714	$y = 1·24x + 16·23$	$0·05 < P < 0·1$

[a] Correlation between transcutaneous bilirubin measurement on forehead and serum total bilirubin concentration measured by American Optical bilirubino-meter.
(Copyright © 1980 American Academy of Pediatrics. Reproduced from *Pediatrics* 65(2), 195–202, with permission.)

5.3.6 *Transcutaneous Bilirubin Measurement and Exchange Transfusions*

Limited testing has been carried out in infants who have received exchange transfusions. Table VII shows that the initial results from pre- and post-exchange data demonstrate an unacceptable level of correlation with total serum bilirubin concentration levels and this contra-indicates the use of transcutaneous bilirubin measurements in such cases.

TABLE VII

Relationship between serum bilirubin concentration and reading on transcutaneous bilirubin meter at different sites of body surface before and after exchange transfusion[a]

Age (h)	Serum bilirubin concen-tration (mg $(100 ml)^{-1}$)	Fore-head	Right upper chest	Middle of sternum	Right upper abdomen	Lower back	Sole	Heel
87	27·1	31·7	25·7	28·7	28·7	20·0	14·7	14·3
Exchange transfusion								
89	16·1	29·3	29·0	28·0	26·7	18·3	14·0	12·3
93	18·5	27·3	28·0	26·7	24·3	18·3	14·3	14·0
101	15·5	27·3	26·0	25·7	24·3	19·0	14·0	16·0
111	16·2	28·3	27·7	26·7	24·3	21·3	14·0	13·7
126	18·0	29·3	29·3	28·0	25·7	22·0	14·0	13·7
157	19·5	32·0	31·7	31·7	28·7	27·0	18·0	15·0
174	20·7	30·0	30·0	30·3	27·3	21·7	17·0	15·0

[a] The patient was a female infant weighing 3900 g at birth.

5.3.7. *Transcutaneous Bilirubin Measurement and Phototherapy*

In 1958 Cremer *et al.* reported that the exposure of infants to sunlight or blue fluorescent light produced a fall in serum bilirubin concentration. The ability of bile pigments to absorb light results in the photodecomposition of bilirubin. There are several well-controlled clinical studies concerning the effectiveness of so-called "phototherapy" as a means of preventing or treating moderate hyperbilirubinaemia. Once phototherapy is started in the jaundiced infant, the skin becomes less yellow. The skin colour cannot therefore be used as a means of assessing serum bilirubin levels, either by visual observation or with the use of the transcutaneous bilirubin meter. Limited testing has been carried out to evaluate the instrument's accuracy in cases where hyperbilirubinaemia has been treated with a regimen of phototherapy. Initial test results indicate that differences between transcutaneous measurements and invasive bilirubin measurements vary considerably, thus suggesting the contra-indication of transcutaneous bilirubin measurements during phototherapy. Table VIII shows the actual results obtained at different body sites, and the wide fluctuations during the course of phototherapy can be seen.

TABLE VIII

Relationship between serum bilirubin concentration and reading on transcutaneous bilirubin meter at different sites of body surface before and after phototherapy[a]

Age (h)	Serum bilirubin concentration (mg $(100 \text{ ml})^{-1}$)	Fore-head	Right upper chest	Middle of sternum	Right upper abdomen	Upper back	Lower back	Sole	Heel
72	17·8	27·5	28·0	28·5	24·0	24·0	23·5	16·5	16·0
Phototherapy									
96	16·5	25·0	17·0	15·7	16·3	25·0	22·7	14·0	14·0
120	14·6	24·0	24·3	23·6	23·6	24·0	24·3	17·0	16·6
216	13·2	24·0	26·3	26·0	23·3	25·0	23·0	14·3	15·0

[a] The patient was a male infant weighing 3910 g at birth.
(Copyright © 1980 American Academy of Pediatrics. Reproduced from *Pediatrics* 65(2), 195–202, with permission.)

5.3.8. *Miscellaneous Factors*

The following factors may affect the correlation between the transcutaneous bilirubin meter reading and serum values. (1) Ecchymoses, bruising or petechiae, caused by extravasation of blood into the skin, will give falsely low readings. One should be aware of this and try to take readings on unaffected areas of skin. (2) The pressure applied, to produce blanching of the skin, is important. Application to the forehead, a comparatively firm site, allows a higher pressure to be achieved than on soft tissue. No data are yet available as to the acceptability of a lower operating force, say 150 g or less, but this should be considered. (3) As already mentioned above, skin colour is critical. Unfortunately we have had no chance to use the device on black infants, but others have (Brown *et al.*, 1981; Vangianichyakorn *et al.*, 1981). Changing skin pigmentation during the early neonatal period is likely to complicate the interpretation of transcutaneous measurements in black infants, as it does in white infants. (4) Drugs, ischaemia, cyanosis, and hypoxia: these have not yet been studied enough

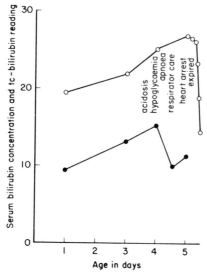

FIG. 12   This shows a delayed fall in transcutaneous readings compared with serum measurements, in a 3240-g term neonate with sacrococcygeal teratoma and klebsiella sepsis. ○, Transcutaneous bilirubin; ●, serum bilirubin concentrations.

to comment upon. However, these factors may have some influence. For example, Fig. 12 shows a marked change of both serum and transcutaneous bilirubin levels in the case of surgically treated sacrococcygeal teratoma with scleredema due to coli-septicaemia. A marked discrepancy between the serum bilirubin level and transcutaneous bilirubin reading is seen. The discrepancy appeared at the onset of apnoea and acidosis. The transcutaneous reading decreased after cardiovascular collapse; it was of interest that the reading continued to decrease after death.

6.0 COMMENT AND CONCLUSIONS

The method described in this chapter is quite different to that reported by Post *et al.* (1976) and by Krauss *et al.* (1976), who used skin-surface reflectance. Our method ensures direct skin contact in order to avoid skin-surface reflectance; thus the yellowish colour intensity is more representative of tissue bilirubin.

The two-wavelength spectrophotometric principle used in the instrument may be associated with certain limitations relating to skin colour. We have tested the device exclusively in Japanese newborns, and acceptable correlation has been found in full-term infants. The device has also been successful when used in Caucasian neonates (Lucey *et al.*, 1980). In this work 48, mostly full-term, infants were studied and a good correlation was found: $r = 0.89$; $P < 0.001$. The authors found that the transcutaneous bilirubin device provided a reliable reflection of serum bilirubin concentrations in newborn infants in the range of 5–20 mg%, and that the device could be relied upon in infants who had been off phototherapy for 24 h or if the measurement site had been shielded from light.

Brown *et al.* (1981) reported that the correlation between transcutaneous bilirubin level and total serum bilirubin concentration was strong in both black ($r = 0.79$) and white ($r = 0.87$) infants. However, they found that in black infants receiving phototherapy there was no longer a strong correlation ($r = 0.36$), although this remained good ($r = 0.79$) in white infants receiving phototherapy. Vangianichyakorn *et al.* indicated a high correlation ($r = 0.80$) in term black infants not receiving phototherapy and a lower correlation ($r = 0.64$) in Spanish infants.

On the basis of these studies one may conclude the following:

(1) The device can be used to estimate serum bilirubin levels in full-term neonates with marked physiological jaundice. It is as accurate as the current laboratory methods.

(2) The method is simple, reproducible and atraumatic. It can be used to follow the rise of serum bilirubin in an infant between 5–15 mg dl^{-1}.

(3) At values over 15 mg dl^{-1}, any infant should have a serum bilirubin determination to confirm the measurement made by the transcutaneous bilirubin meter.

(4) Each hospital should initially perform a study to establish the level of correlation between transcutaneous and serum bilirubin measurements.

(5) The transcutaneous bilirubin meter should not be relied upon entirely in the following situations:

   (a) low birth-weight infants (due to lack of experience at this time);
   (b) black infants;
   (c) infants receiving phototherapy within 24 h before the determination;
   (d) infants with ABO or Rh haemolytic disease whose serum bilirubin level is expected to rise very rapidly;
   (e) seriously ill infants with circulatory collapse or hypoxia.

(6) As a screening method this should be ideal for general paediatric nurseries etc., for following neonates with mild to moderate jaundice.

(7) It is simple to operate. It can be used by nurses, technicians, physicians or trained office aides with reliability.

References

Ballowitz, L. and Avery, M. E. (1970). Spectral reflectance of the skin. *Biol. Neonate* **15**, 348–360.

Bratlid, D. and Winsnes, A. (1971). Determination of conjugated and unconjugated bilirubin by methods based on direct spectrophotometry and chloroform-extraction. A reappraisal. *J. Clin. Lab. Invest.* **28**, 41–48.

Brown, A. K., Kim, M. H., Nachpuckdee, P. and Boyle, G. (1981). Transcutaneous bilirubinometry in infants. *Pediatr. Res.* **15**, 653.

Cremer, R. J., Perryman, P. W. and Richards, D. H. (1958). Influence of light on the hyperbilirubinaemia of infants. *Lancet* **i**, 1094.

Culley, P. E., Waterhouse, J. A. H. and Wood, B. S. B. (1960). Clinical assessment of depth of jaundice in newborn infants. *Lancet* **i**, 88–89.

Gambino, S. R. (1965). Measurement of bilirubin in the newborn utilizing a new optical instrument. *Am. J. Clin. Path.* **44**, 564–565.

Gosset, I. H. (1960). A perspex icterometer for neonates. *Lancet* **i**, 87–88.

Hanley, E. J. and DeWitt, D. P. (1978). A physical model for the detection of neonatal jaundice by multispectral skin reflectance analysis. *Proceedings of the Sixth Annual New England Bioengineering Conf.*, Kingston R. I., 346–349.

Hannemann, R. E., DeWitt, D. P. and Wiechel, J. F. (1978). Neonatal serum bilirubin from skin reflectance. *Pediatr. Res.* **12**, 207–210.

Hannemann, R. E., DeWitt, D. P., Hanley, E. J., Schreiner, R. L. and Bonderman, P. (1979). Determination of serum bilirubin by skin reflectance: Effect of Pigmentation. *Pediatr. Res.* **13**, 1326–1329.

Hertz, H., Dybkaer, R. and Lauritzen, M. (1974). Direct spectrophotometric determination of the concentration of bilirubin in serum. *Scand. J. Clin. Lab. Invest.* **33**, 215–230.

Jackson, S. H. and Hernandez, A. H. (1970). A new "Bilirubinometer" and its use in estimating total and conjugated bilirubin in serum. *Clin. Chem.* **16**, 462–465.

Jendrassik, L. and Cleghorn, R. H. (1936). *Biochem. Z.* **289**, 1.

Krauss, A. N., Post, P. W., Waldman, S. *et al.* (1976). Skin reflectance in the newborn infant. *Pediatr. Res.* **10**, 776–778.

Kubelka, P. and Munk, F. (1931). Ein Beitrag zur Optik der Farbanstriche. *Z. Techn. Physik.* **12**, 593–601.

Lee, K. and Gartner, L. M. (1976). Spectrophotometric characteristics of bilirubin. *Pediatr. Res.* **10**, 782–788.

Levkoff, A. H., Westphal, M. C. and Finklea, J. F. (1970). Evaluation of a direct reading spectrophotometer for neonatal bilirubinometry. *Am. J. Clin. Path.* **54**, 562–565.

Lucey, J. F., Nyborg, E. and Yamanouchi, I. (1980). A new device for trans-cutaneous bilirubinometry. *Pediatr. Res.* **14**, 604–608.

Malloy, H. T. and Evelyn, K. A. (1937). The determination of bilirubin with the photoelectric colorimeter. *J. Biol. Chem.* **119**, 481–490.

Meites, S. and Hogg, C. K. (1960). Direct spectrophotometry of total serum bilirubin in the newborn. *Clin. Chem.* **6**, 421–428.

Michaelsson, M. (1961). *Scand. J. Clin. Lab. Invest.* **13**, Suppl. 56: 1.

Morrison, R. T. and Wilkinson, D. (1962). An evaluation of the icterometer. *Pediatrics* **29**, 740–742.

Peevy, K. J., Mumford, L., Bruce, R. and Gross, S. J. (1978). Estimation of serum bilirubin by spectral reflectance of the skin. *Pediatr. Res.* **12**, 532.

Post, P. W., Krauss, A. N., Waldman, S. *et al.* (1976). Skin reflectance of newborn infants from 25 to 44 weeks gestational age. *Hum. Biol.* **48**, 541–557.

Schreiner, R. L., Hannemann, R. E., DeWitt, D. P. and Moorehead, H. C. (1979). Relationship of skin reflectance and serum bilirubin: Full term caucasian infants. *Hum. Biol.* **51**, 31–40.

Siggaard-Andersen, O. and Komarmy, L. E. (1968). A simple bilirubinometer for ultramicro determination of total and conjugated plasma bilirubin. *Am. J. Clin. Path.* **49**, 863–870.

Vangianichyakorn, K., Sun, S., Abubaker, A. and Glista, B. (1981). Transcutaneous bilirubinometry in black and hispanic infants. *Pediatr. Res.* **15**, 685.

Wyszecki, G. and Stiles, W. S. (1967). "Color Science", pp. 186–189, John Wiley, New York, London and Sydney.

Yamanouchi, I., Yamauchi, Y. and Igarashi, I. (1980). Transcutaneous bilirubinometry: Preliminary studies of noninvasive transcutaneous bilirubin meter in Okayama National Hospital. *Pediatrics* **65**, 195–202.

9. ULTRASOUND IMAGING

P. N. T. Wells

Bristol General Hospital, Bristol, England

1.0 INTRODUCTION

Ultrasound is a form of energy which consists of mechanical vibrations the frequencies of which are so high that they are above the range of human hearing. The lower frequency limit of the ultrasonic spectrum may generally be taken to be about 20 kHz. Most biomedical applications of ultrasound employ frequencies in the range 1–15 MHz. At these frequencies, the wavelength is in the range 1·5–0·1 mm in soft tissues, and narrow beams of ultrasound can be generated which propagate through such tissues without excessive attenuation.

This chapter begins with brief reviews of the physics of diagnostic ultrasound (Sections 2.0–4.0), pulse–echo imaging methods (Section 5.0) and Doppler imaging methods (Section 6.0). For more detailed discussion of these topics, refer to Wells (1977a). The remainder of the chapter is a résumé of the applications of ultrasonic imaging to physiological measurement.

2.0 PROPAGATION OF ULTRASONIC WAVES

2.1 Relationship of Wave Parameters

Ultrasonic energy travels through a medium in the form of a wave. Although a number of different wave modes are possible, almost all biomedical applications involve the use of longitudinal waves. The

NON-INVASIVE MEASUREMENTS: 2
ISBN 0 12 593402 5

particles of which the medium is composed vibrate backwards and forwards about their mean positions, so that energy is transferred through the medium in a direction parallel to that of the oscillations of the particles. The particles themselves do not move through the medium, but simply vibrate to and fro. Thus, the energy is transferred in the form of a disturbance in the equilibrium arrangement of the medium, without any actual transfer of matter.

In an ultrasonic field, cyclical oscillations occur both in space and in time, as illustrated in Fig. 1. The oscillations here are continuous and of constant amplitude, and the particles are in simple harmonic motion; when a particle is displaced from its equilibrium position it experiences a restoring force which is proportional to its displacement. The wavelength, λ, is the distance in the medium between consecutive particles where the displacement amplitudes are equal. Similarly, the wave period, τ, is the time which is occupied by the wave moving forward through a distance, λ, in the medium. The frequency, f, of the wave is equal to the

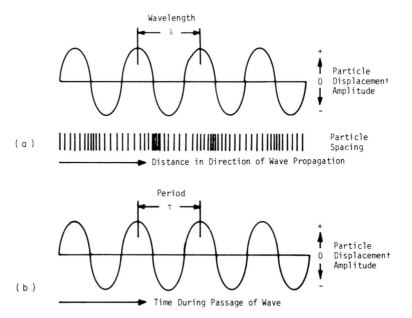

FIG. 1   Diagrams illustrating longitudinal wave motion. (a) Particle displacement amplitude and particle spacing at a particular instant in time in the ultrasonic field: these diagrams represent the distribution of the wave in space. (b) Particle displacement amplitude at a particular point in space in the ultrasonic field: this diagram represents the distribution of the wave in time.

number of cycles which pass a given point in the medium in unit time (usually 1 s); thus,

$$f = 1/\tau. \qquad (1)$$

The wavelength and the frequency are related to the propagation speed, c, by the equation

$$c = f\lambda. \qquad (2)$$

For example, at a frequency of 1 MHz, the period is 1 μs and the wavelength in water ($c = 1500$ m s^{-1}) is 1·5 mm. Values of speed in biological materials are given in Table I. For practical purposes, the value of the speed in any given material may be taken to be independent of the ultrasonic frequency.

2.2 Reflection, Refraction and Scattering

When a wave meets the boundary between two media at normal incidence, it is propagated without deviation into the second medium. At oblique incidence, as shown in Fig. 2, the wave is deviated by refraction unless the speeds in the two media are equal. The relationship is

$$(\sin \theta_i)/(\sin \theta_t) = c_1/c_2. \qquad (3)$$

TABLE I

Ultrasonic properties of some materials of importance in diagnostic imaging

Material	Propagation speed (m s^{-1})	Density (g ml^{-1})	Characteristic impedance (kg m^{-2} s^{-1})	Absorption coefficient at 1 MHz (db cm^{-1})	Frequency dependence of absorption coefficient
Air	330	0·0012	0·0004	1·2	f^2
Aluminium	6300	2·7	17	0·018	f
Blood	1530	1·06	1·6	0·1	$f^{1.3}$
Bone	2700–4100	1·38–1·81	3·7–7·4	10	$f^{1.5}$
Fat	1460–1470	0·92	1·4	0·6	f
Lead zirconate titanate	4000	7·7	30	—	—
Lung	650	0·40	0·26	40	$f^{0.6}$
Muscle	1540–1630	1·07	1·7	1·5–2·5	f
Polyethylene	2000	0·92	1·8	—	—
Water	1520	1·00	1·5	0·002	f^2

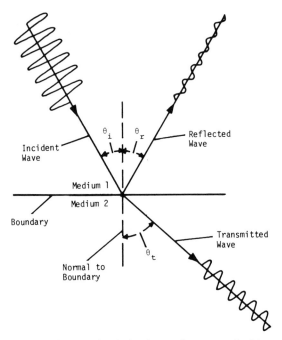

FIG. 2 Diagram illustrating the behaviour of a wave incident on the plane boundary between two media.

In any given medium, the ratio of the instantaneous values of particle pressure and velocity is constant. This constant value is called the *characteristic impedance*, Z, of the medium, and it is equal to the product of the corresponding density and speed; thus

$$Z = \rho c. \tag{4}$$

Typical values of densities and characteristic impedances in materials of interest in diagnostic ultrasound are given in Table I.

If the characteristic impedances of the media on each side of the boundary are equal, the wave travels across the boundary unaffected by the change in the supporting medium (apart from deviation by refraction, if the velocities differ, and the incidence is not normal). If the characteristic impedances are unequal, however, the incident energy is shared between the waves reflected and transmitted at the boundary. At normal incidence, the fraction, R, of the incident energy which is reflected is given by the equation

$$R = [(Z_2 - Z_1)/(Z_2 + Z_1)]^2. \tag{5}$$

As has already been pointed out, $R = 0$ when $Z_1 = Z_2$. There is only a small reflection at the boundary between soft tissues, which, as indicated in Table I, have similar characteristic impedances. On the other hand, $R = 1$, either if $Z_2 \ll Z_1$ (e.g. at the interface between soft tissue and air), or if $Z_2 \gg Z_1$ (which is approximately the situation at the interface between soft tissue and bone). When $R = 1$, complete reflection occurs and no energy is transmitted beyond the boundary.

Reflection at a plane boundary is called *specular reflection*. It is important to realize that specular reflection calculations do not apply to a similar characteristic impedance discontinuity at a rough interface or small obstacle. The specular component of reflection is replaced, by an amount that depends on the geometric characteristics of the discontinuity, by components of scattered energy. This effect becomes important when the dimensions of the discontinuity are in the order of λ or less. For an isolated scatterer that is very much less than the wavelength in size, as illustrated in Fig. 3, or an ensemble of such scatterers (e.g. blood at megahertz frequencies), the intensity of the wave that returns to the source varies inversely as the fourth power of the wavelength; this behaviour is independent of the shape of the scatterer. If the scattering dimensions are of wavelength order, however, the situation is much more complicated, and the intensity of the wave scattered in any particular direction depends on the orientation and shape of the obstacle. This is the situation at many of the boundaries in biological tissues.

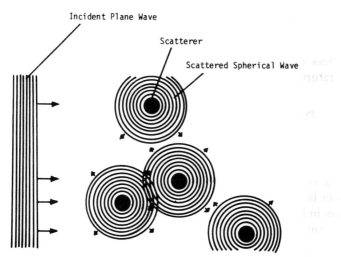

FIG. 3 Scattering. When an ultrasonic wave meets an object which is small compared with the wavelength, the scattered energy is radiated uniformly in all directions in the form of spherical waves.

2.3 Absorption and Attenuation

When a plane ultrasonic wave travels through a medium such as biological soft tissue, it is exponentially attenuated, so that, for a given thickness of a particular material, the ratio of the output and input powers at any given frequency is a constant. Mathematically, the attenuation of a propagating wave can be expressed in terms of an *attenuation coefficient*, α, measured in units of decibels per centimetre.

Attenuation refers to the total reduction in the energy travelling in the direction of propagation; absorption refers only to that component of attenuation by which ultrasonic wave energy is converted into heat. Thus, attenuation mechanisms include reflection, scattering, beam divergence and absorption.

Published data for absorption in materials of interest in ultrasonic diagnosis are given in Table I. Some of these data are actually values of attenuation, but often the literature is not clear enough for the distinction between attenuation and absorption to be made. Furthermore, because the data are sparse, similar mammalian tissues are grouped together regardless of species. Moreover, account is not taken either of tissue temperature or "freshness".

2.4 The Doppler Effect

The frequency of the reflected wave is equal to that of the incident wave if the reflecting boundary is stationary. As shown in Fig. 4, movement of the reflector (or scatterer, or ensemble of scatterers) towards the source, however, results in a compression of the wavelength of the reflected wave and vice versa. Since the speed of ultrasonic propagation in the intervening medium is constant, these changes in wavelength produce corresponding changes in frequency. The phenomenon is called the *Doppler effect*.

At normal incidence, if f is the frequency of the incident wave, and v is the velocity of the reflecting boundary towards the source, the Doppler shift, f_D, in the reflected frequency which occurs in the reflected wave

Fig. 4 Diagrams illustrating the Doppler shift in frequency of an ultrasonic wave reflected by a moving surface. (a) With a stationary target (such as an ensemble of scatterers), the reflected wave has the same frequency as the incident wave. (b) With an approaching target, the reflected waves are compressed, and so shifted upwards in frequency. (c) With a receding target, the reflected waves are shifted downwards in frequency.

$(f_D = f' - f,$ where f' is the received frequency) is given by

$$f_D = 2vf/c, \tag{6}$$

provided that $v \ll c$, as is generally the case in diagnostic applications. In these applications, it often happens that the direction of the motion of the reflecting boundary is at an angle, γ, with the incident wave, although the incident and reflected waves are effectively coincident. This is commonly the situation in the transcutaneous measurement of blood flow velocity in peripheral vessels. Then

$$f_D = 2v \, (\cos \gamma) \, f/c \tag{7}$$

so that it is necessary to know γ in order to calculate v.

3.0 GENERATION AND DETECTION OF ULTRASONIC WAVES

3.1 Piezo-electricity

At megahertz frequencies, such as are employed in diagnostic applications, ultrasound is both generated and detected by the piezo-electric effect. Piezo-electric materials are called "transducers" because they provide a coupling between electrical and mechanical energies. The electric charges bound within the crystalline lattice of the material are arranged in such a way that they react on the application of an electric field to produce a mechanical effect and vice versa. In the undeformed state, the centre of symmetry of the positive charges coincides with that of the negative charges and, because the positive and negative charges are equal in magnitude, there is no effective potential across the transducer. If the transducer is compressed, the centres of symmetry no longer coincide, so that a charge difference appears between the electrodes. The opposite charge difference appears if the transducer is extended. The converse piezo-electric effect occurs because the application of an electric field tends to move the centres of symmetry of the positive and negative charges in opposite directions, causing the transducer to deform. Thus, piezo-electric transducers can act both as generators and detectors of ultrasonic waves.

Although there are many natural crystals which are piezo-electric (the best-known is quartz), the most commonly used transducer material is the synthetic ceramic lead zirconate titanate. This material is polarized during manufacture to make it strongly piezo-electric. It belongs to the group of materials, called *ferro-electrics*, in which there are many tiny electric charge domains which are preferentially orientated in a particular direction by the polarization process.

3.2 Transducers for Diagnostic Applications

Most diagnostic applications of ultrasound employ narrow beams of energy. Such a beam is often best generated (see Section 2.1) by a disc of piezo-electric material electrically excited through two electrodes, one on each parallel surface. The transducer resonates at the frequency at which its thickness is equal to half the wavelength. For example, the ultrasonic speed in lead zirconate titanate is about 4000 m s^{-1}, and the transducer thickness is 1 mm at 2 MHz.

In pulse–echo diagnostic systems (see Section 5.0) the transducer is required to respond to energy pulses of very short duration. This requires that the transducer response should be "damped" by its mounting, as illustrated in Fig. 5.

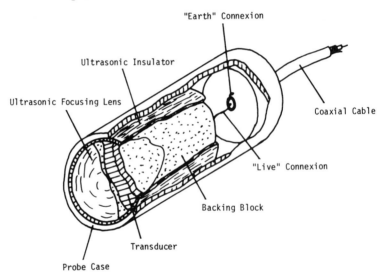

FIG. 5   Construction of a typical probe for pulse–echo operation. The electrical connections are made to thin metal electrodes bonded to the flat surfaces of the transducer. The matching layer may be replaced by a lens, if it is desired to focus the ultrasonic beam.

4.0 THE ULTRASONIC FIELD

4.1 Steady State Conditions

As mentioned in Section 3.2, disc transducers are commonly used in diagnostic applications. The steady state ultrasonic field produced by such a source (and the sensitivity distribution of such a transducer operating as a receiver) may be calculated from the application of Huygens's principle. The theory predicts that, if I_0 is the intensity at the surface of the transducer, and I_z is the intensity at a distance z from the transducer along the central axis, then

$$I_z/I_0 = \sin^2\{(\pi/\lambda)[(a^2 + z^2)^{\frac{1}{2}} - z]\} \qquad (8)$$

where a is the radius of the disc. This relationship is illustrated for a typical example in Fig. 6. Moving along the central axis towards the source, the intensity increases until a maximum is reached at distance z'_{max} from the source given by

$$z'_{max} = a^2/\lambda \qquad (9)$$

provided that $a^2 \gg \lambda^2$. The region between the source and z'_{max} is called the "near field" (or "Frésnel zone"), and the region beyond this is the "far field" (or "Fraunhofer zone"). The beam in the near field is roughly cylindrical. Deep in the far field, the directivity function is given by

$$D_s = \frac{2\mathcal{J}_1\,(ka\,\sin\theta)}{ka\,\sin\theta}, \qquad (10)$$

where θ is the angle relating D_s to the central axis of the beam, and $\mathcal{J}_1$ is the first order Bessel function, and the wave number $k = 2\pi/\lambda$. Thus

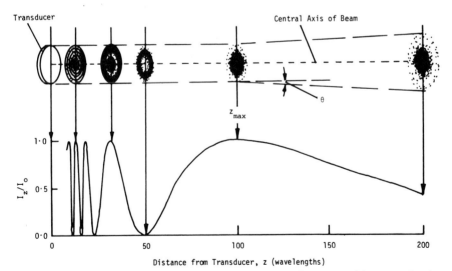

FIG. 6 The ultrasonic field of a typical disc transducer; in this example the distribution shown is for a transducer of 20 wavelengths diameter (i.e. 20 mm at 1·5 MHz in water). The ultrasonic beam normal to the central axis is circular in section, and the elliptical diagrams represent oblique views of such sections.

the main lobe of the beam diverges at angle $\pm \theta$ about the central axis, given by

$$\theta = \sin^{-1}(0 \cdot 61 \ \lambda/a). \qquad (11)$$

In summary, the length of the near field increases, and the divergence in the far field decreases, with increasing diameter of the transducer and with increasing frequency.

4.2 Transient Conditions

If the transducer produces a transient ultrasonic disturbance, as distinct from a steady state continuous wave such as is dealt with in Section 4.1, the ultrasonic field is modified because, at any particular point in the field, the contributions from the different elementary parts of the source surface may not be equal. As a result, the sharply defined inhomogeneities of the steady state become increasingly smeared and homogeneous as the pulse length is reduced.

4.3 Focusing

In the near field (see Section 4.1), an ultrasonic beam may be focused over a limited depth of field. The materials, such as plastics, from which ultrasonic lenses may be constructed generally have higher propagation speeds than water or soft tissues, so that converging lenses are concave.

In considering the phenomenon of focusing, a useful concept is that focusing occurs at the point in the field at which the contributions from the entire surface of the transducer all arrive together, or "in phase". Thus, lenses function by introducing appropriate thicknesses of material in which the speed differs from that in the medium, so that the transit times along ray paths of different lengths to the source are all equal. Similarly, focusing may be achieved by the use of a concave transducer giving equal length ray paths from the transducer surface to the focus. Synthetic focusing (and beam steering) can likewise be achieved with appropriate time grading across an array of small-element transducers; this is discussed in detail in Section 5.7.

5.0 PULSE–ECHO IMAGING METHODS

5.1 Basic Principles

An ultrasonic pulse is reflected when it strikes the boundary between two media of differing characteristic impedances, and the time delay which occurs between the transmission of the pulse and the reception of its echo depends on the propagation speed and the path length. The propagation speeds in different soft tissues are so closely similar (approximately equal to that in water, and around 1500 m s^{-1}), that a constant relationship between time and distance can usually be assumed. Ultrasound travels 10 mm in about 6.7 µs at this speed.

The ultrasonic pulse–echo method depends on the estimations of the ranges and directions of echo-producing targets within the tissue volume interrogated by the ultrasonic beam. Instruments range in complexity from the simple range-finding A-scope with hand-held probe, through the time-position recording system and the static two-dimensional B-scope to real-time systems gathering data from two-dimensional planes within three-dimensional volumes.

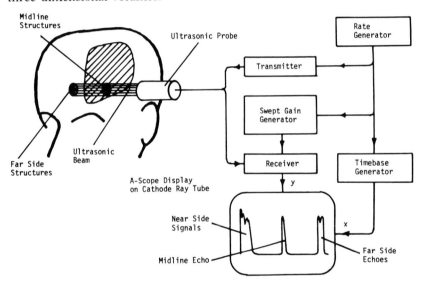

FIG. 7   Basic elements of the A-scope. The output from the receiver is connected to the vertical (y) deflection plates of the cathode ray tube, and that from the time-base generator, to the horizontal (x) plates.

5.2 The A-scope

The basic elements of the simplest type of pulse–echo system for medical diagnosis, called an "A-scope", are illustrated in Fig. 7. The rate generator (or "clock") simultaneously triggers the transmitter, the swept gain generator and the time-base generator. The voltages which appear across the transducer in the probe are amplified by the receiver, and the output from the receiver is arranged to deflect the time-base line on the display. Thus, vertical deflections of the horizontal time-base occur at positions corresponding to echo-producing targets along the ultrasonic beam within the patient. Rapid repetition of the process (typically 2000 times per second) results in a flicker-free display.

5.3 Swept Gain

A substantial improvement in the usefulness of the displayed information is obtained if the echo signals from deeper structures are amplified more than those which originate closer to the probe. This is because deeper echoes are more attenuated by the greater tissue path length; swept gain compensates for this. Ideally, swept gain should lead to similar deflection amplitudes on the display for similar surfaces, irrespective of their distances from the probe. In practice, however, accurate swept gain is difficult to achieve for two main reasons. Firstly, there is a variation in the attenuation rates of different tissues, so that compensating on the basis of 1 dB cm^{-1} MHz^{-1} (e.g. setting the swept gain rate at 4 dB cm^{-1} for operation at 2 MHz, taking the go-and-return path length into account) is at best only a compromise. Secondly, the energy in the ultrasonic pulse is distributed over quite a wide frequency spectrum, and the higher frequency components of the pulse are increasingly attenuated with increasing penetration, since the attenuation coefficient in soft tissues is roughly proportional to the frequency.

5.4 Resolution in Pulse–Echo Systems

Within the limitations imposed by noise and by the maximum permissible transmitted power, the maximum useful dynamic range of the echoes received in conventional medical diagnostic pulse–echo systems is about 100 dB. This dynamic range is shared between the variations in echo amplitude at particular ranges, and the attenuation of echoes which increases with distance. In practice, at any particular range, an echo amplitude variation of about 30 dB is the maximum which may usefully be employed, since the azimuthal resolution is unlikely to be acceptable with a larger dynamic range. Therefore, around 70 dB is available to provide swept gain compensation for attenuation. An attenuation of 1 dB cm^{-1} MHz^{-1} corresponds to 0·15 dB per wavelength, or to 0·3 dB per wavelength of penetration (taking account of the go-and-return path). With 70 dB of swept gain, a penetration of 233 wavelengths would thus seem to be possible; 200 wavelengths is a more realistic figure.

The resolution of any imaging system may be defined in several different ways. The usual definition is that the resolution is equal to the reciprocal of the minimum distance (in range or in azimuth) between two point targets, at which separate registrations can just be distinguished on the display. An alternative definition, equivalent in concept but usually more convenient in practice, is that the resolution is equal to the reciprocal of the distance which appears on the display to be occupied by a point target in the field. Measurements based on this definition avoid problems which arise due to interference between waves scattered by two closely spaced point (or line) targets.

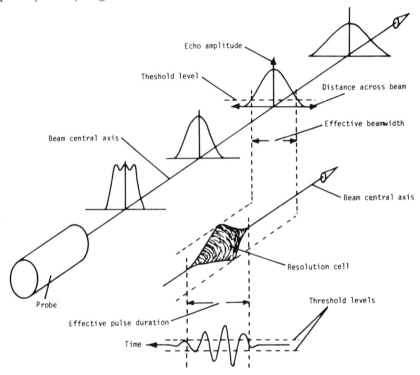

FIG. 8   Factors determining the dimensions of the resolution cell in an ultrasonic pulse–echo system. The axial length of the resolution cell corresponds to the duration of the echo pulse envelope which exceeds the threshold level of the system (this is illustrated in the bottom right-hand side of the diagram). The diameter of the resolution cell is equal to the effective beam-width, which is determined by the system threshold (as illustrated in the top left-hand side of the diagram) and which depends on the space-position along the central axis and the time-position within the pulse.

The resolution cell is the volume of material within which the inter-action providing the data takes place. Except in simple and idealized situations, the dimensions of the resolution cell depend on the distance of the target. As shown in Fig. 8, the length of the resolution cell depends on the duration of the ultrasonic pulse, and its width on the diameter of the ultrasonic beam. The effective values of the pulse duration and the beam-width are determined by the dynamic range lying between the maximum echo amplitude and the detection threshold of the diagnostic system.

5.5 The B-scope

The information obtained with a pulse–echo system is a combination of range and amplitude data which can simply be presented as an A-scan (Section 5.2). The same information, however, may alternatively be displayed on a brightness-modulated time-base in such a way that the brightness increases with echo amplitude; this type of display is called a B-scan. The B-scope is the basis of the time-position (M-mode) recording technique, and the two-dimensional scanning technique.

5.6 Time-position Recording

A time-position recording of structure position along the ultrasonic beam may be generated from a B-scan, as shown in Fig. 9. Most instruments based on this principle generate time and distance (i.e. ultrasonic time-base) markers on the recording to assist in interpretation.

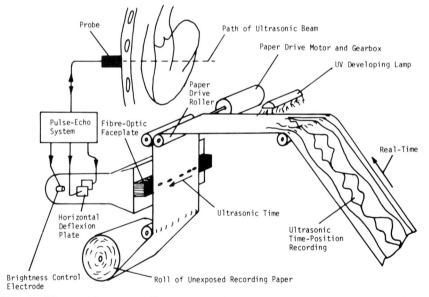

FIG. 9 Time-position recording system using a continuous strip of photographic paper sensitive to ultraviolet light. The B-scan is displayed on a cathode ray tube with a fibre-optic face-plate. This display is extremely bright, and a continuous image of the time-position trace is produced (and developed within a few seconds) as the paper is driven at constant speed past the cathode ray tube.

A time-position recording is composed of many separate B-scan lines lying side by side. Conventionally, increasing distance into the patient is represented by more downward deflection on the recording, and earlier time, horizontally towards the right. The time required to form a single B-scan line depends on the depth of penetration; e.g. a time of 133 μs corresponds to a depth of 100 mm. Structures within the body do not move a significant distance during so short a time, so that (provided that the pulse repetition rate is fast enough) the movements of structures such as heart valves may be studied.

5.7 Two-dimensional B-scanning

The production of an image of a cross-section through soft tissue structures of the body may be accomplished by relating the positions of registrations on the display to the positions of the corresponding echo-producing structures within a defined two-dimensional plane in the patient.

The first type of ultrasonic two-dimensional scanner to come into widespread clinical use was the so-called "static" variety. This type of scanner, the principles of which are illustrated in Fig. 10, is still the most commonly used instrument in the visualization of the relatively static contents of the abdominal cavity. The place of this type of scanner in obstetrics, however, is being taken over by the so-called real-time scanners, which are also particularly suited to cardiological studies. Real-time imaging systems have image frame rates which are sufficiently fast to allow movement to be followed. The actual frame rate necessary to satisfy this definition depends on the circumstances of the particular clinical investigation. For example, if the movment to be followed is that of the heart, a frame rate of at least 40 s^{-1} is necessary to avoid missing details of valve action.

Ultimately the frame rate is limited by the speed of ultrasound in tissue. If, for example, a penetration of 150 mm is required, the time which elapses between the transmission of the ultrasonic pulse and reception of the echo from the maximum range is equal to 200 μs (taking the speed to be 1500 m s^{-1}). The corresponding maximum pulse repetition rate is 5000 s^{-1} (although in practice 2000–3000 s^{-1} would be more likely to be used). Thus the maximum image line rate is 5000 s^{-1}, equal to the product of the number of lines per frame and the number of frames per second. Again, for example, at 40 frames s^{-1}, this corresponds to 125 lines per frame.

Some of the many methods of rapid scanning to produce real-time images are illustrated in Fig. 11. In the first technique, a conventional single-element transducer, or group of single-element transducers, is

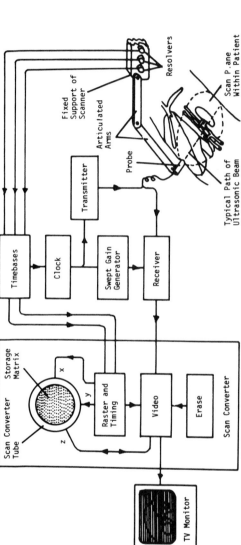

FIG. 10   Basic elements of a typical hand-operated two-dimensional static B-scanner. The ultrasonic part of the instrument produces an intensity-modulated signal that is written on the storage matrix of the scan converter. (The scan converter illustrated here is of the analogue type. Nowadays digital image storage is commonly used, but the analogue method is easier to understand in an initial explanation of the principles involved.) The positions in which the echo signals are recorded are determined by the two time-base generators, which are controlled by the three resolvers mounted on the fixed support of the scanner. The probe is constrained by the two articulated arms so that it can be moved only in a defined two-dimensional plane, in contact with the patient. Coupling is ensured by means of oil or proprietary gel smeared on the skin. The resolvers measure the horizontal and vertical positions of the probe and the direction of the ultrasonic beam. This arrangement results in the time-base being driven across the storage matrix from the corresponding position and in the same direction as the ultrasonic beam travels across the patient. Changes in beam orientation are followed immediately by corresponding changes in that of the time-base. In this way, echoes from any particular reflecting point in the patient are registered at the same position on the storage matrix, independent of the direction of the ultrasonic beam. As the probe is scanned across the patient, echo signals produce registrations on the storage matrix so as to create a charge pattern that is a two-dimensional representation of the anatomical cross-section through the patient in the plane of the scan. This charge pattern is displayed on the television monitor by raster-scanning the scan converter storage matrix. Time-sharing of the scan converter operation allows the image to be viewed whilst scanning is in progress, and the stored image can subsequently be studied on the monitor, or the image may be photographed, either directly from the monitor or on a multiformat camera.

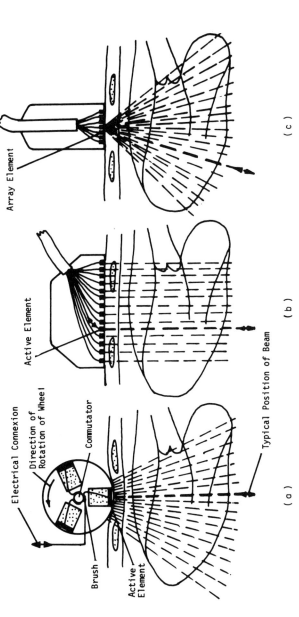

FIG. 11   Methods of real-time scanning. (a) Fast mechanical scanner: in this example, three single-element transducers are mounted on a wheel which rotates continuously, and as each transducer in turn comes into contact with the patient, it sweeps out a new frame made up of an image sector. (b) Electronically scanned linear array transducer array: the transducer elements are addressed sequentially, to sweep out image frames with a rectangular format. (c) Transducer array with electronic beam steering: delays are introduced into the separate signal paths associated with the individual elements in the array, to sweep out image frames with a sector format.

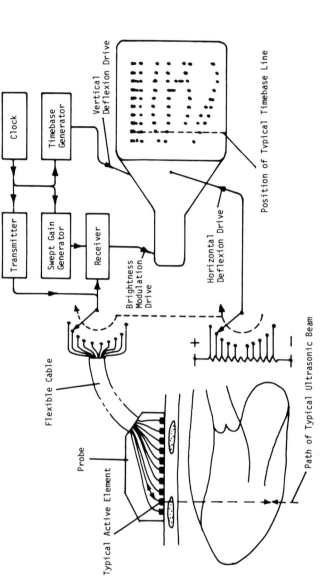

FIG. 12 Linear array real-time scanning system. In this diagram, the probe contains ten separate transducer elements (in a practical system, the number is usually between 20 and 128). The clock, typically operating at a p.r.f. of 2000 s⁻¹, triggers the transmitter. In this simple example, the transmitter pulse is applied to one of the transducer elements through a sequencing switch. (Again, in a practical system, this sequencing switch, and the second switch operating synchronously with it, are usually electronic and not mechanical.) Simultaneously, the clock triggers the time-base generator connected to the vertical deflection plates of the cathode ray tube display. Echoes returning from within the patient are detected by the transducer element which emitted the original pulse, fed through the sequencing switch, and amplified (under swept gain control, triggered by the clock), to brightness-modulate the display. Each element is addressed rapidly in sequence, and a two-dimensional image is built up by the second sequencing switch applying appropriate horizontal deflection voltages to the display.

mechanically driven to form images in real-time but which are otherwise similar to those made by conventional "static" two-dimensional scanners.

The principles of the electronically addressed linear array type of real-time scanner are illustrated in Fig. 12. In the earliest linear array systems

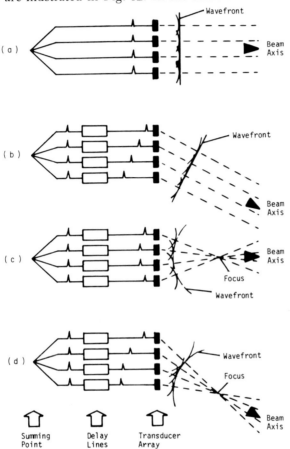

FIG. 13   Principles of electronically steered array scanning. These diagrams illustrate the transmission of ultrasound by an array of four long narrow elements; the same principles apply to the reception of ultrasound. The long axes of the elements are normal to the plane of the diagram, and viewed in this direction each element emits a cylindrical wavelet in response to electrical excitation. (a) When the elements are excited simultaneously (as indicated by the simultaneous "blips" on the connecting wires), the wavelets combine to form a wavefront of which the corresponding beam travels directly away from the array. (b) When the elements are excited in sequence, the beam is deviated off the central axis. (c) When spherical time grading is used, the beam is focused. (d) When the time grading consists of combined linear and spherical distributions, the beam is both deviated and focused.

the transducer elements were addressed one at a time. In this situation a compromise is necessary. On the one hand, it is desirable to have a large number of lines in the image, and this requires a large number of transducers. On the other hand, it is desirable to have good resolution, which depends on having a non-divergent (or even focused) beam of ultrasound. Unfortunately, the beam divergence in the far field increases as the transducer is made narrower. The need to compromise can be avoided, however, by having an array of many narrow transducers (so that the objective of high line density can be achieved) operated in groups to ensure an adequate aperture (so that the resolution is acceptable). A common arrangement is to have 64 elements operated in groups of four, stepped one element between lines, thus giving 61 lines of ultrasonic information.

The second type of electronically controlled real-time scanner makes use of the beam-steering capability of an array. Consider the array, consisting of four long narrow elements, illustrated in Fig. 13. The introduction of appropriate time delays in the signal paths allows the beam to be directed through any desired angle (within certain limitations), and also to be focused. Moreover, on reception the position of the focus may be swept continuously to coincide with the instantaneous range of the target; this technique is also valuable with linear array scanners. Typically, electronically steered arrays have around 20 elements, and the external dimensions of the probe are similar to those of a conventional single-element transducer probe.

5.8 Frame Freeze

At any moment during scanning the operator may see an image which merits detailed study. In this case, a "frame freeze" capability is most useful. Basically there are two methods by which this may be provided. One method uses a conventional analogue scan converter, such as that commonly used as an image store in static scanners; but this does have the disadvantages that the selective erasure of part of an image is difficult and time-consuming (in relation to the frame time of the image). These problems are avoided in the second method of frame freeze, which uses a digital image store.

5.9 Dimensional Measurements

In many clinical applications it is important to be able to make measurements of distance from the ultrasonic two-dimensional images. It is not satisfactory to measure directly from the cathode ray tube display, even when the image is stationary for long enough to make this possible, because the image is distorted by non-linearities in the deflection system. Electronic calipers may be provided to eliminate this problem and several arrangements have been devised: generally the choice depends on the personal preference of the operator.

6.0 DOPPLER IMAGING METHODS

6.1 The Scope of Ultrasonic Doppler Methods

Nowadays, ultrasonic Doppler methods are both widely used and of established value in the study of movement in clinical diagnosis. In most applications, the Doppler shift in frequency of a continuous wave ultrasonic beam (see Section 2.4) reflected from a moving structure (such as the fetal heart) or from a moving ensemble of scatterers (such as flowing blood) is used to provide information about the velocity of the movement, either for interpretation by ear, or for analysis by instrument. Two-dimensional Doppler scanning is becoming accepted. Pulsed Doppler systems, which combine the range-measuring capability of the pulse–echo method with the velocity-measuring capability of Doppler, are beginning to demonstrate their potential value in scanning and analysis.

Ultrasonic Doppler velocimetry (Roberts and Sainz, 1979) has an established role in the study of arterial disease. The method is based on the measurement of blood flow velocity by the Doppler shift in back scattered ultrasound, according to Eqn (7) in Section 2.4. The maximum frequency envelope of the Doppler frequency spectrum is analysed to provide indices of pulsatility, damping and transit time. The analysis has been refined by Skidmore and Woodcock (1980), who have used the Laplace transform method. These analytical techniques, although they provide important physiological data, do not come within the normal meaning of "ultrasonic imaging". For this reason, detailed discussion is beyond the scope of the present chapter, and only a brief discussion of the principles involved is included in Section 7.2.

6.2 Ultrasonic Doppler Detectors

The same restrictions and limitations (such as the necessity to maintain good ultrasonic coupling, and the inability to operate successfully through gas) which apply to ultrasonic pulse–echo methods, also apply to ultrasonic Doppler methods.

The choice of the ultrasonic frequency depends on the clinical application. A compromise is necessary between the penetration, the variation

of Doppler shift frequency for a given variation in target velocity, the
sensitivity to small reflectors, and the size and shape of the ultrasonic
field. In obstetrics and cardiology, the optimal frequency is generally
2–3 MHz, but in blood flow studies it may be as high as 10 MHz.

Substitution of typical values, including a typical target velocity of
100 mm s^{-1}, in Eqn (7) reveals that a 2 MHz ultrasonic beam is shifted in
frequency by about 260 Hz if the beam and target motions are coincident.
For practical purposes, the Doppler shift frequency may be taken to be
proportional both to the ultrasonic frequency and to the reflector velocity.
Thus, in medical diagnostic applications, the ultrasonic Doppler shift
frequency generally lies in the audible range.

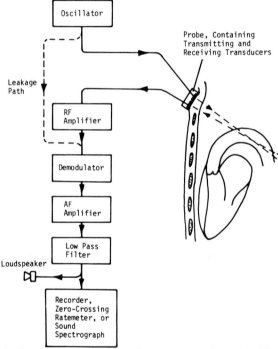

FIG. 14   Block diagram of typical continuous wave ultrasonic Doppler system;
a blood flow velocity detector is shown here.

Figure 14 is a block diagram of a continuous wave Doppler system.
The transmitter is an oscillator which operates continuously, providing an
output of constant amplitude and frequency. The ultrasonic probe contains
separate transmitting and receiving transducers. (These are generally
necessary, because it is important to minimize the direct transfer of energy
from the transmitter to the radio frequency amplifier in order to avoid
overloading the receiver.) The output from the r.f. amplifier consists of a

mixture of signals, some of frequency equal to that of the transmitter (these are due to reflections from stationary structures in the ultrasonic field and electrical leakage) and some of frequencies shifted by the Doppler effect (due to reflections from moving structures). These signals are mixed in the demodulator, the output from which contains the difference frequencies between the transmitted ultrasonic wave and the Doppler shifted received waves. The output from the demodulator is filtered to allow these difference frequencies to pass, whilst unwanted (higher) frequencies are stopped. The difference frequencies are amplified, and either an operator listens to them, or they are analysed electronically. In clinical applications, the Doppler shifted signals do not consist of a single frequency, but they extend over a frequency spectrum (since the beam simultaneously interrogates structures moving at different velocities). For this reason measurements of Doppler shift signals made by rate-meters, such as the zero-crossing frequency meter, although generally satisfactory for fetal heart rate studies, need to be interpreted with caution in cardiovascular applications. In most cardiological investigations with Doppler techniques, it is very much safer to subject the Doppler signals to frequency spectrum analysis; this may be done by on-line or off-line instruments.

Continuous wave Doppler systems give information about the velocities of targets, but not about their distances from the transducer. This information can be obtained, however, by pulsing the transmitted ultrasound to allow both Doppler signals and range delays to be measured simultaneously. A block diagram of a typical pulsed Doppler system is shown in Fig. 15. In practice, the upper frequency limit (which is related to the maximum value of target vector velocity that can be measured) which can be detected without ambiguity depends upon the sampling rate. The maximum sampling rate is limited by the ultrasonic transit time to and from the target of interest, and by the reverberation decay time. It is well known in information theory that if a signal waveform has frequencies in its spectrum extending from zero to an upper frequency f_{max}, it is possible to convey all the information in the signal provided that the sampling frequency is at least $2f_{max}$. This sets an upper limit to the maximum unambiguously measurable velocity vector at any given penetration for a given ultrasonic frequency. For example, the chosen frequency might be 2 MHz, and the necessary penetration, 200 mm. The corresponding maximum pulse repetition rate would be 3750 s^{-1}. Theoretically the maximum ultrasonic Doppler shift frequency corresponding to this sampling rate would be 1875 Hz; and substitution in the Doppler equation (Section 2.4) reveals that the maximum vector velocity would be 720 mm s^{-1} (with $\gamma = 0°$). It turns out that, at any given frequency and

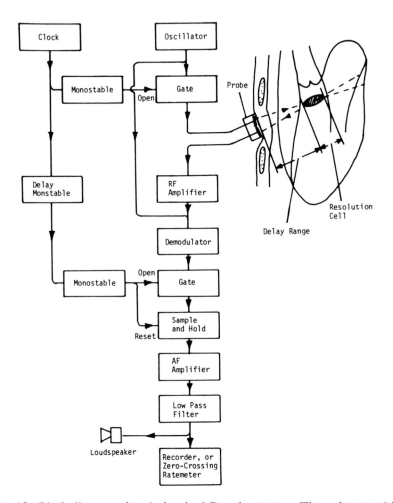

FIG. 15 Block diagram of typical pulsed Doppler system. The pulse repetition rate is controlled by the clock, which triggers the monostable to open the gate to allow the transmitting transducer to be excited for a period corresponding to the width of the target volume which it is desired to study. Echoes returning from within the patient are amplified, and mixed in the demodulator with the signal from the oscillator (equal in frequency to that which was transmitted). The delay monostable triggers the monostable controlling the receiver gate, so that the gate opens to allow a voltage, which is in effect a sample corresponding to the Doppler shift due to motion in the target volume, to be stored in the sample-and-hold circuit. The sample-and-hold is reset immediately prior to being updated by a new sample resulting from the following ultrasonic pulse. The output from the sample-and-hold is thus a rectangular wave with a long "mark" and a short "space", the envelope of which is an audible signal representing the Doppler shifted information from the target volume.

geometry, the product of the maximum vector velocity and the maximum target range is equal to a constant.

Simple Doppler systems, whether continuous wave or pulsed, merely measure the magnitude of the frequency difference between the transmitted and received ultrasonic signals, and not the sign of the difference. This sign carries the information about the direction of the movement of the target, either towards or away from the probe. This directional information is vital in some diagnostic situations.

If the signal always consists, at any instant in time, of movement (or flow) in only one direction, a detector capable of switching a logic circuit to indicate whether the movement is in the forward or reverse direction would be satisfactory. (This is the method employed in many commercially manufactured systems.) Almost invariably, however, in practice the Doppler signal consists, at least for some of the time in each periodic cycle, of simultaneous signals from targets moving in opposite directions. Logic circuits are then inappropriate, and there is really no substitute for the sound spectrograph as a display device. Other directionally sensitive detection arrangements include single side-band and super-heterodyne techniques.

6.3 Resolution in Doppler Systems

The lateral resolution of a Doppler system at any particular distance from the transducer depends on the effective width of the ultrasonic beam. This in turn depends on the beam profile and the signal processing arrangements, including the detector threshold level.

The range resolution of a continuous wave Doppler is, in effect, such that any detectable target gives a signal; the penetration is limited by attenuation. Pulsed Doppler systems have range resolution determined by the effective length of the ultrasonic pulse, and in principle the situation resembles that in a conventional pulse–echo system.

6.4 Doppler Imaging

In many clinical situations, adequate diagnostic information can be obtained with a hand-held Doppler probe. It is necessary, however, for the investigator to know the anatomy of the structures being studied in order to interpret the results.

In studies of the vascular system it is often useful also to have a two-dimensional map showing the position of blood vessels. The ultrasonic Doppler shifted signals from flowing blood are sufficiently characteristic to allow their presence to be identified by logic circuitry. This capability

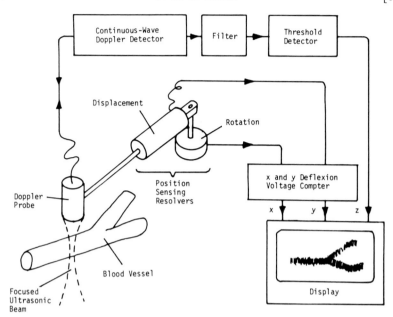

Fig. 16   Continuous-wave Doppler system for two-dimensional visualization of blood vessel distribution.

is exploited in the two-dimensional scanner shown in Fig. 16. The probe is mounted on a two-dimensional co-ordinate measuring scanner, the resolvers of which provide data enabling computation of the x and y voltages that control the deflection circuits of a direct-view electronic storage tube. The probe is arranged so that the ultrasonic beam is at least slightly inclined to the direction of flow in the vessels to be visualized. When the beam passes through moving blood the Doppler detector generates an output which is filtered (to remove artefacts due to low velocity movements such as those of the probe over the skin) and, provided that the output exceeds a pre-set threshold level, it switches on the electron beam of the display. A two-dimensional map showing those regions in which flow has been detected is constructed on the display by scanning the probe over the area of skin overlying the vessel. Since arteries and veins often lie close together, a directionally sensitive circuit is arranged to inhibit the display when flow is detected in the opposite direction to that in the vessel under study.

A two-dimensional scan of a blood vessel made with a continuous-wave Doppler instrument is essentially a plan view, representing the projection of the blood vessel onto the skin surface along the line-of-sight of the ultrasonic beam. The same type of image can be obtained with a pulsed

Doppler scanner range-gated to a constant depth (or even to a variable depth, under the control of the operator) within the blood vessel. Because the pulsed Doppler scanner is capable of measuring range (as well as velocity), and thus of displaying the depth of detected flow, cross-sectional and longitudinal images of the blood vessel lumen can also be produced. Typically a pulsed Doppler scanner has 32 serial gates, each representing flow in a 1 mm increment along the ultrasonic beam, and the system operates at 5 MHz and is directionally sensitive. Furthermore, in appropriate anatomical situations, it is possible to determine the orientation, lumen cross-sectional area and flow velocity profile by means of pulsed Doppler scanning, and thus to estimate blood flow volume.

6.5 Combined Pulse–Echo Imaging and Doppler Measurement

The combination of pulse–echo two-dimensional imaging of blood vessels (or of structures suspected of being blood vessels, or of the heart) with pulsed Doppler measurement of blood velocity in defined volume elements in the image is becoming established as a powerful technique. It may be expected to be further refined.

7.0 RÉSUMÉ OF APPLICATIONS IN NON-INVASIVE PHYSIOLOGICAL MEASUREMENT

It is not possible to present here anything more than a rather superficial review of the applications of ultrasonic imaging which have particular relevance to non-invasive physiological measurement. The wider subject of ultrasonic diagnosis as a whole is not dealt with here at all, and for information on this the reader is referred to the medical ultrasonic literature (see, e.g. de Vlieger *et al.*, 1978; Taylor, 1979; Wells, 1977b; White, 1973).

7.1 Applications in Cardiology

Except for a small triangular area in the left parasternal region, the anterior of the heart is normally covered by lungs and pleura. The intercostal spaces in this area provide the most commonly used access route to the heart in echocardiography. Occasionally, a suprasternal or subxiphoid route is used and, even more rarely, the examination may be made from the oesophagus, or with a catheter-mounted transducer. These latter routes are not usually justifiable, however, except in patients precluded from the usual approach by emphysema or by previous surgery.

The best-established ultrasonic examination of the cardiovascular system is based on time-position recording of the movements of the cardiac valves (Roelandt, 1977). In order of increasing difficulty, the functions of the mitral, aortic, tricuspid and pulmonary valves can be studied. The normal atrioventricular valves open rapidly at the beginning of ventricular diastole, and the leaflets then move closer together until they momentarily flip apart during atrial systole. They remain closed throughout ventricular systole. The normal aortic and pulmonary valves, on the other hand, exhibit box-like motion waveforms, being open during systole and closed in diastole.

In patients who have combined mitral stenosis and regurgitation, the shape of the echocardiogram depends on which lesion is dominant. If stenosis is dominant, the tracing resembles that in pure mitral stenosis. On the other hand, if regurgitation is dominant, the early diastolic slope is greater than that in pure stenosis and a more gradual slope continues for the next phase of diastole.

Mitral valve prolapse may be diagnosed from the mitral echocardiogram by the appearance of an abrupt posterior displacement of the anterior leaflet during systole.

In patients with torn anterior chordae, the anterior leaflet of the mitral valve moves erratically in diastole; the fluttering is much coarser than that seen in aortic regurgitation. With torn posterior chordae, the posterior mitral valve leaflet remains posterior throughout systole and returns anteriorly during ventricular diastole. Prolapse of the mitral valve, on the other hand, does not affect the posterior leaflet motion in early systole.

Hypertrophic cardiomyopathy, with or without obstruction, is a condition in which there is asymmetrical hypertrophy of the interventricular septum and the left ventricular wall. Mitral valve echocardiography reveals that the normal gradual anterior systolic movement is replaced by a more pronounced deflection which begins after the onset of ejection from the ventricle and continues to the end of systole. This is because the anterior mitral valve leaflet and the interventricular septum come into contact during systole in patients with hypertrophic obstructuive cardiomyopathy (HOCM). Clinically it is often difficult to distinguish these patients from those with coronary artery disease. In addition, there may be a reduced diastolic slope due to the reduced rate of filling resulting from the low compliance of the ventricle. Even in the presence of fixed left ventricular outflow obstruction, abnormal systolic anterior movement of the anterior mitral valve leaflet may be found on retrospective examination of the echocardiograms of patients who subsequently proved to have HOCM.

A myxoma of the left atrium is a pendunculated soft tissue mass which arises from the interatrial septum. It may enlarge sufficiently either to grow, hour-glass fashion, through the mitral valve orifice, or it may move into the left ventricle during diastole, returning to the left atrium during systole. The tumour may be diagnosed from its echocardiogram which is reminiscent in shape of that which occurs in mitral stenosis, and which is characterized by multiple echoes in the left atrium, particularly in diastole.

In the presence of aortic regurgitation, an apical diastolic rumbling (the "Austin–Flint murmur") is often heard. In the presence of this murmur it is not possible to determine, using conventional methods of diagnosis, whether or not the mitral valve is diseased. Echocardiography of the mitral valve can quickly resolve this problem.

M-mode studies of mitral valve prostheses can often be clinically useful, although the results are not always definite. The Starr–Edwards valve gives rise to a pattern which resembles that seen from a natural valve with mitral stenosis. The ball remains in the fully open position during diastole, and any movement which may be seen during this phase is that of the cage. There seems always to be some regurgitation at the beginning of systole. Malfunction of the prosthesis may be due to fibrous overgrowth causing the ball to stick, and this can usually be seen on the M-mode tracing. Malfunction of other types of prostheses usually also have characteristic M-mode tracings, but in practice the difficulty is that satisfactory recordings cannot often be obtained because of the orientation of the ultrasonic beam in relation to the valve movement.

The echocardiographic examination of the aortic valve may also provide useful physiological data. The normal aortic valve echocardiogram is seen within the aortic root as slender cusp echoes producing a box-like configuration during systole (when the valve is open), and a single, nearly central, line in diastole. In disease, the density of cusp echoes correlates well with the assessment of valve calcification found at operation, but the presence of these echoes makes it impossible to measure the movements of the cusps. Consequently, aortic echocardiography is not very helpful in grading the severity of aortic stenosis. In the rare absence of calcification, the expected reduction in the separation of the aortic valve cusps in stenosis is not seen if the valve is domed so that the orifice is displaced superiorly beyond the ultrasonic beam, as with bicuspid aortic valve. In aortic regurgitation, however, the valve separation is generally greater than in the normal; if the separation is increased, regurgitation is almost certain to be present. False negatives, however, do occur.

Dissecting aortic aneurysm may be diagnosed from the appearance on the echocardiogram of normal aortic valve leaflets within two anterior and two posterior echoes which correspond to the dilated aortic root and the false lumen of the aneurysm.

Qualitatively, echocardiograms of the tricuspid valve are similar to those of the mitral valve. Moreover, the tricuspid valve may flutter during diastole in patients with pulmonary regurgitation, just as the mitral valve may do in aortic regurgitation. The diastolic slope of the tricuspid valve echocardiogram may be significantly decreased in the absence of stenosis, by restrictive processes involving the right ventricle and pericardium, and so caution in interpretation is necessary.

The M-mode recording of the movement of the pulmonary valve may be of value in assessing pulmonary hypertension. In comparison with the normal, in hypertension the pulmonary valve opens more rapidly, the pre-ejection period is longer, and the posterior displacement following atrial systole (the A-wave) is smaller or even absent. Some caution is necessary, because the A-wave amplitude is also affected by respiration, so that its amplitude must be measured during the inspiratory phase of quiet respiration. Mid-systolic fluttering may occur in patients with pulmonary hypertension, and it is interesting that the mitral valve motion may also mimic mitral stenosis in this condition.

It has been shown experimentally that there is good correlation between the stroke volume and the product of the amplitudes of the mitral ring movement and the distance between the anterior and posterior heart walls. In clinical practice, it is easier to measure the left ventricular internal dimension (minor axis) in diastole ($LVID_d$) and in systole ($LVID_s$), and to calculate the stroke volume (LVSV) by substitution in an empirical formula (Feigenbaum et al., 1969):

$$(LVSV) \simeq (LVID_d)^3 - (LVID_s)^3. \tag{12}$$

It should be pointed out, however, that the method is not usable in all patients (because of access difficulties), and its accuracy is further reduced in some cardiac abnormalities.

Echocardiographic diagnosis of hypertrophic cardiomyopathy may generally be made by the observation of abnormal mitral valve motion. These patients, however, represent only one subgroup of a cardiac disease in which the characteristic anatomical abnormality is asymmetric septal hypertrophy (ASH). In most patients with ASH, left ventricular outflow is unobstructed and cardiac dysfunction is presumably due to widespread left ventricular myocardial abnormality; a few patients exhibit HOCM. The interventricular septum is thickened in obstructive ASH. Moreover the thickening of the free wall behind the posterior mitral leaflet appears to regress after surgery for the relief of outflow obstruction.

In favourable circumstances, it is possible echographically to measure the thickness of the posterior left ventricular wall. This has been used to measure stress–strain relationships and to estimate preload and after-load.

The velocity of circumferential fibre shortening (VCF) may be measured from ultrasonic recordings of the posterior left ventricular wall during systole. Mean VCF is often depressed in patients with non-localized impaired left ventricular function. Related studies are described by Gibson (1979).

Some of the basic problems of assessing left ventricular function from M-mode recordings, which arise because the technique is limited to one spatial dimension, can be avoided by the use of real-time two-dimensional imaging. Dyskynesia is then generally quite easy to detect, if not to quantify. The anterior and posterior ventricular walls, and the inter-ventricular septum, can all be visualized in suitable patients. The medial lateral walls of the left ventricle, which are difficult to study by M-mode echocardiography, are accessible by two-dimensional scanning. The examination can also be carried out during stress testing.

In the study of the heart, three main approaches with Doppler techniques are promising. Firstly, measurements of flow in the thoracic aorta reflect left heart function. These data can be obtained with a continuous wave Doppler system operating at a frequency of around 2 MHz, by positioning the probe in the suprasternal notch (Sequeira et al., 1976). The orientation of the aortic arch is such that the ultrasonic beam can, when directed from the suprasternal notch, intersect the direction of blood flow tangentially. Other angles of attack occur, due to the curvature of the vessel, but the highest Doppler shift frequency corresponds to the highest velocity within the beam. The use of a real-time sound spectrograph to display direction-ally detected Doppler shift signals allows the operator to obtain optimal orientation, and to recognize flow signals from branch arteries which, since they serve the head and neck, are in the opposite direction to the flow in the aortic arch. The spectral display allows turbulence to be identified, and can be interpreted even if the signal-to-noise ratio is poor.

The clinical usefulness of transcutaneous aortovelography is still being assessed. The part of the aorta in which the flow velocity is monitored is close to the heart, so that information on left heart action is obtained. In any particular individual, instantaneous cardiac output is likely to be proportional to the measured velocity, provided that the systolic cross-sectional area of the aorta, the velocity flow profile, and the fraction of flow in the branches of the aortic arch, all remain constant. Preliminary observations bear out the validity of these assumptions, and therefore the method may be useful in critical care situations. The waveform of the envelope of the frequency spectrum also seems to reflect the cardiac performance in other respects. Thus, it is possible to estimate indices of early systolic acceleration, peak-velocity, and durations of acceleration and deceleration phases of the systolic period. Because of the difficulty of absolute measurement, in clinical practice it is likely that changes in these indices will prove to be more useful than their actual values. In addition, the frequency spectrum may reflect functional abnormalities in the heart and aorta, such as aortic regurgitation, coarctation of the aorta, and so on.

Secondly, the waveforms of blood flow detected with a continuous wave Doppler probe placed transcutaneously over the jugular vein depend on the right heart function (Kalmanson et al., 1974).

Thirdly, potentially valuable information can be obtained by measuring blood flow within the heart. Thus, the continuous wave Doppler method can be used to measure the instantaneous maximum blood flow velocity within the cardiac chambers, and especially in the region of the valves. Using present techniques, this is not easy to do because of the problems of structure identification in the absence of real-time two-dimensional imaging for guidance. Continuous wave Doppler instruments do not suffer from the range–velocity limitation of pulsed systems, however, and the technique is a sensitive way of detecting regurgitation and turbulence. Generally, however, this disadvantage of pulsed Doppler is not a serious problem at least for qualitative diagnosis, and the great advantage of range selection deserves to be emphasized (Baker et al., 1977). The combination of pulsed Doppler flow measurement with real-time two-dimensional visualization is emerging as an extremely powerful diagnostic tool. The small size of the resolution cell allows turbulent volumes to be detected: in this way, murmurs can be identified which relate to the diastolic rumbles of mitral and tricuspid stenoses, mitral regurgitation, left ventricular outflow obstruction, aortic stenosis, aortic regurgitation, augmented right ventricular filling sound in atrial septal defects, pulmonary stenosis, pulmonary regurgitation, and high velocity flow through the obstruction in co-arctation of the aorta. The analyses generally depend on inspection of the frequency spectrum.

7.2 Applications in Vascular Physiology

Indirect meaurement of blood pressure depends on the equality of the external pressure applied to the arterial wall to that in a pneumatic cuff wrapped around the limb within which the artery lies. When the pressure in the cuff is greater than the systolic arterial pressure no blood flows through the artery. As the pressure is decreased, blood begins to flow for a changing proportion of the cardiac cycle, until it flows without interruption when the cuff pressure is less than the diastolic arterial pressure.

There are various ways in which the arterial pulse distal to the cuff may be detected in order to deduce the blood pressure from measurement of the cuff pressure. This is usually measured by a mercury manometer in a sphygmomanometer, but other types of pressure gauge may be used. The clinician usually detects the pulse either by palpation, or by ausculation, or simply by looking at the range of pressure over which the height of the mercury column in the manometer exhibits small pulsations whilst the cuff pressure is steadily reduced. The ultrasonic Doppler method may be used to detect these pulsations. It is certainly possible to detect the arterial pulsations and to measure the blood pressures in patients in whom other non-invasive methods would fail (Kirby et al., 1969).

Automatic blood pressure recorders using Doppler ultrasound to detect the presence or absence of the pulse have been developed. Whilst these instruments are better than many others—particularly those using microphones—reservations about their performances have been expressed (Labarthe et al., 1973).

In generalized arterial disease, the characteristics of the arteries in the leg reflect the progress of atherosclerosis throughout the arterial system. Arterial disease may modify the shape of the blood flow pressure pulse at the ankle. This may be determined non-invasively by using the ECG R-wave as a timing reference, and measuring the delays in the arrivals of different parts of the pressure wave beyond a cuff with decreasing pressures, at a Doppler probe positioned over the posterior tibial artery. Another approach to arterial characterization depends on measurements of the transit times of the arterial pulse past consecutive segments of artery, and from measurements of the pulsatility of the arterial pulse waveforms at the input and output sites of the arterial segments. Measurement of the pulsatility depends on obtaining the waveform of the maximum Doppler shift frequency, and this is usually done manually by tracing from the frequency spectrum, although reliable maximum frequency followers have recently been developed. This approach allows the collateral circulation to be graded into one of four classes, according to its status (Gosling, 1976). More recent studies have introduced the concept of characterizing the arterial segment between the heart and measurement site (e.g. at the common femoral artery) in terms of the Laplace transform of the blood flow velocity/time signal. Thus it is possible to obtain numerical indices of arterial stiffness, proximal lumen size and distal peripheral impedance (Skidmore et al., 1980).

Two-dimensional Doppler imaging of blood vessels gives further important data on localized peripheral arterial disease, and is a valuable guide for selecting sites for the monitoring of flow waveforms. The relatively inexpensive continuous wave instruments are only capable of imaging the projection of the blood vessels, but pulsed Doppler systems can also make cross-sections and longitudinal sections, thus increasing the reliability of lesion detection (Lusby, 1980).

Studies of venous flow are limited by the general absence, except in vessels close to the heart, of natural pulsation. Deep vein thrombosis can sometimes be detected by changed flow characteristics in the superficial femoral vein in response to squeezing the calf or foot. Doppler imaging has a valuable place in assessing the local site of deep vein thrombosis (Day *et al.*, 1976).

There seems to be no doubt that the clinical value of a method of transcutaneous measurement of blood flow volume would be immense. Two approaches aimed at achieving this goal are showing promise. The first depends on establishing the geometry of the blood vessel and its flow profile, using by now conventional pulsed Doppler instrumentation (Fish *et al.*, 1978). The second method, still very much in the experimental stage, uses a wide beam to measure the power backscattered from the blood; this gives an estimate of the cross-sectional area of the vessel. The power density spectrum reflects the flow profile shape; and the other information required is the vector velocity at the centre of the vessel, which can be determined by a narrow pulsed beam (Hottinger and Meindl, 1979).

Although the applications of ultrasound to blood flow studies are naturally dominated by those based on the Doppler effect, high resolution pulse–echo imaging of blood vessels and associated anatomy and pathology can provide complementary data. The instruments used are normally small real-time devices, which may have integral Doppler capability (Phillips *et al.*, 1980).

These techniques of measuring and visualizing blood flow have an important role in monitoring patients in the post-operative period, in addition to their presently better-accepted place in diagnosis and pre-operative assessment.

7.3 Applications in Obstetrics

The ultrasonic measurement of fetal biparietal diameter is of established value in the assessment of maturity (Campbell, 1969), and nowadays many obstetricians routinely carry out the test early in pregnancy and at 34–36 weeks gestation (Neilson *et al.*, 1980). The fetal head is first identified on a two-dimensional scan (real-time scanning is most convenient), and the scan plane is then adjusted until the head is seen as an ovoid with the midline echoes bisecting its longer axis. The biparietal diameter (the distance between the skull echoes at the opposite sides of the skull along the shorter axis) is then calculated by assuming the value of the speed of ultrasound. In the normal fetus, the biparietal diameter increases from about 27 mm at 14 weeks gestation to about 80 mm at 30 weeks. The rate of growth slows as pregnancy progresses, and the biparietal diameter is about 95 mm at term (38 weeks).

It is not possible to measure the biparietal diameter in very early pregnancy. Between 6 and 14 weeks gestation, however, the maturity of pregnancy may be estimated from measurement of the fetal crown–rump length, since at this stage the rate of normal growth is around 10 mm per week (Robinson, 1973).

Thus, according to the stage in pregnancy, appropriate ultrasonic techniques can provide an accurate indication of the size of the fetus. If the size does not match the dates, there are two possible explanations: either the dates are wrong, or the fetus is growing at an abnormal (usually slow) rate.

After about the twelfth week of gestation, the motion of the heart of a viable fetus can almost always be detected as an audible signal by means of a Doppler system with the probe placed on the maternal abdomen (Bishop, 1966). Earlier in pregnancy, when a large proportion of threatened abortions present, the motion of the heart can usually be seen as an oscillation on a pulse–echo time-base, when the ultrasonic beam is appropriately positioned under guidance from a two-dimensional scan (Robinson, 1972). Although complete absence of fetal pulsations is clinically important, the heart-rate itself is not affected in threatened abortion. Fetal limb and breathing movements, however, may be more sensitive indications of fetal distress (Dawes, 1974; Trudinger et al., 1979). Limb movements can be quantitated by continuous two-dimensional real-time imaging. Fetal breathing can be detected both by pulse–echo studies of fetal abdominal and thoracic movements (McDicken et al., 1979), and by Doppler signals of blood flow in the fetal vena cava modified by intrathoracic respiratory pressure changes (Gough and Poore, 1979).

Towards the end of pregnancy, it is possible to estimate fetal blood flow either using continuous wave (FitzGerald and Drumm, 1977) or pulsed (Eik-Nes et al., 1980; Gill and Kossoff, 1979) Doppler techniques. Provided that the fetal ECG can be detected, it is also possible to extract fetal cardiac valve timings from real-time two-dimensional ultrasonic images (Cousin, 1980).

Also in mid to late pregnancy, urine in the fetal bladder may be visualized as an echo-free area low in the abdominal cavity. Measurements of three orthogonal dimensions, when multiplied together, give an estimate of fetal bladder volume, and hence it is possible to determine the urine-production rate (Wladimiroff and Campbell, 1974).

In the perinatal period, the normal fetal heart rate is in the range 120–160 min^{-1}, there is no significant change in rate during uterine contractions, and there is a beat-to-beat variability of not more than about 5 min^{-1}. Unfavourable patterns (Thomas and Blackwell, 1975) include bradycardia without beat-to-beat variations, or with decelerations during contractions, and tachycardia. On the other hand, accelerations at the beginning of contractions are a good sign. Loss of beat-to-beat variations alone is not worrying (it may be due to maternal medication). The ultrasonic Doppler measurement of the fetal heart rate during labour is such a simple procedure (Thomas *et al.*, 1973) that many obstetricians routinely monitor all deliveries for signs of fetal distress.

Also during labour, it has been shown experimentally that it is possible to monitor the extent of cervical dilatation using an ultrasonic pulse transit-time technique (Zadorf *et al.*, 1976). Transmitting and receiving trans-ducers are attached on diametrically opposite regions of the cervix.

7.4 Other Applications

7.4.1 *Detection of Gas Bubbles*

Gas bubbles may occur in circulating blood (and represent a potential hazard as emboli) during certain surgical procedures, and in soft tissues during decompression of divers. Some anaesthetists monitor blood flow in suitable major arteries by means of Doppler systems, and listen for abnormal sounds due to passing bubbles (Hills and Grulke, 1975). Bubbles forming in soft tissues may be detected by changes in backscatter with a pulse–echo system (Rubissow and MacKay, 1974).

7.4.2 *Diagnosis of Brain Death*

As more effective life-support systems are developed, and as the need for organ donors becomes greater, there is a growing requirement for more rigorous tests of brain death. In addition to the usual criteria, the absence of ultrasonically detected intracerebral pulsations, and the reduction in the amplitude of the midline echo, are useful indices (Oka, 1978).

7.4.3. *Kidney Function*

Two-dimensional ultrasonic scanning may be the first examination of patients presenting with renal failure of unknown cause, or it may follow excretory urography. Sanders and Jeck (1976) have demonstrated that ultrasonic scanning can identify those patients with small or absent kidneys, and with hydronephrosis or polycystic disease.

An indication of reduction in blood flow to a failing transplanted kidney may be obtained by serial studies with a continuous wave Doppler flowmeter (Sampson, 1969). Conventional ultrasonic imaging is also helpful in such cases, because it can lead to the diagnosis of structure abnormalities, and guide renal biopsy (Bartrum *et al.*, 1976).

Urine production rate may be measured by serial studies of bladder volume based on appropriately chosen two-dimensional ultrasonic scans (Alftan and Mattson, 1969). Urine velocity can be measured during voiding (in the male) by the Doppler method, if small phosphate crystals are first allowed to form in the urine by giving the patient a drink of milk with sodium bicarbonate tablets (Albright and Harris, 1975). In this way, turbulence and strictures in the urethra can be located.

7.4.4 *Physiological Activity*

The level of activity of an individual may reflect his psychiatric status. Haines (1974) has described an instrument using 40 kHz ultrasound, resembling a burglar alarm, which is suitable for measuring the amount of patient movement in a defined volume.

7.4.5. *Speech Production*

Movements of the lateral wall of the pharynx (Kelsey *et al.*, 1969) and the vocal folds (Hamlet, 1972) can be made by means of time-position recording with the pulse–echo ultrasonic probe appropriately positioned on the surface of the neck. Vocal fold motion can also be studied by the Doppler method (Minifie *et al.*, 1968). These data are useful in assessing cleft palate patients prior to therapy, in the rehabilitation of patients following laryngectomy and in speech research.

7.4.6. *Tissue Temperature Measurement*

The speed of 5 MHz ultrasound has been measured in tissue samples by Bowen *et al.* (1979) in order to evaluate the feasibility of non-invasive monitoring of temperature distributions produced during hyperthermia treatments for cancer. The data for kidney, liver and muscle indicate that the rate of speed change is correlated with corresponding speed, but the relationship for fat is different. These encouraging results may lead to the development of a tele-thermometer.

References

Albright, R. J. and Harris, J. H. (1975). Diagnosis of urethral flow parameters by ultrasonic backscatter. *IEEE Trans. Biomed Engng* **BME-22**, 1–11

Alftan, O and Mattson, T. (1969). Ultrasonic method of measuring residual urine. *Annls Chir. Gynaecol. Fenn.* **58**, 300–303.

Baker, D. W., Strandness, D. E. and Johnson, S. L. (1977). Pulsed Doppler techniques: some examples from the University of Washington. *Ultrasound Med. Biol.* **2**, 251–262.

Bartrum, R. J., Smith, E. H., D'Orsi, C. J., Tilney, N. L. and Daniono, J. (1976). Evaluation of renal transplants with ultrasound. *Radiology* **118**, 405–410.

Bishop, E. H. (1966). Obstetric uses of the ultrasonic motion sensor. *Am. J. Obstet. Gynecol.* **96**, 863–867.

Bowen, T., Connor, W. G., Nasoni, R. L., Pifer, A. E. and Sholes, R. R. (1979). Measurement of the temperature dependence of the velocity of ultrasound in soft tissues. *In* "Ultrasonic Tissue Characterisation II" (Ed. M. Linzer), pp. 57–61, N.B.S. Special Publication 525, US Government Printing Office, Washington.

Campbell, S. (1969). The prediction of fetal maturity by ultrasonic measurement of biparietal diameter. *J. Obstet. Gynaecol. Br. Commonw.* **76**, 603–609.

Cousin, A. J. (1980). Processing of Doppler returns from the foetal heart: Acoustic requirements for extraction of valvular timing information. *Med. Biol. Engng Comput.* **18**, 563–568.

Dawes, G. S. (1974). Breathing before birth in animals and man. An essay in developmental medicine. *New Engl. J. Med.* **290**, 557–559.

Day, T. K., Fish, P. J. and Kakkar, V. V. (1976). Detection of deep vein thrombosis by Doppler angiography. *Br. Med. J.* **1**, 618–620.

de Vlieger, M., Holmes, J. H., Kazner, E., Kossoff, G., Kratochwil, A., Kraus, R., Poujol, J. and Strandness, D. E. (Eds) (1978). "Handbook of Clinical Ultrasonics", John Wiley, New York.

Eik-Nes, S. H., Brubakk, A. O. and Ulstein, M. K. (1980). Measurement of human fetal blood flow. *Br. Med. J.* **280**, 283–284.

Feigenbaum, H., Wolfe, S. B., Popp, R. L., Haine, C. L. and Dodge, H. T. (1969). Correlation of ultrasound with angiography in measuring left ventricular diastolic volume. *Am. J. Cardiol.* **23**, 111.

Fish, P. J., Wilson, I. M., Holt, B. and Walters, D. (1978). Multichannel pulsed Doppler imaging: Measurement accuracy and beam vessel angle estimation. *In* "Ultrasound in Medicine" (Eds D. White and E. A. Lyons), Vol. 4, pp. 359–362, Plenum Press, New York.

FitzGerald, D. E. and Drumm, J. E. (1977). Non-invasive measurement of human fetal circulation using ultrasound: A new method. *Br. Med. J.* **2**, 1450–1451.

Gibson, D. G. (1979). The contribution of digitized echocardiography to clinical cardiology. *In* "Echocardiology" (Ed. C. T. Lancée), pp. 29–35, Martinus Nijhoff, The Hague.

Gill, R. W. and Kossoff, G. (1979). Pulsed Doppler combined with B-mode imaging for blood flow measurement. *Contr. Gynecol. Obstet.* **6**, 139–141.

Gosling, R. G. (1976). Extraction of physiological information from spectrum-analysed Doppler-shifted continuous-wave ultrasound signals obtained non-invasively from the arterial system. *In* "Medical Electronics" (Eds D. W. Hill and B. W. Watson), Monographs 18–22, Peter Peregrinus, Stevenage.

Gough, J. D. and Poore, E. R. (1979). A continuous wave Doppler ultrasound method of recording fetal breathing *in utero*. *Ultrasound Med. Biol.* **5**, 249–256.

Haines, J. (1974). An ultrasonic system for measuring activity. *Med. Biol. Engng* **12**, 378–381.

Hamlet, S. L. (1972). Vocal fold articulatory activity during whispered sibilants. *Archs Otolar.* **95**, 211–213.

Hills, B. A. and Grulke, D. C. (1975). Evaluation of ultrasonic bubble detectors using calibrated microbubbles at selected velocities. *Ultrasonics* **13**, 181–184.

Hottinger, C. F. and Meindl, J. D. (1979). Blood flow measurement using the attenuation-compensated volume flowmeter. *Ultrasonic Imaging* **1**, 1–15.

Kalmanson, D., Veyrat, C., Chiche, P. and Witchitz, S. (1974). Non-invasive diagnosis of right heart diseases and left-to-right shunts using directional Doppler ultrasound. *In* "Cardiovascular Applications of Ultrasound" (Ed. R. S. Reneman), pp. 361–370, North-Holland, Amsterdam.

Kelsey, C. A., Crummy, A. B. and Schulman, E. Y. (1969). Comparison of ultrasonic and cineradiographic measurements of lateral pharyngeal wall motion. *Invest. Radiol.* **4**, 241–245.

Kirby, R. R., Kemmerer, W. T. and Morgan, J. L. (1969). Transcutaneous Doppler measurement of blood pressure. *Anesthesiology* **31**, 81–89.

Labarthe, D. R., Hawkins, C. M. and Remington, R. D. (1973). Evaluation of performance of selected devices for measuring blood pressure. *Am. J. Cardiol.* **32**, 546–553.

Lusby, R. J. (1980). Pulsed Doppler assessment of the profunda femoris artery. *In* "Diagnosis and Monitoring in Arterial Surgery" (Eds R. N. Baird and J. P. Woodcock), pp. 39–46, John Wright, Bristol.

McDicken, W. N., Anderson, T., McHugh, R., Row, C. R., Boddy, K. and Cole, R. (1979). An ultrasonic real-time scanner with pulsed Doppler and T-M facilities for foetal breathing and other obstetrical studies. *Ultrasound Med. Biol.* **5**, 333–339.

Minifie. F. D., Kelsey, C. A. and Hixon, T. J. (1968). Measurement of vocal fold motion using an ultrasonic Doppler velocity monitor. *J. Acoust. Soc. Am.* **43**, 1165–1169.

Neilson, J. P., Whitfield, C. R. and Aitchison, T. C. (1980). Screening for the small-for-dates fetus: A two-stage ultrasonic examination schedule. *Br. Med. J.* **280**, 1203–1206.

Oka, M. (1978). Echo pulsations in brain death. *In* "Handbook of Clinical Ultrasound" (Eds M. de Vlieger *et al.*), pp. 807–816, John Wiley, New York.

Phillips, D. J., Powers, J. E., Eyer, M. K., Blackshear W. M. Jr, Bodily K. C. Strandness D. E. and Baker D. W. (1980). Detection of peripheral vascular disease using the Duplex Scanner III. *Ultrasound Med. Biol.* **6**, 205–218.

Roberts. V. C. and Sainz, A. J. (1979). Ultrasonic Doppler velocimetry. *In* "Non-Invasive Physiological Measurements" (Ed. P. Rolfe), Vol. 1, pp. 153–174, Academic Press, London and New York.

Robinson, H. P. (1972). Detection of fetal heart movement in the first trimester of pregnancy using pulsed ultrasound. *Br. Med. J.* **4**, 466–468.

Robinson, H. P. (1973). Sonar measurement of fetal crown-rump length as a means of assessing maturity in the first trimester of pregnancy. *Br. Med. J.* **4**, 28–31.

Roelandt, J. (1977). "Practical Echocardiology". *In* "Ultrasound in Biomedicine Series" (Ed. D. N. White), Vol. 1, Research Studies Press, Forest Grove.

Rubissow, G. J. and MacKay, R. S. (1974). Decompression study and control using ultrasonics. *Aerospace Med.* **45**, 473–478.

Sampson, D. (1969). Ultrasonic method for detecting rejection of human renal allotransplants. *Lancet* **ii**, 976–978.

Sanders, R. C. and Jeck, D. L. (1976). B-scan ultrasound in the evaluation of renal failure. *Radiology* **119**, 199–202.

Sequeira, R. F., Light, L. H., Cross, G. and Raftery, E. B. (1976). Transcutaneous aortovelography. A quantitative evaluation. *Br. Heart. J.* **38**, 443–450.

Skidmore, R. and Woodcock, J. P. (1980). Physiological interpretation of Doppler-shift waveforms—I. Theoretical considerations. *Ultrasound Med. Biol.* **6**, 7–10.

Skidmore, R., Woodcock, J. P., Wells, P. N. T., Bird, D. and Baird, R. N. (1980). Physiological interpretation of Doppler-shift waveforms—III. Clinical results. *Ultrasound Med. Biol.* **5**, 227–231.

Taylor, K. J. W. (General Editor) (1979 *et seq.*). "Clinics in Diagnostic Ultrasound" (continuing series), Churchill Livingstone, New York.

Thomas, D. L., Torbet, T., Hansen, S. and Hay, D. M. (1973). A comprehensive system for monitoring foetal heartrate and uterine contractions. *Med. Biol. Engng* **11**, 703–709.

Thomas, G. and Blackwell, R. J. (1975). The analysis of continuous fetal heart rate traces in the first and second stages of labour. *Br. J. Obstet. Gynaecol.* **82**, 634–642.

Trudinger, B. J., Gordon, Y. B., Grudzinskas, J. G., Hull, M. G. R., Lewis, P. J. and Arrans, M. E. L. (1979). Fetal breathing movements and other tests of fetal wellbeing: A comparative evaluation. *Br. Med. J.* **2**, 577–579.

Wells, P. N. T. (1977a). "Biomedical Ultrasonics", Academic Press, London and New York.

Wells, P. N. T. (Ed.) (1977b). "Ultrasonics in Clinical Diagnosis", (2nd Edn), Churchill Livingstone, Edinburgh.

White, D. N. (Ed.) (1973 *et seq.*). *Ultrasound in Medicine and Biology*.

Wladimiroff, J. W. and Campbell, S. (1974). Fetal urine-production rates in normal and complicated pregnancies. *Lancet* **i**, 151–154.

Zadorf, I., Neuman, M. R. and Wolfson, R. N. (1976). Continuous monitoring of cervical dilatation during labour by ultrasonic transit-time measurement. *Med. Biol. Engng* **14**, 299–305.

10. MEASUREMENT OF ELECTRICAL PROPERTIES OF TISSUE AT MICROWAVE FREQUENCIES: A NEW APPROACH TO DETECTION AND TREATMENT OF ABNORMALITIES

R. L. Magin and E. C. Burdette

University of Illinois at Urbana-Champaign, Urbana, USA
Georgia Tech Engineering Experiment Station, Atlanta, USA

1.0 INTRODUCTION

1.1 Potential of the Method

Recent developments promise the detection of locally abnormal tissues and the measurement of tissue blood flow from non-invasive electrical measurements made with small diameter probes. These probes (Burdette *et al.*, 1980a; Stuchly *et al.*, 1981) measure the electrical properties of tissues over a wide frequency range: 1–10 000 MHz. Since the electrical properties of tissues are sensitive to surrounding tissue organization, blood perfusion and the development of pathology, these measurements provide a new and unique way to study local tissue physiology. In addition, they also allow the electrical properties of tissues to be measured *in vivo*. Areas of biomedical and clinical practice which could benefit from this electrical property information are electromagnetic-induced hyperthermia for

NON-INVASIVE MEASUREMENTS: 2
ISBN 0 12 593402 5

cancer treatment, electrogmagnetic thawing of frozen organs, detection of pathological conditions, and clinical diagnostic monitoring of parameters such as lung water and blood flow. The contributions that a knowledge of tissue electrical properties can make in these areas will be described in this chapter after a presentation of the theory, procedure and results of dielectric probe measurements. The potential which this technique offers as a new method for measuring local blood flow, tissue concentration of polar drugs, and composition of whole blood will be illustrated by the presentation of recent experimental results.

1.2 Background and Previous Work

Several early investigators (Cook, 1951; Herrick *et al.*, 1950; Schwan, 1957) measured the *in vitro* dielectric properties of biological tissues using short-circuited transmission line and wave-guide techniques. These previous studies and more recent measurements in the microwave frequency range (Schwan and Foster, 1980; Stuchly and Stuchly, 1980; Schepps and Foster, 1980) establish a useful base of *in vitro* dielectric data for various normal tissues over an extremely wide frequency range, from 10 kHz to 10 GHz. Examination of these data reveals the existence of three distinct plateau regions in tissue dielectric properties. These plateau regions are illustrated in Fig. 1 by measurements of the relative dielectric constant of muscle (Schwan and Foster, 1980). Each plateau is separated by areas of rapid change in the dielectric properties as a function of frequency. Schwan (1957) described these regions of change as the α, β and γ dispersion regions, and related the observed phenomena to different tissue components. These changes are important because each is related to a different mechanism of interaction between the electromagnetic fields and the constituents of biological tissues. The α dispersion occurs at relatively low frequencies, around 80 Hz. It arises from the interaction between charges on the surface of cellular membranes and ions in solution. The β dispersion, which occurs at approximately 50 kHz, is due to the presence of cellular membranes (enclosing cell water) which act as insulating structures. As the frequency is increased above 50 kHz, the cellular membranes are effectively short-circuited. Thus, the cellular membrane effect decreases with increasing frequency; and, at frequencies greater than 10 GHz, the tissue electrical properties are largely determined by the water and electrolyte content of the tissue. Because of its water content, biological tissue exhibits another dispersion phenomenon, the γ dispersion, in the microwave region near 25 GHz. This is the frequency at which water molecules at 37°C interact most strongly with the rapidly changing microwave field. Between the β and γ dispersions a broad spectral disper-

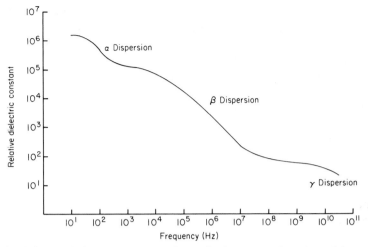

FIG. 1   Relative dielectric constant of muscle tissue as a function of frequency. Three distinct plateaus are separated by the α, β and γ dispersion regions. The electrical conductivity of muscle increases in a similar step-like manner.

sion of small magnitude known as the δ dispersion is often observed. It extends from a few hundred megahertz to approximately 4 GHz and is probably caused by a partial rotation of large polar molecules along with that of protein-bound water.

Several groups of investigators have carefully examined electrical properties of normal animal and human tissues (Tinga and Nelson, 1973; Johnson and Guy, 1972; Geddes and Baker, 1967; Schwan and Foster, 1980). These studies have proven to be extremely useful in understanding the mechanism of interaction between electromagnetic energy and biological tissues. However, because these data were obtained from *in vitro* measurements, they do not reflect the actual physiological conditions of living tissues, and cannot describe changes in electrical properties due to physiological changes such as blood flow or the development of local pathology.

2.0 BASIC PRINCIPLES

2.1 Definitions

The physical properties of biological tissues to be presented and discussed are the electrical conductivity, σ, and the relative dielectric constant, K (Von Hippel, 1954). The electrical conductivity of a material is defined as the conductance per unit length (mho m^{-1}). It is simply a measure of the ease with which an electric current flows in the material. The relative dielectric constant is defined as the capacitance of a material relative to that of empty space. Capacitance indicates the ability of the material to store electrical charges. These two properties are frequently combined in a parameter known as the complex permittivity, $\varepsilon^* = \varepsilon' - j\varepsilon''$ where $j = \sqrt{(-1)}$ designates the imaginary part of a complex number. The relative dielectric constant K is defined by the ratio $\varepsilon'/\varepsilon_0$, where ε_0 is the permittivity of empty space. The electrical conductivity, σ, is related to the imaginary part of the complex permittivity by the expression $\sigma = 2\pi f \varepsilon''$, where f represents the frequency, in hertz, of the time varying electromagnetic field. Magnetic properties are not to be considered because the magnetic permeability of most biological material is equal to that of free space and magnetic losses are very small (Von Hippel, 1954). The frequency range of interest in these studies extends from 1 to 10 000 MHz. This includes part (1–300 MHz) of the radiofrequency spectrum and a significant portion (300–10 000 MHz) of the microwave frequency range.

2.2 Relaxation Phenomena at Microwave Frequencies

The electrical properties of a material determine its ability to interact with an electromagnetic field. This interaction results from the presence of components within the material that can be affected by electric and magnetic forces generated by electric and magnetic fields (Harrington, 1961). In non-magnetic materials such as tissues, an electromagnetic field will primarily act upon components within the material that have either a net electrical charge or an electric dipole moment (Von Hippel, 1954). The electric field generates forces which impart motion to these components resulting in a flow of electric current within the material. In biological tissues, the components possessing a net electrical charge are predominantly ions (Na^+, Cl^-, K^+, Ca^{2+}), while polar molecules (water, amino acids, sugars, proteins) are the main source of electric dipole moments. Because the electrical properties of a tissue are determined by such a wide variety of components, these properties exhibit significant variations as a function of frequency, polarization, tissue type, water content and temperature. For the interested reader, a more complete description of the concepts and terminology of tissue electrical properties and relaxation phenomena is provided by Salter (1979) in Volume 1 of this series and by Schwan and Foster (1980).

2.3 Tissue Electrical Measurements *In Vitro* and *In Vivo*

An issue of concern is the validity of *in vitro* tissue electrical properties obtained from measurements of excised tissues as opposed to data collected from *in vivo* experiments. As early as 1922, Osterhout described significant changes in the dielectric characteristics of biological materials following death. He attributed the observed changes to the loss of membrane function and the breakdown of cellular structure (Osterhout, 1922). In 1938, Rajewsky made low frequency measurements (300 kHz to 30 MHz) which showed a deterioration of the dielectric properties concomitant with a significant decrease in metabolic rate (Rajewsky, 1938). However, the reported changes did not begin until approximately a day after death of the organism. Recent *in situ* low frequency measurements (Burdette *et al.*, 1980b) revealed changes in the electrical conductivity of the liver within an hour of death, while one day was required for changes in the dielectric permittivity. The observed changes in these low frequency dielectric properties are caused by a breakdown of cellular membranes. Schwan and Foster (1980) state that the high frequency data are relatively unaffected by death of the tissue because at microwave frequencies the dielectric characteristics of tissue are due predominantly to the water and protein contents of the tissue. However, it is recognized that blood loss and changes in the water content of tissue do occur upon excision. These factors affect the tissue's dielectric properties and may be responsible for the reported changes following excision. On the other hand, recent measurements have demonstrated changes in tissue dielectric properties at microwave frequencies immediately following death that are not due to moisture loss (Burdette *et al.*, 1980b).

Tissue dielectric property measurements have traditionally been made using electrical circuit apparatus such as impedance bridges, resonant circuits and short-circuited transmission lines or wave-guides (Cook, 1951; Von Hippel, 1954). However, these techniques generally require tissue excision and large sample volumes (Schwan, 1957; Johnson and Guy, 1972), and therefore are not well-suited for *in vivo* measurements. The recent development of coaxial line probes (Burdette *et al.*, 1980a; Athey *et al.*, 1981) provides a more appropriate method for measuring the *in situ* dielectric properties of biological tissues at radio and microwave frequencies. A recent review by Stuchly and Stuchly (1981) compares these new coaxial line techniques. The reader is referred to this paper as a guide to selecting appropriate *in vivo* measurement techniques for specific applications.

3.0 PROBE DIELECTRIC MEASUREMENT SYSTEM

3.1 Description of the Apparatus

A small diameter (2·2 mm) cylindrical probe uniquely suited to the measurement of the *in vivo* dielectric properties of tissues has been developed and tested (Burdette *et al.*, 1980a). The basic concept of this probe system was first described by Magin and Burns (1972). The technique relies on an antenna modelling theorem (Deschamps, 1962) which relates a change in the terminal impedance of an antenna inserted into a lossy medium to the dielectric properties of that medium. If the impedance of the antenna is known in both a reference medium (air) and in the medium under study (tissue), this modelling theorem may be used to determine the dielectric properties of that medium. Because the impedance of a very short monopole antenna is known, this type of antenna may be used as a small coaxial probe that can be brought into contact with living tissue and thus permit measurements of the tissue's dielectric properties. The volume of tissue whose dielectric properties are measured consists only of the region in the immediate vicinity of the probe tip. The extent of the measured volume is determined by the fringing electric fields near the probe as described by Burdette *et al.* (1980a). In the limiting case where the monopole antenna length approaches zero, the probe becomes an open-circuited transmission line and the analysis reduces to that for the fringing field at the end of the line (probe tip). The equipment required for these measurements consists of a signal source, reflectometer, network analyser and data recorder. Semi-automation of the data acquisition procedure is accomplished using a minicomputer, as shown in Fig. 2a, which depicts the overall system block diagram. The complete *in vivo* probe dielectric measurement system is pictured in Fig. 2b. This system measures probe impedance in the form of a complex reflection coefficient which is converted by a program in the minicomputer into the dielectric constant and conductivity information.

The accuracy of the coaxial probe system was determined through measurements on standard dielectric materials and on normal biological tissues (Burdette *et al.*, 1980a). De-ionized water, 0·1 M saline, methanol and ethylene glycol were studied in the frequency range 1–10 000 MHz. The relative dielectric constant and electrical conductivity determined for these liquids agreed to within $\pm 5\%$ of the values reported by other researchers (Buckley and Maryott, 1958; Cook, 1952; Hasted, 1972; Von Hippel, 1954) and within $\pm 3\%$ of the theoretical values (Von Hippel, 1954). The *in vivo* dielectric properties of normal tissues (skeletal muscle, kidney, fat, brain and blood) from dog and rat generally agreed with previously published *in vitro* results (Cook, 1951; Lin, 1975; Schwan,

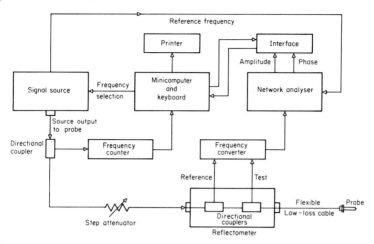

FIG. 2 (*top*) Block diagram of the system used for determining the electrical properties of tissues *in vivo*. (*bottom*) Photograph of probe dielectric measurement system showing probe, network analyser and desk-top computer.

1957; Foster *et al.*, 1979; Stuchly *et al.*, 1981; Brady *et al.*, 1981), but in several tissues differences between *in vivo* and *in vitro* dielectric properties were observed. These measurements on standard liquid dielectrics and normal tissue samples established the reliability and accuracy of this new dielectric property measurement technique from 1–10 000 MHz.

3.2 Probe Design and Construction

A variety of probe configurations have been investigated. The length of the extended centre conductor of these probes has ranged from a cut-off open-ended coaxial line to 1·0 cm, and probe outside diameters have ranged from the size of an 18 gauge hypodermic needle (approximately 1 mm) to 3·6 mm. Four probe designs that have been used for tissue

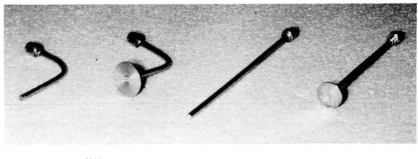

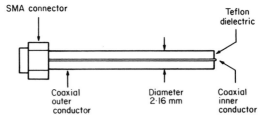

SMA connector

Teflon dielectric

Coaxial outer conductor

Diameter 2·16 mm

Coaxial inner conductor

FIG. 3   (*top*) Photograph of four coaxial dielectric measurement probes. (*bottom*) Dimensioned illustration of the probe used to obtain most of the tissue dielectric measurements described in this chapter.

dielectric measurements are pictured in Fig. 3 (top). The dielectric measurement probe used in most of the reported research is illustrated in Fig. 3 (bottom). It was fabricated from 2·2-mm diameter semi-rigid coaxial cable. An SMA type connector was attached to the probe by first removing the centre conductor and Teflon dielectric material. The connector was soldered to the outer conductor and the probe was then reassembled using the centre conductor as the centre pin of the connector, thus avoiding additional soldering. In this manner, it was possible to attach the SMA type connector without deforming the Teflon dielectric insulation within the coaxial cable. While disassembled, the centre and outer conductors of the probe were gold plated. Plating the probe with this inert metal greatly reduced chemical reactions between the probe and the electrolytes in the tissue. The process virtually eliminated oxidation of the probe's metallic surfaces and thus helped to improve tissue/probe electrical contact without contributing to electrode polarization effects at low frequencies.

3.3 Control of Measurements

The most important factor that must be controlled while performing tissue dielectric measurements is the contact between tissue and probe. The following conditions can also affect the accuracy and/or repeatability of probe measurements:

(1) tissue dehydration;
(2) accumulation of excess fluid in the measurement area;
(3) formation of a dried film of tissue fluids on the probe;
(4) variation in probe contact pressure;
(5) changes in tissue temperature;
(6) tissue inhomogeneity in the measurement area.

Each one of the above factors constitutes a potential source of error in the performance of the dielectric measurements, and therefore steps must be taken to minimize these factors.

3.4 Tissue Measurement Procedure

Impedance measurements are performed by positioning the probe antenna in direct contact with the tissue under study (Burdette *et al.*, 1980a). For the measurement of surface and subdermal tissues, it is not necessary to insert the probe into the tissue. However, in order to measure the dielectric properties of internal organs, the tissue must be surgically exposed. A geared assembly is used to position the probe on the tissue. This allows accurate repositioning of the probe for repeated measurements. For measurements of deeper tissues, a small-diameter probe can be inserted into the tissue being studied. A temperature- and humidity-controlled chamber (37°C, 90% RH) is used during measurements on small animals (rats, mice) to prevent changes in the temperature and moisture content of exposed tissues (Burdette *et al.*, 1983). Dielectric data are collected by recording the probe impedance while changing the frequency of the signal source. Measurements are repeated four to eight times with the probe being repositioned for each set of measurements and the results averaged. Methanol and distilled water are used to clean the probe between each set of measurements to prevent an insulating film from developing on the probe.

Multiple measurements at a single tissue site gave reproducible results ($\pm 2\%$) which were relatively insensitive to contact pressure as long as there was complete contact between the probe and the tissue. The measurements were sensitive, however, to the location of the probe on the tissue. Measurements performed with the probe located in well-vascularized regions differed by as much as 20% in dielectric constant from those obtained in poorly vascularized regions.

When normal tissues and neoplastic surface lesions were studied in human cancer patients, the contact pressure between probe and tissue was found to influence the measured probe impedance. The contact pressure was standardized in those studies by applying increasing probe pressure against the tissue until the phase angle of the complex reflection coefficient remained constant. This probe pressure was then maintained for the duration of the measurement. Using this procedure, small standard errors ($\pm 3\%$) were obtained for multiple measurements of the same tissue.

4.0 TISSUE ELECTRICAL PROPERTIES

4.1 Normal Tissues

Using the probe dielectric measurement system (Burdette *et al.*, 1980a) both *in vivo* and *in vitro* dielectric measurements have been performed on several types of tissue: skeletal muscle (canine and rat), kidney cortex and medulla (canine), visceral fat (canine), brain cortex (rat) and whole blood (rat). The relative dielectric constant and electrical conductivity of *in vivo*

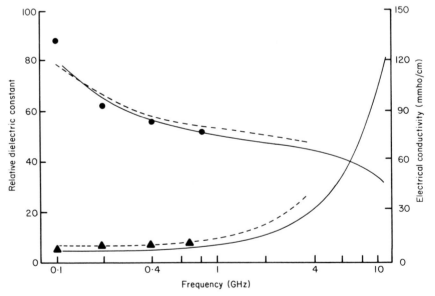

FIG. 4   Relative dielectric constant ($\bullet$) and electrical conductivity ($\blacktriangle$) of *in vivo* (— — —) and *in vitro* (———) canine kidney cortex compared to reference data (Schwan, 1957).

TABLE I

The relative dielectric constant, K, of a variety of solid murine tumours and of selected normal tissues. The values reported at each frequency are the average of three separate measurements

Tissue	Frequency (MHz)				
	13·56	27·12	433	918	2450
Tumours (mouse)					
Mendecki	201·8	112·9	34·7	31·4	27·0
C3HBA	206	121·5	38·8	37·4	37·0
16C	—	—	46·5	43·6	42·2
Lewis Lung	268·8	154·6	57·2	50·4	52·0
Glioblastoma	—	—	60·3	58·6	58·5
Ependymoblastoma	261·6	136·3	65·0	65·0	65·0
B16	234·7	125·8	46·5	41·8	40·0
Normal tissues					
Muscle (canine)	222·9	107·4	49·3	48·8	46·8
Fat (canine)	46·9	23·4	16·3	14·6	11·2
Kidney (canine)	330·0	181·0	58·5	54·4	50·1
Brain (mouse)	267·8	192·8	49·4	46·9	44·6
Inflated lung (rat)	3·2	2·1	1·5	1·47	1·45

TABLE II

The electrical conductivity, σ (mmho cm^{-1}) of a variety of solid murine tumours and of selected normal tissues. The values reported at each frequency are the average of three separate measurements

Tissue	Frequency (MHz)				
	13·56	27·12	433	918	2450
Tumours (mouse)					
Mendecki	3·3	4·1	6·6	7·8	12·0
C3HBA	3·4	3·9	6·6	7·8	11·8
16C	—	—	8·3	9·7	15·6
Lewis lung	6·0	6·8	10·1	10·8	22·4
Glioblastoma	—	—	12·1	13·4	26·3
Ependymoblastoma	8·6	9·0	10·2	11·5	25·0
B16	4·6	5·2	9·8	11·9	21·0
Normal tissues					
Muscle (canine)	7·0	7·7	10·6	11·0	21·0
Fat (canine)	2·0	2·0	2·6	3·3	6·3
Kidney (canine)	8·0	8·8	10·9	14·3	21·6
Brain (mouse)	3·7	4·8	7·5	9·9	18·7
Inflated lung (rat)	0·07	0·08	0·11	0·11	0·17

TABLE III

The relative dielectric constant, K, and electrical conductivity, σ (mmho cm^{-1}) of human normal tissues and of tumours. Each value is the average of three separate measurements

Tissue	Frequency (MHz)					
	500		918		2450	
Tumours	K	σ	K	σ	K	σ
Breast carcinoma	56	8·0	50	10·0	44	14·0
Adenocarcinoma	38	5·0	36	6·0	31	10·0
Melanoma	36	3·5	34	4·5	31	9·0
Normal tissues						
Skin	40	5·0	37	7·0	34	10·0
Breast (through skin)	43	4·0	38	6·0	32	10·0

canine kidney cortical tissue are compared to *in vitro* kidney data and to data from Schwan (1957) in Fig. 4. This comparison between the *in vivo* and *in vitro* dielectric property data shows the *in vivo* values to be larger than the corresponding *in vitro* results. Similar differences between *in vivo* and *in vitro* tissue measurements of relative dielectric constant and electrical conductivity were observed in muscle, fat and brain (Burdette *et al.*, 1980a). A summary of these *in vivo* data at selected frequencies (13·56, 27·12, 433, 918 and 2450 MHz) is presented in Tables I and II. *In vivo* dielectric properties of human skin and breast tissues at 500, 918 and 2450 MHz are presented in Table III.

4.2 Abnormal Tissues

In vivo dielectric property measurements of seven mouse tumour lines have been performed over the frequency range 10–4000 MHz (Burdette *et al.*, 1983). Tables I and II summarize these data and compare them with those of several normal animal tissues. The dielectric properties of the three mouse mammary adenocarcinoma lines (C3HBA, Mendecki, 16C) differ from each other, although they all lie intermediate in value between muscle and fat. The dielectric constant and electrical conductivity of Lewis lung carcinoma and B16 melanoma are approximately equal to those of canine muscle at frequencies above 300 MHz. The large difference between the dielectric properties of the Lewis lung carcinoma and normal lung tissue is due to the fact that normal lung tissue consists almost entirely of air.

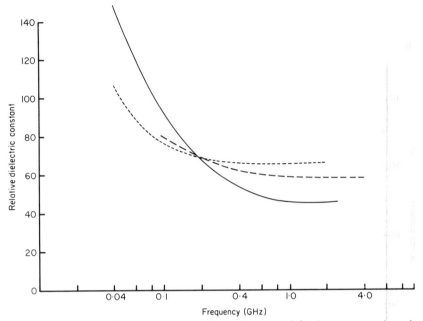

FIG. 5   Relative dielectric constant of two implanted brain tumours in mice compared to mouse brain cortex. ———— Brain cortex; — — — glioblastoma; — — — — — ependymoblastoma.

A comparison of *in vivo* dielectric properties of two different types of brain tumours (glioblastoma, ependymoblastoma) in mice with normal brain tissue from the mouse is presented in Figs 5 and 6. The two model mouse tumours have dielectric constants that are greater than those of normal mouse brain cortex only in the 200–400 MHz frequency range, while the tumour electrical conductivities are higher than that for the normal rat brain cortex over the entire frequency range measured.

The *in vitro* dielectric properties of three human tumours, skin and breast tissue have been measured in the frequency range 500–4000 MHz (Burdette *et al.*, 1983). Table III summarizes these data at 500, 918 and 2450 MHz. Breast carcinoma exhibited a larger dielectric constant and electrical conductivity than either normal skin or breast tissue over the 500–2450 MHz frequency range. The dielectric properties of the adenocarcinoma were very similar to those of normal human skin throughout this frequency range. Melanoma, which usually occurs within the skin, has a smaller dielectric constant and electrical conductivity than normal skin in the same frequency range.

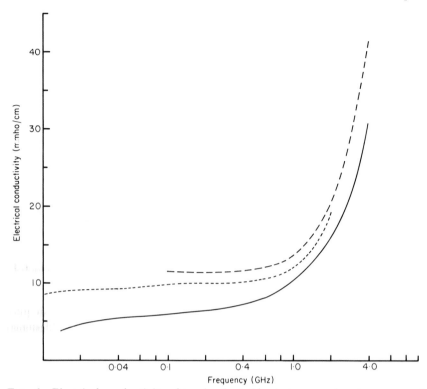

FIG. 6   Electrical conductivity of two implanted brain tumours in mice compared to mouse brain cortex. ———— Brain cortex; — — — glioblastoma; — — — — — ependymoblastoma.

5.0 EFFECTS OF PHYSICAL AND PHYSIOLOGICAL CHANGES ON TISSUE DIELECTRIC PROPERTIES

5.1 Physical Factors

Tissue water content, composition and temperature are the three physical parameters that have the most influence on tissue dielectric properties at microwave frequencies. In the 1–100 MHz frequency range, the composition of the tissue (cellular structure, electrolytes, proteins) is an important determinant of the observed changes in dielectric properties. However, of these three factors, tissue water content is the most significant in characterizing tissue dielectric properties. In fact, for the purpose of describing their general dielectric characteristics, all tissues can be divided into two classes: those of high water content and those of low water content.

Johnson and Guy (1972) provide a convenient table which describes the basic dielectric properties of these two classes of tissues over an extended frequency range (1–10 000 MHz). In general, tissues of low water content have dielectric constants and electrical conductivities that are smaller by approximately a factor of ten compared to those of tissues with high water content. For example, Johnson and Guy (1972) show that, at 2450 MHz, tissues of high water content have values of $K = 47$ and $\sigma = 2 \cdot 21$ mhom^{-1}, while for tissues of low water content $K = 5 \cdot 5$ and $\sigma = 0 \cdot 15$ mhom^{-1}. However, by classifying tissues into only two categories, no allowance is made for physiological or biochemical differences between various tissues which can contribute more subtly to the tissue dielectric properties. Temperature changes in the range of biological interest (20–40°C) do not have a large effect on tissue dielectric properties at frequencies away from relaxation. Both K and σ vary by less than 2% per degree Celsius in this temperature range at microwave frequencies. A more complete description of the temperatuie behaviour of tissue dielectric properties at microwave frequencies is given by Schwan and Foster (1980).

The significance of water content in describing tissue dielectric properties was illustrated in the work of Schepps and Foster (1980) which demonstrated that K and σ for many normal and tumour tissues can each be described by a relatively simple equation that depends only on the frequency and the tissue water content. These empirical equations allow prediction of the approximate dielectric properties at 37°C of soft tissues containing over 60% water content by volume over the frequency range 10–17 000 MHz.

Although tissue dielectric properties do not change much as a function of temperature within the normal range of biological temperature variations, they are greatly affected when a phase change between the liquid and solid states occurs. Consider the change in the dielectric constant of a suspension of human granulocytes at 2·45 GHz as a function of temperature as shown in Fig. 7. The small value of the relative dielectric constant for this cell suspension below its solid–liquid phase transition is due to the restricted motion of the water molecules in the frozen state. Above the solid-to-liquid phase transition temperature the relative dielectric constant of the cell suspension increases to its characteristic liquid state value of approximately 60. The electrical conductivity of this suspension shows a similar dramatic increase with increasing temperature as the phase transition temperature is encountered (Popovic et al., 1977). This type of dielectric behaviour was also seen in kidney tissue in experiments designed to thaw frozen canine kidneys using intense microwave radiation (Burdette and Karow, 1978; Burdette et al., 1980c). Note also in Fig. 7 that

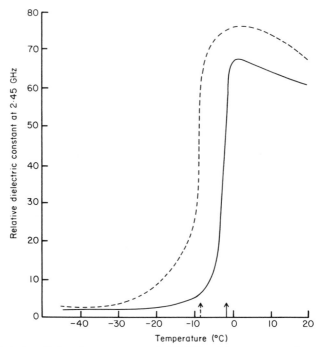

FIG. 7   Relative dielectric constant at 2·45 GHz of human granulocyte suspensions as a function of temperature with and without 6% dimethylsulfoxide (DMSO). The small arrows indicate the solid-to-liquid phase change temperature. ——— No DMSO; – – – – – – 6% DMSO.

the addition of the polar cryoprotectant dimethylsulfoxide (DMSO) at a final concentration of 6% decreases the solid-to-liquid phase transition temperature and increases the relative dielectric constant.

5.2   Physiological Factors

A wide variety of physiological events can alter the dielectric properties of tissues to the extent that measurable changes are observed. Significant changes in dielectric properties would be expected to follow the development of a physiological or pathological condition that alters the tissue water content. For example, the occurrence of pulmonary oedema should greatly increase the dielectric constant and conductivity of lung tissue. This effect was examined by Pederson et al. (1976) in an attempt to develop a non-invasive system for the detection of pulmonary disease. In a similar manner, conditions which result in the local accumulation of fibrotic or fatty tissue should cause a marked decrease in the dielectric properties of the affected tissue. Jacobi et al. (1979) have shown that such changes in local tissue dielectric properties may allow one to obtain images of regions of the body which exhibit anomalous dielectric behaviour. Recent experiments (Larsen and Jacobi, 1979; Jacobi and Larsen, 1980) have demonstrated the feasibility of electromagnetic imaging techniques on an isolated canine kidney. In addition, Larsen et al. (1978) have reported dielectric property changes in the 3–30 MHz frequency range which were associated with physiological, as well as pathophysiological, states in suspensions of living cells.

In fairly recent experiments Burdette et al. (1980d) demonstrated a change in the dielectric properties of an isolated perfused kidney as a function of organ blood flow, as shown in Fig. 8. Stated simply, an increase in total renal blood flow produces a decrease in relative dielectric constant and electrical conductivity, while a decrease in renal blood flow produces the opposite effect. These changes may be due to variations in the tissue's blood volume. If increased renal flow caused an increased presence of tissue water, one would expect a corresponding increase in dielectric property values rather than the observed decrease. Thus, these results obtained using the probe measurement technique suggest that the pressure–flow–volume relationship is not a simple one, and that redistribution of flow within the renal vasculature influences local renal dielectric properties. Also, tissue water shifts may not be the only significant factor in the measured dielectric changes. The dielectric probe also permitted measurements of dielectric property changes from different anatomical regions (cortex, medulla) of the kidney. Note that although the dielectric constant and electrical conductivity of cortex and medulla were different, the effect of blood flow on the dielectric properties was similar for both regions with perhaps the greatest effect on the medullary dielectric constant.

It is also possible to use tissue dielectric measurement techniques *in vivo* to monitor variations in the concentrations of the blood components (red blood cells, white blood cells, platelets) and of drugs. The data presented in Fig. 9 show the changes in tissue dielectric properties seen when the composition of blood is varied. Note that the dielectric constant of plasma is greater than that of packed red blood cells or of granulocytes plus plasma, which is to be expected, since the cellular components have a much lower dielectric constant than plasma. The addition of 6% DMSO increases the dielectric properties of the granulocyte suspension to above those for the plasma alone. In Fig. 10 the electrical properties at 2·45 GHz of kidney tissue perfused at 25°C with four concentrations of DMSO are displayed (Burdette et al., 1978). These data show a linear increase in the dielectric constant with increasing DMSO concentration. This result demonstrates the potential that *in vivo* dielectric measurements offer in

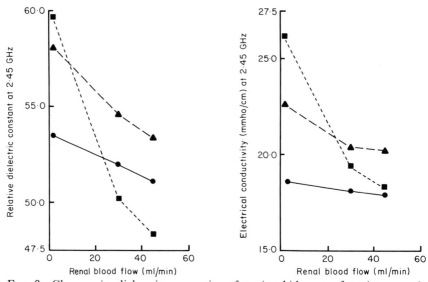

FIG. 8 Changes in dielectric properties of canine kidney surface (●———●), cortex (▲ — — ▲) and medulla (■ – – – – ■) corresponding to different total renal blood flow rates. These measurements were performed at 2·45 GHz and 37 °C on one experimental animal.

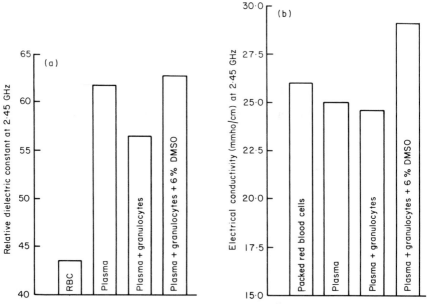

FIG. 9 Measurements of the dielectric properties of human blood components (plasma, packed red blood cells and granulocytes) at 2·45 GHz and 20 °C. (a) Relative dielectric constant. (b) Electrical conductivity.

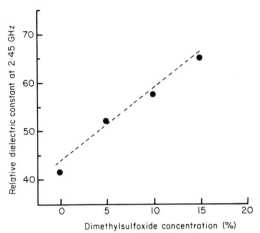

FIG. 10   Relative dielectric constant of rabbit kidney tissue measured at 2·45 GHz
and 25 °C for increasing dimethylsulfoxide (DMSO) concentrations.

providing a method for measuring the concentration of polar substances
within a tissue.

6.0 EXPERIMENTAL DEVELOPMENTS AND CLINICAL APPLICATIONS

6.1 Electromagnetic Techniques for Medical Diagnosis

A wide variety of electromagnetic techniques may be potentially useful in
medical diagnosis (Iskander and Durney, 1980). The relatively well-
established methods of electrical-impedance plethysmography and
electromagnetic blood flow measurement have been accepted in clinical
medicine. However, many additional bio-electromagnetic applications
are now being tested experimentally. These new techniques include the
microwave measurement of lung water (Pederson *et al.*, 1978), electromagnetic imaging (Jacobi and Larson, 1980), an impedance camera
(Henderson and Webster, 1978), microwave radiometry (Myers and Barrett
1980) and cancer treatment by localized microwave hyperthermia (Har-
Kedar and Bleehen, 1976). Each of these applications requires an understanding of the interaction of electromagnetic waves with specific tissues.
This interaction is largely determined by the anatomy and the electrical
properties of the tissue under study. In this section, we will examine the
role that *in vivo* tissue dielectric measurements play in the design of
microwave systems that treat or interrogate regions of the body.

6.2 Absorption of Electromagnetic Radiation by the Body

Information on dielectric properties is essential for the determination of the potential hazards to personnel of electromagnetic radiation (Michaelson, 1980). Realistic models of the amount of electromagnetic radiation absorbed by the body require accurate knowledge of the tissue geometry and dielectric properties. This information allows calculation of the power absorbed per gram of tissue if the appropriate equations describing the absorption, scattering and reflection of the incident electromagnetic waves can be solved. The recent application of sophisticated mathematical and computer techniques to these problems has resulted in the characterization of the dose distribution of electromagnetic energy within a variety of animal and human forms (Gandhi, 1980). These results offer the possibility of establishing a more meaningful correlation between the microwave exposure conditions, the amount of energy absorbed and the observed biological effects that may present a hazard to the body.

6.3 Selection of Suitable Frequencies for Tumour Treatment

In vivo dielectric constant and electrical conductivity measurements of tumours provide essential information for the design of microwave hyperthermia systems. The interpretation of this information will be illustrated by considering a few specific examples. The design goal in these systems is the selection of a frequency at which a significant difference in absorbed power exists between the target tumour tissue and the surrounding normal tissues. In this process, proper consideration must be given to the attenuation of microwave energy by any intervening normal tissues.

Comparing the dielectric constant and conductivity of the three mammary adenocarcinomas (C3HBA, Mendecki and 16C) to similar data for canine fat over the frequency range 13·56 MHz to 2·45 GHz (Tables I and II), the tumour values are found to be between 100–300% greater. This result indicates that a breast tumour would absorb energy more readily at these frequencies than the surrounding normal, fat-like breast tissue. These data illustrate the potential for electromagnetic radiation to induce differential hyperthermia for the treatment of breast cancer.

The electrical properties of Lewis lung carcinoma are also compared to those of several normal tissues in Tables I and II. Throughout the frequency range examined, the dielectric properties of the tumour are more nearly comparable to those of muscle than to those of the inflated lung or fat. Thus, at these frequencies, the tumour tissue would absorb significantly more electromagnetic energy than the surrounding lung.

The dielectric properties of two implanted brain tumours (ependymo-blastoma and glioblastoma) are compared in Figs 5 and 6 to the properties of normal rat brain cortex. The electrical conductivity of both tumours from 100 to 2000 MHz is 50–100% higher than normal brain tissue. These data suggest that frequencies in the range 500–1500 MHz may be used for inducing differential brain tumour hyperthermia. Applicators at these frequencies could be designed to radiate energy selectively into brain tumours.

The data in Table III compare the dielectric properties of a human breast carcinoma to those of normal skin and breast tissue. These data illustrate that the electrical conductivity and dielectric constant of breast tumour tissue are elevated in comparison with normal breast tissue. Superficial breast tumours of the type measured might, therefore, be successfully treated by microwave radiation-induced hyperthermia at these frequencies. The second human tumour listed in Table III is an adenocarcinoma in the neck of a patient. Its properties should be compared to those of the normal skin and muscle surrounding it. These data indicate that this frequency band is unsuitable for inducing differential hyperthermia because of the similarity of the values for the dielectric properties of these tissues. Finally, the entries in Table III can be used to compare the dielectric properties of a superficial melanoma to those of the patient's normal skin. Both the conductivity and dielectric constant of this tumour are lower than those of the normal skin surrounding it. This indicates that heating at lower frequencies with direct contact electrodes might be successfully employed in treating these tumours, but additional dielectric property information at frequencies of less than 500 MHz is needed.

The results of the *in vivo* measurements of the dielectric properties of animal and human tumours can be summarized as follows: (1) measurable differences exist between tumours; (2) significant differences exist between tumours and normal tissues; and (3) these differences can be used to guide the selection of suitable frequencies for the selective heating of tumour tissue.

6.4 Future Developments

Several applications of *in vivo* tissue dielectric measurement techniques should be considered more completely in future research. The limited data shown in Fig. 8 indicate that dielectric property measurements have the sensitivity necessary to detect changes in the local blood flow to an organ. Additional research is necessary to establish the usefulness of this technique. Tissue dielectric measurements are also a sensitive indicator of the composition of the blood and can be used to measure non-invasively the tissue concentration of polar drugs. Since these measurements are sensitive to the normal physiological function of tissues, it is not unreasonable to speculate that many abnormal conditions (oedema, fibrosis, tumours, impaired blood flow) could be detected by examination of the appropriate tissue electrical parameter at an optimum frequency. The problem for the future is to establish an experimental correlation between these pathological conditions and changes in tissue dielectric properties. With this knowledge in hand, diagnosis of abnormal tissue conditions could be made from non-invasive *in vivo* measurements of the tissue's electrical properties.

Acknowledgements

We wish to thank F. Dunn, S. M. Michaelson, T. W. Athey, B. Oakley and K. R. Foster for critical reading of the manuscript.

References

Athey, T. W., Stuchly, M. A. and Stuchly, S. S. (1982). Measurement of radio frequency permittivity of biological tissues with an open-ended coaxial line: Part I. *IEEE Trans Microwave Theory Tech.* **MTT-30**, 82–86.

Brady, M. M., Symons, S. A. and Stuchly, S. S. (1981). Dielectric behavior of selected animal tissues *in vitro* at frequencies from 2 to 4 GHz. *IEEE Trans. Biomed. Engng* **BME-28**, 305–307.

Buckley, F. and Maryott, A. A. (1958). Tables of dielectric dispersion data for pure liquids and dilute solutions. NBS Circular 589.

Burdette, E. C. and Karow, A. M. (1978). Kidney model for study of electromagnetic thawing. *Cryobiology* **15**, 142–151.

Burdette, E. C., Karow, A. M. and Jeske, A. H. (1978). Design, development, and performance of an electromagnetic illumination system for thawing cryopreserved kidneys of rabbits and dogs. *Cryobiology* **15**, 152–167.

Burdette, E. C., Cain, F. L. and Seals, J. (1980a). *In vivo* probe measurement technique for determining dielectric properties at VHF through microwave frequencies. *IEEE Trans. Microwave Theory Tech.* **MTT-18**, 414–427.

Burdette, E. C., Cain, F. L. and Seals, J. (1980b). *In situ* permittivity measurements: perspective, techniques, results. *In* "Proc. Microwave Dosimetric Imagery Symposium", IEEE Press, New York.

Burdette, E. C., Wiggins, S., Brown, R. and Karow, A. M. (1980c). Microwave thawing of frozen kidneys: A theoretical-based experimentally-effective design. *Cryobiology* **17**, 393–402.

Burdette, E. C., Friederich, P. G., and Cain, F. L. (1980d). *In vivo* techniques for measuring electrical properties of tissues. Annual Technical Report No. 2, U.S. Army Medical Research and Development Command Contract No. DAMD17-78-C-8044, Sept. 1980 (available through Defense Technical Information Center).

Burdette, E. C., Seals, J., Magin, R. L. and Auda, S. P. (1983). Dielectric properties of animal and human tumors determined from *in vivo* measurements. *Cancer Res.* **43**, in press.

Cook, H. F. (1951). The dielectric behaviour of some types of human tissues at microwave frequencies. *Br. J. Appl. Phys.* **12**, 295–304.

Cook, H. F. (1952). Comparison of the dielectric behaviour of pure water and human blood at microwave frequencies. *Br. J. Appl. Phys.* **3**, 249–255.

Deschamps, G. A. (1962). Impedance of an antenna in a conducting medium. *IRE Trans. Antennas Propagat.* **AP-10**, 648–650.

Foster, K. R., Schepps, J. L., Stoy, R. D. and Schwan, H. P. (1979). Dielectric properties of brain tissue between 0·01 and 10 GHz. *Phys. Med. Biol.* **24**, 1177–1187.

Gandhi, O. P. (1980). State of the knowledge for electromagnetic absorbed dose in man and animals. *Proc. IEEE* **68**, 24–32.

Geddes, L. A. and Baker, L. E. (1967). The specific resistance of biological material: A compendium of data for the biomedical engineer and physiologist. *Med. Biol. Engng* **5**, 271–293.

Har-Kedar, I. and Bleehen, N. M. (1976). Experimental and clinical aspects of hyperthermia applied to the treatment of cancer with special reference to the role of ultrasonic and microwave heating. *In* "Advances in Radiation Biology" (Eds J. T. Lett and H. Adler), Vol. 6, pp. 229–266, Academic Press, New York and London.

Harrington, R. F. (1961). "Time Harmonic Electromagnetic Fields", McGraw-Hill, New York.

Hasted, J. B. (1972). Water—A comprehensive treatise. *In* "The physics and Physical Chemistry of Water" (Ed. F. Franks), Vol. 2, pp. 255–305, Plenum Press, New York and London.

Henderson R. P. and Webster, J. G. (1978). An impedance camera for specific measurements of the thorax. *IEEE Trans. Biomed. Engng* **BME-25**, 250–254.

Herrick, J. F., Jelatis, B. G. and Lee, G. M. (1950). Dielectric properties of tissues important in microwave diathermy. *Fedn Proc.* **9**, 60–71.

Iskander, M. F. and Durney, C. H. (1980). Electromagnetic techniques for medical diagnosis: A review. *Proc. IEEE* **68**, 126–132.

Jacobi, J. H. and Larsen, L. E. (1980). Microwave time delay spectroscopic imagery of isolated canine kidney. *Med. Phys.* **7**, 1–7.

Jacobi, J. H., Larsen, L. E. and Hast, C. T. (1979). Water-immersed microwave antennas and their application to microwave interrogation of biological targets. *IEEE Trans. Microwave Theory Tech.* **MTT-27**, 70–78.

Johnson, C. C. and Guy, A. W. (1972). Nonionizing electromagnetic wave effects in biological materials and systems. *Proc. IEEE* **60**, 692–718.

Larsen, L. E. and Jacobi, J. H. (1979). Microwave scattering parameter imagery of an isolated canine kidney. *Med. Phys.* **6**, 394–403.

Larsen, L. E., Jacobi, J. H. and Key, A. K. (1978). Preliminary observations with an electromagnetic method for the noninvasive analysis of cell suspension physiology and pathophysiology. *IEEE Trans. Microwave Theory Tech.* **MTT-26**, 581–595.

Lin, J. C. (1975). Microwave properties of fresh mammalian brain tissue at body temperature. *IEEE Trans. Biomed. Engng* **BME-22**, 74–76.

Magin, R. L. and Burns, C. P. (1972). Determination of biological tissue dielectric constant and resistivity from *in vivo* impedance measurements. *In* "Proc. Region 3 IEEE Conference," H6, pp. 1–3, IEEE Press, New York.

Michaelson, S. M. (1980). Microwave biological effects: An overview. *Proc. IEEE* **68**, 40–49.

Myers, P. C. and Barrett, A. H. (1980). Microwave thermography of normal and cancerous breast tissue. *Ann. N. Y. Acad. Sci.* **335**, 443–455.

Osterhout, W. J. Z. (1922). "Injury Recovery and Death, in Relation to Conductivity and Permeability", Lippincott, Philadelphia and London.

Pederson, P. C., Johnson, C. C., Durney, C. H. and Bragg, D. G. (1976). An investigation of the use of microwave radiation for pulmonary diagnostics. *IEEE Trans. Biomed. Engng* **BME-23**, 410–412.

Pederson, P. C., Johnson, C. C., Durney, C. H. and Bragg, D. G. (1978). Microwave reflection and transmission measurements for pulmonary diagnosis and monitoring. *IEEE Trans. Biomed. Engng* **BME-25**, 40–48.

Popovic, V. P., Popovic, P., Burdette, E. C. and Schaffer, R. E. (1977). Restoration of phagocytosis after freezing and microwave thawing of granulocytes. *Cryobiology* **14**, 698–699.

Rajewsky, B. (1938). "Ultrakurzwellen, Ergebnisse der Biophysikalischen Forschung", Geog Thieme, Leipzig.

Salter, D. C. (1979). Quantifying skin disease and healing *in vivo* using electrical impedance measurements. *In* "Non-Invasive Physiological Measurements" (Ed. P. Rolfe), Vol. 1, Chap. 2, Academic Press, London and New York.

Schepps, J. L. and Foster, K. R. (1980). UHF and microwave dielectric properties of normal and tumor tissues: Variation in dielectric properties with tissue water content. *Phys. Med. Biol.* **25**, 1149–1159.

Schwan, H. P. (1957). Electrical properties of tissue and cell suspensions. *Adv. Biol. Med. Phys.* **5**, 147–209.

Schwan, H. P. and Foster, K. R. (1980). RF-field interactions with biological systems: Electrical properties and biophysical mechanisms. *Proc. IEEE* **68**, 104–113.

Stuchly, M. A. and Stuchly, S. S. (1980). Dielectric properties of biological substances—tabulated. *J. Microwave Power* **15**, 19–26.

Stuchly, M. A. and Stuchly, S. S. (1981). Coaxial line reflection methods for measuring dielectric properties of biological substances at radio and microwave frequencies—A review. *IEEE Trans. Instrum. Meas.* **IM-29**, 176–183.

Stuchly, M. A., Athey, T. W., Samaras, G. H. and Taylor, G. E. (1982). Measurement of radio frequency permittivity of biological tissues with an open-ended coaxial line: Part II—Experimental results. *IEEE Trans. Microwave Theory Tech.* **MTT-30**, 87–92.

Stuchly, M. A., Athey, T. W., Samaras, G. H. and Taylor, G. E. (1981). Dielectric properties of biological substances at radio frequencies. Part II. Experimental results. *IEEE Trans. Microwave Theory Tech.* **MTT-** (in press).

Tinga, W. R. and Nelson, S. O. (1973). Dielectric properties of materials for microwave processing—tabulated. *J. Microwave Power* **8**, 29–61.

Von Hippel, A. R. (1954). "Dielectric Materials and Applications", MIT Press, Boston.

SUBJECT INDEX

Page numbers in italics indicate references to tables or figures.

A

Absorption of light by haemoglobin, 252–253, 254–256, 290–291, 292

Accelerations in fetal heart rate, 21, 24–26

Action potential, 109–110, 113, 133

Aerospace medicine, oximetry in, 277

Aliasing in Fourier analysis, 201–202

Altitude, oximetry at, 276

Amplifier,
  for diaphragm electromyograph, 141–144
  for electrogastrography, 198, *199*

Anaemia, fetal, and electrocardiography, 28, 37

Anaesthesia,
  oximetry in, 274–275
  ultrasound scanning in, 348

Analysis,
  in electrogastrography,
    autocorrelation, 190–191, 207–208, 210–212
    computer, 203
    EMP statistics, 204–205
    Fourier-type, 201–203
    intensity modulated raster scanning, 205, *206*, 210, *211*
    phase lock techniques, 203–205
    visual, 200–201
  of fetal breathing movements, autocorrelation, 77, *78*
  in fetal cardiac assessment, 22–24, 44
  in laser Doppler blood flow techniques
    autocorrelation, 235
    Fourier-type, 234
  in transcutaneous bilirubin measurement, 289, 297–306

Anatomy,
  fetal cardiovascular, 3–5, 7–9
  skin, 221

Apnoea,

diaphragm electromyographic investigation of, 171

Arrhythmias, fetal, 9, 30–1
  and breathing movements, 79

Artefacts,
  in electrogastrography, 189–190, 200–201
  in electromyography: diaphragm, 140–141, 149–153
    uterine, 107–108, 121–122
  in laser Doppler blood flow measurement, 243–244
  in measurement of fetal limb movements, 92
  in transcutaneous bilirubin measurement, 308
  in ultrasonic measurements of fetal breathing movements, 72–73

Arterialization in oximetry, 257, 262, 269

Asphyxia,
  and fetal breathing movements, 62
  and fetal electrocardiography, 36–37, 40, 47–48

Audio output in Doppler blood flow measurement, 243

B

Behaviour,
  and gastric activity, 195
  and muscle activity, 155–158, 160–164

Bilirubin, transcutaneous measurement, 287–290, 309–310
  clinical trials, 296–309
  Minolta transcutaneous bilirubin meter, *293*, 294–296
  principles, 290–294

Blanking technique in eliminating maternal electrocardiogram, 15

377